Packaging for Sustainability

Karli Verghese · Helen Lewis
Leanne Fitzpatrick
Editors

Packaging for Sustainability

Dr. Karli Verghese
Centre for Design
RMIT University
GPO Box 2476
Melbourne VIC 3001
Australia
e-mail: Karli.Verghese@rmit.edu.au

Dr. Leanne Fitzpatrick
Birubi Innovation
5 Brooklyn Ave
Dandenong VIC 3175
Australia
e-mail: leanne@birubi.com.au

Dr. Helen Lewis
Centre for Design
RMIT University
GPO Box 2476
Melbourne VIC 3001
Australia
e-mail: lewis.helen@bigpond.com

ISBN 978-0-85729-987-1   e-ISBN 978-0-85729-988-8
DOI 10.1007/978-0-85729-988-8
Springer London Dordrecht Heidelberg New York

Library of Congress Control Number: 2011941769

British Library Cataloguing in Publication Data
A catalogue record for this book is available from the British Library

© Springer-Verlag London Limited 2012
Capilene is a registered trademark of Patagonia, Inc., PO Box 150, Ventura, California 93002, United States of America
EASYTAB is a registered trademark of THE GILLETTE COMPANY, One Gillette Park, Boston, Massachusetts, United States, 02127
Apart from any fair dealing for the purposes of research or private study, or criticism or review, as permitted under the Copyright, Designs and Patents Act 1988, this publication may only be reproduced, stored or transmitted, in any form or by any means, with the prior permission in writing of the publishers, or in the case of reprographic reproduction in accordance with the terms of licenses issued by the Copyright Licensing Agency. Enquiries concerning reproduction outside those terms should be sent to the publishers.
The use of registered names, trademarks, etc., in this publication does not imply, even in the absence of a specific statement, that such names are exempt from the relevant laws and regulations and therefore free for general use.
The publisher makes no representation, express or implied, with regard to the accuracy of the information contained in this book and cannot accept any legal responsibility or liability for any errors or omissions that may be made.
The authors do not accept any liability to any person or organisation for the information or advice provided in this book or incorporated into it by reference. Nor do the authors accept any liability for any loss or damages incurred as a result of reliance placed on the content of the book.

Printed on acid-free paper

Springer is part of Springer Science+Business Media (www.springer.com)

Foreword

Other than packaging, it is doubtful if there is anything in this world linked to so many diverse business sectors, which improves the daily lives of billions, prevents waste and yet is regarded negatively by most people. Responding to all of this places immense demands on packaging professionals.

And now, with sustainability one of the biggest buzz words resonating in our global village, the packaging value chain faces an additional demand: to produce 'sustainable packaging'. Every member of the chain, from packaging designers through to end users and the waste recovery sector, is being challenged to embrace sustainability in their respective link with the business of packaging. Help is needed and this book, from the founders of The Sustainable Packaging Alliance, provides it.

There is a growing need to facilitate and communicate an understanding of the relative sustainability of packaging and that need is growing. It is needed internally within company departments, externally for business-to-business communication in the packaged goods supply chain and, to meet consumer expectations, companies increasingly need to communicate publicly about their packaging. Additionally, growing calls for mandatory corporate sustainability reporting coming both from governments and market regulatory authorities is placing new responsibilities on corporate reporting, which inevitably will include references to how a company packages the products it sells.

The absence of a common understanding of what is really meant by the term 'sustainable packaging' coupled nevertheless with growing calls for packaging to 'be sustainable' has produced a great deal of misinformation, confusion and even panic within the packaging world. Often, conflicting demands and expectations from consumers, regulators and packaged goods supply chain partners have resulted in companies responding to this pressure in different ways. Without the benefit of a common industry language to enable informed discussions between stakeholders about the relationship between packaging and sustainability, suboptimal and counterproductive measures have sometimes resulted.

Although there is an accepted consensus among packaging experts that, in absolute terms, 'sustainable packaging' cannot be defined, all recognise that packaging does make a valuable contribution to economic, environmental and social sustainability by protecting products, preventing waste, enabling efficient business conduct and providing end users with the benefits of the products it contains. Choosing optimum packaging for each product is therefore a critical component of every company's sustainability strategy.

This book identifies and explains the building blocks which are essential to move towards more sustainable packaging. It also provides a guide to using the tools needed to begin the construction: a life cycle thinking approach and life cycle assessment methodologies. Properly used, these methods enable a data-based analysis of all the selected parameters necessary to assess the performance of packaging in a holistic way. This cannot be done in isolation from the product the packaging will contain or from its overall supply chain. Looking only at the packaging will not produce optimum results and could generate an inefficient, more costly and ultimately less sustainable outcome.

With this understanding and these tools, successful packaging policy will also necessitate a strategic corporate approach to realise the desired result. Learning to work with supply chain partners, with trade and industry associations and other stakeholder groups including regulators is considered an indispensable element of successful corporate packaging strategies. This will require the establishment of agreed goals, a commitment of management support and a plan to carry all of this through.

By this process, delivering packaging which is more sustainable will take time and investment but the benefits can ultimately include cost reductions, reduced environmental impact, enhanced consumer perception, improved decision making and ultimately extended influence within the packaging value chain and the broader corporate world. It is a journey every company needs to take and on which the authors of this book, Karli Verghese, Helen Lewis and Leanne Fitzpatrick, will be your guides.

27 April 2011

Julian Carroll
EUROPEN

Acknowledgments

We would like to thank a number of people who have been involved in the writing, reviewing and production of this book.

Professor Ralph Horne for his support of this project and the financial contribution from the Centre for Design at RMIT University for copy-editing of the final manuscript.

Co-authors of individual chapters: Simon Lockrey, Dr. Enda Crossin, Andrew Carre, Professor Margaret Jollands and Helaine Stanley for their time and valuable contributions.

Fran Macdonald for her valuable contribution to the copy-editing of the final manuscript.

Emeritus Professor Harry Lovell and Dr. Juanita Day who reviewed an early version of Chap. 6.

To all who gave permission for the use of figures and case studies throughout the book and the companies that supplied photos of products.

Julian Carroll for preparing the foreword to this book.

Anthony Doyle, Claire Protherough and Grace Quinn from Springer for their support and guidance over the past three years, from the idea generation through to publishing of the book.

Our families for their patience and support.

29 April 2011 Karli Verghese
Melbourne Helen Lewis
  Leanne Fitzpatrick

Contents

1 **Developing the Strategy** 1
Leanne Fitzpatrick, Karli Verghese and Helen Lewis

2 **Designing for Sustainability** 41
Helen Lewis

3 **Marketing and Communicating Sustainability** 107
Helen Lewis and Helaine Stanley

4 **Complying with Regulations** 155
Helen Lewis

5 **Applying Life Cycle Assessment** 171
Karli Verghese and Andrew Carre

6 **Packaging Materials** 211
Karli Verghese, Enda Crossin and Margaret Jollands

7 **Selecting and Applying Tools** 251
Karli Verghese and Simon Lockrey

8 **Implementing the Strategy** 285
Leanne Fitzpatrick, Helen Lewis and Karli Verghese

Editors' Biographies 329

Appendix A: Application of a Sustainable Packaging Framework 331

Appendix B: Labels and Logos 335

Appendix C: Matrix of International Regulations, Policies and Standards 341

Glossary 373

Index 379

Abbreviations

AS	Australian Standards
ASTM	American Society for Testing and Materials
BBP	Benzyl butyl phthalate
BPA	Bisphenol A
CDL	Container Deposit Legislation
CO_2	Carbon dioxide
COMPASS	Comparative Packaging Assessment
CPET	Crystalline polyethylene terephthalate
DBP	Dibutyl phthalate
DEHP	Diethylhexyl phthalate
DIDP	Diisodecyl phthalate
DINP	Diisononyl phthalate
DRCs	Display-ready containers
ECF	Elemental chlorine-free
EDIT	Eco-Design Indicator Tool
EN	European standard
EPS	Expanded polystyrene
eq.	Equivalent
EU	European Union
EUROPEN	European Organisation for Packaging and the Environment
FDA	Food and Drug Administration
FSC	Forest Stewardship Council
GPP	Global Packaging Project
GPPS	General purpose polystyrene
HDPE	High density polyethylene
HIPS	High impact polystyrene
ISO	International Organisation for Standardisation
LCA	Life cycle assessment
LCI	Life cycle inventory
LCIA	Life cycle impact assessment
LCM	Life cycle management

LCT	Life cycle thinking
LDPE	Low density polyethylene
LLDPE	Linear low density polyethylene
LPB	Liquid paperboard
LOHAS	Lifestyles of health and sustainability
MRF	Materials recovery facility
MSW	Municipal solid waste
NGO	Non Government Organisation
NZ	New Zealand
PC	Polycarbonate
PE	Polyethylene
PET	Polyethylene terephthalate
PCF	Processed chlorine-free
PIQET	Packaging Impact Quick Evaluation Tool
PLA	Polylactic acid
PP	Polypropylene
PS	Polystyrene
PVC	Polyvinyl chloride
rPET	Recycled polyethylene terephthalate
RPCs	Reusable plastic containers
SPA	Sustainable Packaging Alliance
SPC	Sustainable Packaging Coalition
SPI	Society of the Plastic Industry
TCF	Totally chlorine free
TPS	Thermoplastic starch
UK	United Kingdom
US	United States
VOCs	Volatile organic compounds
WRAP	Waste and Resources Action Programme
+ve	Positive
−ve	Negative

Chapter 1
Developing the Strategy

Leanne Fitzpatrick, Karli Verghese and Helen Lewis

Abstract The commercial operating environment is increasingly demanding more sustainable products and improved environmental performance. How a business responds to this demand should be embedded within its corporate strategy. For many years, packaging has been at centre stage in political and consumer campaigns to address environmental issues. Packaging does generate environmental impacts in all stages of its life cycle however these cannot be isolated from the impacts of the product it protects. Packaging's role in the corporate strategy should be clearly identified, by understanding the environmental life cycle of products and their packaging and relevant current or emerging environmental regulations. Packaging sustainability initiatives should optimise the product-packaging system and reduce specific environmental impacts including those of the packaging itself. This chapter outlines issues to consider in constructing the business case for investing in packaging for sustainability and, setting goals and targets for packaging's contribution to a business's more general sustainable development goals.

Contents

1.1	Introduction	3
1.2	Sustainable Development	11
1.3	Corporate Sustainability	15
	1.3.1 Stage of Evolution Towards Sustainable Development	16
	1.3.2 Sustainable Product Development	19

L. Fitzpatrick (✉)
Birubi Innovation, 5 Brooklyn Avenue, Dandenong, VIC 3175, Australia
e-mail: leanne@birubi.com.au

K. Verghese · H. Lewis
Centre for Design, RMIT University, GPO Box 2476, Melbourne, VIC 3001, Australia
e-mail: Karli.Verghese@rmit.edu.au

H. Lewis
e-mail: lewis.helen@bigpond.com

1.4 Packaging's Role in Sustainable Development.. 22
1.5 Developing the Business Case 25
    1.5.1 Corporate and Brand Positioning.. 25
    1.5.2 Supply Chain Requirements.. 26
    1.5.3 Solid Waste Reduction 29
    1.5.4 Resource Efficiency... 33
    1.5.5 Climate Change 34
1.6 Conclusion 35
References... 37

Figures

Figure 1.1	The corporate report card..	4
Figure 1.2	Integrating sustainable development into the strategic and corporate planning process..	6
Figure 1.3	Life cycle map of a product-packaging system	7
Figure 1.4	Eco-efficiency versus Eco-effectiveness.............................	14
Figure 1.5	Two metabolisms outlined in the cradle-to-cradle model ...	15
Figure 1.6	Product life cycle management..	20
Figure 1.7	Simplified product environmental life cycle map...............	21
Figure 1.8	The evolution of packaging concerns and regulations........	22
Figure 1.9	Packaging waste generation as a proportion of municipal solid waste (MSW), United States, 1960–2008	29
Figure 1.10	Waste hierarchy	30
Figure 1.11	Waste disposal and recovery, United States, 1960–2008 ...	31
Figure 1.12	Categories of products generated in Municipal Solid Waste, United States, 2008..	31
Figure 1.13	Sustainability and packaging...	36

Tables

Table 1.1	Packaging system terminology...	8
Table 1.2	Triple bottom line issues for business..................................	12
Table 1.3	Competencies of businesses that lead in sustainable development	16
Table 1.4	Sustainability phase model...	17
Table 1.5	Sustainability challenges, competencies and opportunities phase model	18
Table 1.6	Examples of packaging initiatives driven by retailers.........	27
Table 1.7	Cost of packaging waste	33

Photos

Photo 1.1	Packet of corn chips (Photo: Cathy Kaplan, Istock photo)....	10
Photo 1.2	Packaging litter in Yarra River, Melbourne (Photo: Karli Verghese)	32

Case Studies

Case Study 1.1	A strategic approach to sustainability - Marks and Spencer's 'Plan A' ..	5
Case Study 1.2	The corn chips or their packet?..	9
Case Study 1.3	The wine or the glass bottle?...	11

Case Study 1.4	Evolution of sustainability at Procter & Gamble	18
Case Study 1.5	Sustainable Packaging Alliance's definition of sustainable packaging	23
Case Study 1.6	Sustainable Packaging Coalition's criteria for sustainable packaging	24
Case Study 1.7	Benchmarking sustainability performance: The Dow Jones Sustainability Indexes	25
Case Study 1.8	Packaging Restricted Substances List and Design Requirements: Nike	28
Case Study 1.9	Packaging sustainability performance expectations: McDonald's	28
Case Study 1.10	Economic and environmental benefits: Packaging waste reduction at Walmart	34

1.1 Introduction

This book explains:
- the business case for investing in packaging for sustainability
- the role of packaging in sustainable development
- strategies, tools and case studies to help a business develop more sustainable products and improve environmental performance through packaging.

It contains valuable resources for packaging and product developers, marketers, sustainability and environment managers, procurement and supply chain managers, senior executives and corporate boards.

Businesses Must Address Sustainability

Contributing to sustainability is now a pre-requisite for a business's own sustainability. Increasingly, market, regulatory, supply chain and social forces require businesses to:
- provide more sustainable products and services
- publicly report on their sustainability performance
- respond to climate change by reducing energy consumption and greenhouse gas emissions
- improve efficiency in the use of limited resources such as raw materials, land and water
- increase consumption of renewable resources and use land appropriately
- reduce waste associated with the production, distribution, use and disposal of their products.

Sustainable Development Creates New Business Challenges

> 'The key is to build sustainability into the business, rather than present it as an additional activity; to have it owned by the business rather than by a corporate "CSR group"'. [1, p. 386]

A business must respond to these forces whilst continuing to achieve performance goals such as profitability, market share and revenue growth, and meet other regulatory requirements for product safety, occupational health and safety, trade practices and so on (see Fig. 1.1). For many businesses, a fundamental shift in strategy and the development of new skills and knowledge are required before sustainable development is seriously practised throughout their operations. The magnitude and nature of these changes are often significant, creating barriers to organisational change and constraining the rate of improvement in sustainable development.

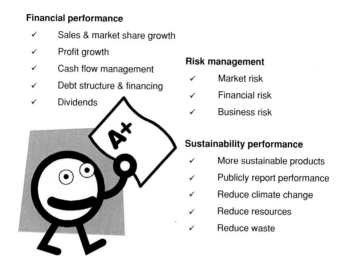

Fig. 1.1 The corporate report card

Build Sustainable Development into the Corporate Strategy

Many businesses are making the changes necessary to incorporate sustainable development into their business model (see the Marks & Spencers case study (Case Study 1.1) and Case Study 8.1 (VIP Packaging).

The first step is to develop a coherent corporate sustainability strategy that clarifies:
- the business's specific case for sustainable development
- the environmental life cycle impacts of its products and services
- the impacts and requirements of current and future environmental regulations.

The strategy should outline how the business will:
- create growth from more sustainable products and services
- improve the efficiency of its operations and supply chains through reductions in material and energy consumption and waste
- meet or exceed its regulatory requirements
- report and communicate the environmental benefits of its products and achievements.

It should also include goals/targets and performance indicators or metrics that inform the strategies and operating plans of relevant corporate functions such as marketing, product and packaging development, procurement, operations and manufacturing (see Fig. 1.2).

1 Developing the Strategy

> **Case Study 1.1 A Strategic Approach to Sustainability—Marks and Spencer's 'Plan A'**
>
> Marks & Spencer (M&S) is a leading retailer in the United Kingdom with over 21 million customers visiting their stores every week. In 2007 the company announced an ambitious sustainability strategy called Plan A—'committing to change 100 things over five years, because we have got only one world and time is running out'. These commitments covered five 'pillars': climate change, waste, natural resources, fair partner, and health and wellbeing. M&S has committed to reducing the weight of non-glass packaging by 25% and ensuring that all packaging can be easily recycled or composted by 2012.
>
> In 2010 the company updated Plan A with a new set of commitments and announced its ambition to become 'the world's most sustainable major retailer by 2015'. There is a strong business case: M&S reports that by using less energy, reducing packaging and waste, and creating new markets such as M&S Energy, it has generated an additional profit of £50m for 2009–2010. Chairman Sir Stuart Rose stressed that sustainability is now an integral part of M&S's business model:
>
>> 'We aim to engage every one of our 21 million customers by building Plan A qualities into all of the 2.7 billion M&S products we sell and helping customers to develop their own Plan A eco-plans. We also aim to accelerate the transition of Plan A from 'Plan' to 'How We Do Business' by integrating it into processes and giving our people the skills, tools and motivation required to make a difference'.
>
> Chief Executive Marc Boland also noted that: 'Plan A is a good fit with the M&S brand ... an excellent way to stay close to customers and their concerns and it's also where society is heading in the future'.
> *Source*: Marks and Spencer [14]

Packaging can contribute to achieving the business's sustainable development goals/targets and should be:
- integral to the corporate sustainability strategy
- reflected in the short, medium and long term goals for corporate activities such as marketing, communication and sales, product and packaging development, procurement and supply chain management, manufacturing processes and environmental improvements.

> 'An indicator is used as a proxy for an issue or characteristic an organisation wants to measure. An indicator describes a concept and can express movement—whether positive or negative—toward a goal. Generally, an indicator focuses on a piece of a system that can provide a sense of the bigger picture. For example, the indicator "small business survival rate" provides information about the overall economic health of a region.
> A metric is the method used to express an indicator. Metrics are often computational or quantitative, but can also be qualitative. Metrics are typically expressed as a numerator and a denominator; 'A per B'. For example, a metric to quantify the indicator 'virgin material content' could be expressed as '% of total virgin material used per tons of packaging component'.
> *Source*: The Consumer Goods Forum [12, p. 15]

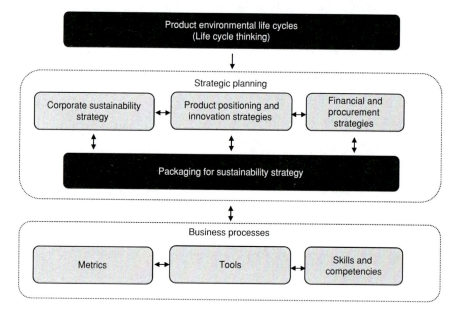

Fig. 1.2 Integrating sustainable development into the strategic and corporate planning process

Confirm the Role of Packaging in the Sustainability Strategy

> 'The packaging industry should not only aim to improve the production process of their packages, but also provide packages whose functionality helps to reduce other more relevant environmental impacts in the life cycle such as, for example, [food] losses.' *Source*: Busser and Jungbluth [5, p. S81]

1 Developing the Strategy

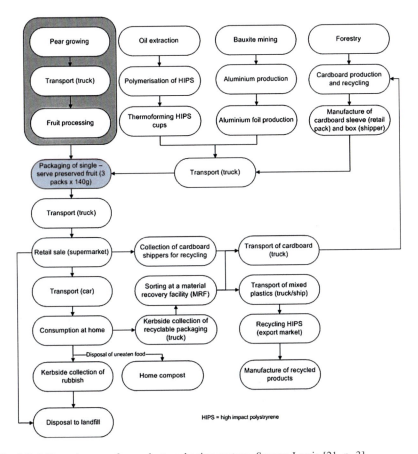

Fig. 1.3 Life cycle map of a product-packaging system. *Source*: Lewis [21, p. 3]

For many years, packaging has been at centre stage in political and consumer campaigns to address perceptions of unsustainable consumerism in Western societies (see Sect. 1.4). Its use, disposal and recovery generate environmental impacts by consuming materials, energy and water, and generating wastes and emissions—all of which must be addressed in the sustainability strategy. However, packaging may also provide hidden environmental benefits when its primary function is considered; that is, product protection. Packaging when used effectively:

- enables the safe and efficient supply of products
- minimises the environmental impacts of producing, transporting, using and disposing of those products
- contributes to achieving sustainable development goals/targets.

The environmental impacts of a product and its packaging are therefore interlinked (see Fig. 1.3), and the links need to be understood by any business operating in the packaging supply chain. To achieve this requires new levels of collaboration and communication between businesses in the packaging supply chain (brand owners, packaging suppliers, material producers, waste recovery)—which in itself is a significant challenge. If not achieved it may lead to ill-informed decisions that:

- increase environmental impacts by shifting the burden from one impact area, such as packaging waste, to another, such as product waste
- hinder the uptake of better strategies; for example, by sending mixed messages to consumers.

Put Packaging's Environmental Impacts and Benefits into Context

A number of packaging components are used to contain and distribute products and make up a 'packaging system'. The packaging system is also often described in terms of its role at different stages within the product supply chain (see Table 1.1).

Table 1.1 Packaging system terminology

Label	Other terms	Definition	Functions
Primary	Sales, Consumer, Retail	The sales unit at the point of purchase	Protection, promotion, convenience, information, handling, safety
Secondary	Display, Merchandising	Packaging used at the point of purchase to contain or present a number of sales units; it can be removed from the product without affecting its characteristics	Protection, promotion, convenience, information, utilisation, handling, safety
Tertiary	Distribution, Traded, Transport	Used to facilitate handling and transport of a number of sales units or grouped packages in order to prevent physical handling and transport damage; does not include road, rail, ship and airfreight containers	Protection, information, handling, safety
Industrial	Business-to-business	Used for transport and distribution of products for industrial use	Protection, information, handling, safety

Source: Definitions of primary, secondary and tertiary packaging are based on the European Packaging and Packaging Waste Directive [22]

The environmental impacts of the product are generally but not always greater than those of its packaging system. This 'generally accepted fact' often leads to a view that it is not necessary or productive to focus on the packaging as well as the product, but this is naive. The packaging impacts cannot be separated from those of the product, so it is the product-packaging system as a whole that must be optimised (Case Studies 1.2, 1.3 and 2.10 on Omo Surf 'Small & Mighty' detergent in Chap. 2) also demonstrate that:

- The impacts may arise from secondary or even tertiary packaging rather than the product's primary packaging, which is often the focus of attention when considering environmental impacts of packaging
- Decisions about the product can lead to reductions in the environmental impacts of packaging
- Decisions about the product-packaging system have flow-on impacts for transportation, storage, distribution and so on.

1 Developing the Strategy

Case Study 1.2 The Corn Chips or their Packet?

A life cycle assessment was conducted in Victoria, Australia on producing and supplying 400 gram packets of corn chips to a retail outlet to assess greenhouse gas impacts. In this case study, packaging is not the major driver of impacts although it does play a significant role. The packaging, consisting of the aluminium foil (retail bag) and a corrugated box (secondary packaging) for transporting the retail bags through the supply chain, accounted for 21% of the impacts. Supply chain transportation accounted for 9% of the impacts, and these are strongly influenced by the choice and design of the product and its packaging. The overall breakdown was 6% of greenhouse emissions pre-farm, 36% were on farm and 58% were from off-farm processes and activities.

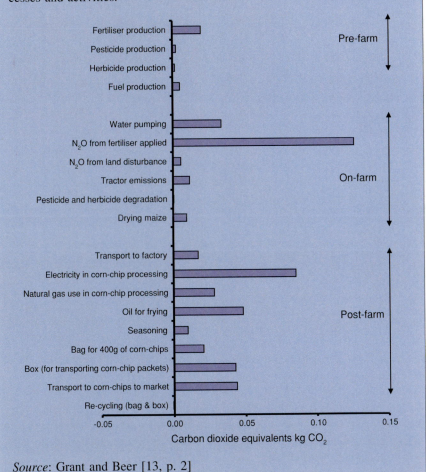

Source: Grant and Beer [13, p. 2]

(continued)

Case Study 1.2 (continued)

Photo 1.1 Packet of corn chips (Photo: Cathy Kaplan, Istock photo)

Packaging Design becomes More Complex

Packaging design is a complex process that considers many aspects of marketing (such as brand positioning and shelf presence), packaging function (such as product protection and manufacturing capability) and cost (such as product cost and margin, and capital investment and returns). Many corporate functions and stakeholders—internal and external—are involved or affected directly and indirectly by packaging design (see Tables 8.1 and 8.8).

⇒ See Chap. 8 for how to create a packaging for sustainability action plan.

Designing for sustainability adds further complexity to the packaging design process. It requires changes to existing product and packaging development and therefore new capabilities in those areas. It also affects production, distribution, use and recovery of packaging.

1 Developing the Strategy

Case Study 1.3 The Wine or the Glass Bottle?

A life cycle assessment was conducted on wine production in the Spanish region of La Rioja, including: cultivation of grapes, wine making and bottling, distribution and sales, and disposal of empty bottles. In this case study, packaging and distribution had a significant impact on the overall environmental impacts. The bottle was the largest contributor to impacts in two of the four indicators evaluated (acidification and photochemical oxidant potentials) and a major contributor to the third (global warming potential). If the distribution system is added (including secondary packaging and transport systems) then the contribution is even more significant. The product—viticulture—dominates eutrophication potential and contributes more than 40% to global warming potential.

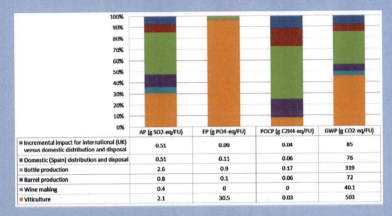

Source: Gazulla et al. [17]

1.2 Sustainable Development

First of all, it is important for a business to understand sustainable development and its implications. This understanding should be shared with customers, suppliers and other stakeholders.

Sustainable Development

The term '*sustainable development*' entered the public debate after the World Commission on Environment and Development published its landmark report, *Our Common Future*, in 1987. It was defined as 'development that meets the needs of the present without compromising the ability of future generations to meet their own needs' [23, p. 43]. This work established many of the basic sustainability principles that have been explored since that time. For example, it highlighted the need to address economic development, social equity and environmental protection as equally important goals of sustainability.

Triple Bottom Line

The concept of the *'triple bottom line'* (TBL)—also referred to as a concern for 'people, profit and the planet'—was introduced to translate the principle of sustainable development into something meaningful for business. The term was popularised by John Elkington in 1997 [24] through the concept of TBL accounting and accountability and encouraged a more complete approach to sustainability (see Table 1.2). As Elkington noted, '[s]ociety depends on the economy—and the economy depends on the global ecosystem, whose health represents the ultimate bottom line' [24, p. 73].

The three elements of sustainability interact and open up new opportunities and challenges for business. In Chap. 2 we provide a framework for designing packaging for sustainability that takes a triple bottom line approach.

Table 1.2 Triple bottom line issues for business

Bottom line	Issues for business
Economic	Cost competitiveness
	Demand for products and services
	Level of innovation
	Human and intellectual capital
	Profit margin
Social and ethical	Animal testing
	Community relations
	Human rights
	Working conditions
	Irresponsible marketing
	Impacts on indigenous people
	Employment of minorities
Environmental	Environmental compliance
	Use and protection of natural capital
	Environmental management costs
	Material, energy and water consumption
	Solid waste and pollution
	Life cycle impacts of products and services
	Performance against best practice standards

Source: Based on Elkington [24, pp. 69–94]

Sustainable Consumption and Production

In 2002 the United Nations world summit on sustainable development identified the promotion of *sustainable patterns of consumption and production* as one of three overarching objectives of sustainable development. In the following year

> **Sustainable consumption and production** is…….., the use of services and related products which respond to basic needs and bring a better quality of life while minimising the use of natural resources and toxic materials as well as the emissions of waste and pollutants over the life-cycle so as not to jeopardize the needs of future generations' *Source*: United Nations Department of Economic and Social Affairs (UNDESA) and the United Nations Environment Programme (UNEP) [6, p. 4]
>
> It requires a:
>
> ………, product and service lifecycle perspective to ensure sustainable management of natural resources from the extraction to the production, distribution, consumption and disposal/reuse phases…[Sustainable consumption and production] aims at 'doing more and better with less', by providing policies, tools, measures, infrastructure, and supporting behaviour changes leading to green, resource efficient economies that ensure well-being, quality of life, and social development for all, while minimising environmental degradation along the whole life cycle, *Source*: United Nations Department of Economic and Social Affairs (UNDESA) and the United Nations Environment Programme (UNEP) [6, p. 3]

the United Nations launched the 'Marakesh process' to develop a ten year program of sustainable consumption and production activities. The draft program emphasises the need for long term, systemic change to achieve increased resource efficiency.

The European Commission has also developed an action plan to promote sustainable consumption and production. Their plan links the broad concept of sustainable development to products and services, noting that 70–80% of all environmental impacts arise from the consumption activities of eating, drinking, housing and travel [25]. It includes policies to:

- promote 'eco-design' and 'eco-labelling'
- promote 'green procurement' (buying sustainable goods and services) by government agencies
- encourage retailers to reduce environmental impacts in their supply chains
- promote leaner production.

In the Commission's view the challenge is to create a '*virtuous circle*' by:
- 'improving the overall environmental performance of products throughout their life-cycle'
- 'promoting and stimulating the demand for better products and production technologies'
- 'helping consumers to make better choices through more coherent and simplified labelling'. [25, p. 3].

Eco-efficiency

> **Eco-efficiency** is… "a management philosophy that encourages business to search for environmental improvements that yield parallel economic benefits. It promotes activities that 'create more value with less impact [2]".

A similar philosophy underpins the work of the World Business Council for Sustainable Development, which has established a business case for sustainable development based on eco-efficiency [2]. Businesses are encouraged to measure and report on the following metrics [26]:

- value provided by the business; for example, the quantity of goods or services produced or net sales
- environmental impacts of operations, such as energy consumption, materials consumption, water consumption, greenhouse gas emissions and ozone depleting substances emissions
- eco-efficiency ratios for each impact category; the measure of value divided by the measure of a specific environmental indicator. For example, the eco-efficiency ratio for energy consumption could be reported as 'kilograms per gigajoule.'

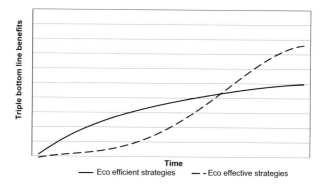

Fig. 1.4 Eco-efficiency versus Eco-effectiveness. *Source*: based upon Alston [27]

Eco-effectiveness

> **Eco-effectiveness** is… taking an eco-effective approach to design that might result in an innovation so extreme that it resembles nothing we know, or it might merely show us how to optimise a system already in place. It's not the solution itself that is necessarily radical but the shift in perspective with which we begin, from the old view of nature as something to be controlled to a stance of engagement. *Source*: McDonough and Braungart [3, p. 84].

1 Developing the Strategy

McDonough and Braungart in their influential book *Cradle to Cradle; Remaking the Way We Make Things* [3] argue that a focus on 'eco-efficiency' will not address the challenges presented by sustainability. They argue that eco-efficiency itself is not sustainable, as reducing the amount of a material used ('dematerialisation') or chemical emitted doesn't stop depletion or destruction; it only slows it down [3, p. 54].

Instead they promote an 'eco-effective' approach that challenges how things are done, encourages step-change innovation, and complements eco-efficiency, although eco-effectiveness may take longer in the initial phase to produce results (see Fig. 1.4).

⇒ See Chap. 5—Applying LCA—for an explanation of the difference between cradle-to-cradle and cradle-to-grave.

An eco-effective approach avoids waste by designing products for recovery and continuous cycling of materials through one of two 'metabolisms' without cross-contamination: a biological metabolism, such as composting, or a technical metabolism, such as an industrial recycling process [3, p. 104] (see Fig. 1.5).

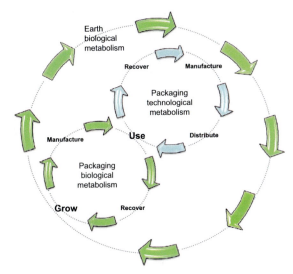

Fig. 1.5 Two metabolisms outlined in the cradle-to-cradle model. *Source*: based upon Alston [27]

1.3 Corporate Sustainability

> **Corporate sustainability** is... a business approach that creates long term shareholder value by embracing opportunities and managing risks deriving from economic, environmental and social developments. Corporate sustainability leaders achieve long term shareholder value by gearing their strategies and management to harness the market's potential for sustainable products and services while at the same time successfully reducing and avoiding sustainability costs and risks. [4]

Businesses that lead in sustainable development are highly competent to address global and industry challenges in an integrated way across corporate activities (see Table 1.3).

Table 1.3 Competencies of businesses that lead in sustainable development

Business activity	Competency
Strategy	Integrating long-term economic, environmental and social aspects in their business strategies while maintaining global competitiveness and brand reputation
Financial	Meeting shareholders' demands for sound financial returns, long-term economic growth, open communication and transparent financial accounting
Customer and product	Fostering loyalty by investing in customer relationship management and product and service innovation that focuses on technologies and systems, which use financial, natural and social resources in an efficient, effective and economic manner over the long-term
Governance and stakeholder	Setting the highest standards of corporate governance and stakeholder engagement, including corporate codes of conduct and public reporting
Human	Managing human resources to maintain workforce capabilities and employee satisfaction through best-in-class organisational learning and knowledge management practices and remuneration and benefit programs

Source: Dow Jones [4]

In addition to normal corporate planning, sustainable development requires a business to think about:

- the extent to which it has already evolved in sustainable development terms (see Sect. 1.3.1)
- how products will be developed, redesigned, improved or deleted to be more sustainable
- specific environmental regulations that require immediate and longer term attention.

⇒ Read more about packaging environmental regulations in Chap. 4.

1.3.1 Stage of Evolution Towards Sustainable Development

The sustainability phase model [28] (see Table 1.4) and variants of it [29, 30] are useful tools to inform the business case for sustainability and a realistic strategy.

Benchmark your Stage of Evolution

Dunphy et al. [28] describe six stages of a business's evolution as it responds to the drivers and challenges of sustainable development. This evolution reflects the

1 Developing the Strategy

Table 1.4 Sustainability phase model

Phase	Actions
Rejection	All resources are exploited for immediate economic gain
	Potential constraints on corporate activities are actively opposed
Non-responsive	Business is unaware or ignorant of corporate ethic beyond financial gain
	Environmental consequences of activities are taken for granted or disregarded
Compliance	Safe, healthy workplace is emphasised
	Environmental issues that could lead to litigation or strong community action are avoided
	Business reacts to growing legal requirements and community expectations
Efficiency	Advantages of sustainable practices are recognised
	Reducing cost and increasing efficiency are emphasised
	First steps are taken to incorporate sustainability as an integral part of the business
Strategic pro-activity	Sustainable development is an important part of the corporate strategy
	Sustainable development is viewed as potential for competitive advantage
	Sustainability is embedded in the goal of long-term profitability
The sustaining corporation	Working for a sustainable world is the business's ideology
	It continues to pursue economic goals
	It pro-actively promotes sustainability values and practices
	It commits to facilitating ecological viability of the planet and contributes to just, equitable social practices

Source: Dunphy et al. [28, pp. 14–16]

business's understanding of, and engagement with, sustainability issues and how it adapts its corporate models and cultures in response.

The first two phases ('Rejection' and 'Non-responsive') have no impact on the corporate strategy. In phase three ('Compliance') and phase four ('Efficiency'), resources are allocated to address environmental issues, and environment is an adjunct to the corporate strategy. In the final phases ('Strategic Pro-activity' and 'The Sustaining Business'), sustainability activities are no longer under challenge, and sustainability thinking drives the corporate strategy and is integrated through all departments and levels of the business.

Position for the Future

In the 'Compliance' phase, the business case would reflect current compliance (for example, through waste disposal licenses, energy and water usage limits), emerging compliance issues, such as a carbon tax, and costs of non-compliance. Strategies would focus on meeting regulatory requirements. However, as a step towards sustainable development, the sustainability strategy would include goals and activities that position the business for the future. A portfolio of activities would be identified to meet current phase goals and reflect strategic intent to evolve to the next phase ('Efficiency'). Resources would be allocated not only to immediate compliance but also to activities that increase efficiency. Ideally, these will include projects that achieve both traditional targets, such as market share growth and cost reduction, and better environmental outcomes, such as carbon reduction. They would also develop the skills, competencies and organisational culture change required for the next phase.

Some of the changes a business must make to facilitate further sustainable development are highlighted in Table 1.5.

Table 1.5 Sustainability challenges, competencies and opportunities phase model

Phase		Actions
Stage 1	Viewing compliance as an opportunity	Ensure that compliance with norms becomes an opportunity for innovation
Stage 2	Making value chains sustainable	Increase efficiencies throughout the value chain
Stage 3	Designing sustainable products and services	Develop sustainable offerings or redesign existing ones
Stage 4	Developing new corporate models	Find new ways to deliver and capture value that change the basis of the competition
Stage 5	Creating next practice platforms	Question through the sustainability lens the dominant logic of business today

Source: Nidumolu et al. [30, pp. 60–61]

As a business succeeds in delivering 'win–win' outcomes for itself and the environment, it is likely to progress further up the evolutionary curve. At this point, sustainable development starts to be perceived not as a cost of doing business but as an enabler of better corporate performance. This process is demonstrated in the Procter & Gamble case study (Case Study 1.4).

> **Case Study 1.4 Evolution of Sustainability at Procter & Gamble**
>
> Procter & Gamble (P&G) are pioneers in the development of life cycle assessment (LCA) and have actively used LCA in their science-based approach to sustainable development. In 1956 P&G produced their first discussion paper on the environmental science of their packaging materials to inform their product development processes.
>
> In 2005, in response to the increased attention on sustainability by governments, non-government organisations and consumers, P&G undertook a review that identified sustainability as a new business opportunity.
>
> In 2007 a sustainability strategy was released. It focused on improving the environmental profile of their products and operations, engaging with employees, reaching out to external stakeholders and increasing social sustainability. Implementing the strategy increased efficiencies in manufacturing operations that reduced energy, greenhouse gas emissions, water and waste and provided significant cost savings.

(continued)

1 Developing the Strategy

Case Study 1.4 (continued)

> P&G also identified an important role for innovation in its sustainability strategy:
>
> '[S]ustainable innovation products … are defined as products with an improved environmental profile, where the improvements are significant and obvious. To qualify, a product needs to deliver at least a 10% improvement, across the lifecycle, in one of the key indicators (energy consumption, water consumption, total materials use for product or packaging, transport, or replacement of non-renewable with renewable resources), with no meaningful deterioration in any of the other indicators [1, p. 388].'
>
> In May 2010, P&G launched its supplier environmental sustainability scorecard to enhance supply chain collaboration, improve environmental performance and encourage the sharing of ideas and capabilities across the supply chain. Suppliers were required to report against four environmental impact indicators: energy, water, waste disposal and greenhouse gases. The information is used to calculate a total environmental footprint that is measured annually.
>
> In late 2010 P&G revised its sustainability strategy and set new goals. These include:
> - 'using 100% renewable or recycled materials for all products and packaging
> - having zero consumer waste go to landfill
> - designing products to delight consumers while maximising the conservation of resources
> - powering [its] plants with 100% renewable energy
> - emitting no fossil-based CO_2 or toxic emissions
> - delivering effluent water quality that is as good as or better than influent water quality with no contribution to water scarcity and
> - having zero manufacturing waste go to landfill' [18, p. 28].
>
> For the past ten years P&G has produced a sustainability report outlining its goals, activities and achievements.
> *Sources*: White [1], P&G [18], Mohan [19], P&G [20]

1.3.2 Sustainable Product Development

Different Approaches for Different Products

'Product life cycle management' is often used to inform marketing, product development and production strategies (see Fig. 1.6).

This approach identifies activities and allocates resources that reflect the age and market penetration of a product. For new products, marketing activities focus on awareness in order to create demand, there is little competition, and profitability is low. As sales increase, profitability improves and competition creates downward pressure on price. For mature products, marketing activities focus on brand differentiation and product/range diversification to maintain or increase market share, and there is intense competition and increased pressure to focus on cost reduction and efficiency. When sales of a mature product decline or stabilise at a lower level, profitability pressure is high, and cost reduction and efficiency strategies dominate.

Sustainability goals and activities also differ with a product's age and contribution to sales and profitability. Sustainable development should be embedded in the development of new products, while existing products should be reviewed and redesigned, improved or even deleted.

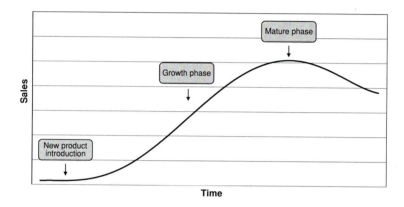

Fig. 1.6 Product life cycle management

Understand the Environmental Life Cycle

In order to set appropriate goals and priorities, it's important to understand a product's 'environmental life cycle'; that is, the environmental impacts associated with its production, use and disposal (Fig. 1.7).

Environmental impacts are the consequences of providing a particular product or service. An environmental life cycle map makes the impacts visible and can be used to estimate future costs and benefits of meeting environmental regulations and initiatives.

⇒ Read more about applying life cycle thinking in Chap. 5.

Adopt Life Cycle Thinking

Life cycle management and associated tools for life cycle assessment (commonly known by its acronym, LCA) are used to generate product environmental life cycle maps and identify improvement strategies. Their purpose is optimisation of the 'system' as a whole, which requires supply chain partnerships to achieve better

1 Developing the Strategy

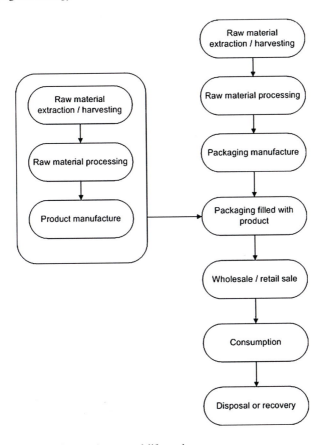

Fig. 1.7 Simplified product environmental life cycle map

and long term environmental benefits that avoid creating new impacts or 'burden-shifting'.

> **Burden shifting** occurs when there is a transfer of environmental impacts from one point or business in the supply chain to another. This may happen for example, if a package is light-weighted by replacing a recyclable material with a non-recyclable material that has no waste collection and reprocessing facilities. In this case, the benefits of the material-saving must be evaluated against the increased amount of post-consumer waste generated that becomes landfill.

As a business develops life cycle management, it progresses to a prevention-oriented approach that leads to innovation right across the business and supply chain [31, p. 56]. A better understanding of the product-packaging system means that environmental impacts or 'hotspots' are routinely identified and addressed through preventative action. In leading sustainability businesses, life cycle thinking leads to consideration of environmental alongside social and economic impacts (the triple bottom line).

1.4 Packaging's Role in Sustainable Development

The application of life cycle thinking is still not widespread or uniformly applied within the business community, let alone the broader community. Packaging regulations have tended to focus on waste reduction and recycling at the expense of other impacts, although this is changing (see Chap. 4). With a better understanding of the role that packaging can play in sustainable development, government policies and industry initiatives are now more focused on the use of LCA and sustainability metrics to inform the use and design of packaging (Fig. 1.8).

Fig. 1.8 The evolution of packaging concerns and regulations

First There was Litter and Waste

Regulatory and political action for many years responded to strong public opinion that packaging is 'bad for the environment', particularly due to litter, waste generation, marine life impacts and examples of over-packaging. Pressures on the availability of land for waste disposal and increasing costs of waste management also led to regulations that targeted packaging waste reduction, reuse and recovery.

Then There was 'Sustainable Packaging'

A new approach is becoming more evident globally in government policy and industry self-regulation. Initiatives over the past decade have promoted 'sustainable packaging' rather than waste reduction, through the development of guidelines, standards and scorecards that:

⇒ Read more about designing packaging for sustainability in Chap. 2

- recognise the role of packaging and its interaction with the product

- encourage the application of LCA to inform packaging design
- promote the inclusion of high level design principles for packaging such as efficiency, renewable materials and recovery after use.

One of the first attempts to define sustainable packaging was made by the Sustainable Packaging Alliance in Australia. They identified four main principles (see Case Study 1.5) and suggested that the packaging design process should consider:

- the entire life cycle of the package from raw materials through to disposal
- interactions between the package and the product so that the environmental impacts of the product-packaging system as a whole are minimised
- triple bottom line impacts and benefits of packaging.

The latest development of this work into a decision-support tool for packaging design is described in Chap. 2.

Case Study 1.5 Sustainable Packaging Alliance's Definition of Sustainable Packaging

Four principles for packaging design were identified—'effective', 'efficient', 'cyclic' and 'clean'—and key performance indicators proposed that were expressed in terms such as 'reduces product waste' and 'improves functionality' to highlight the fact that sustainability is a process of continuous improvement rather than a pre-determined endpoint.

Source: Lewis et al. [8]

A similar definition that included new elements, particularly about renewable energy and materials, was developed by the Sustainable Packaging Coalition in the United States in 2005 to ensure that 'all parties are working towards the same vision' [32]. This has recently been developed into a series of indicators and metrics [33] to help businesses measure progress against their sustainable packaging criteria (see Case Study 1.6).

⇒ Read more about packaging sustainability metrics in Sect. 8.2.8

Some retailers and brand owners have also developed criteria and metrics for sustainable packaging. The most influential of these is Walmart, which in 2009 announced a supplier sustainability assessment, and plans to measure the environmental impacts of products.

⇒ Read more about Walmart's packaging evaluation tool in Sect. 7.3.2

> **Case Study 1.6 Sustainable Packaging Coalition's Criteria for Sustainable Packaging**
>
> Sustainable packaging:
> - is beneficial, safe and healthy for individuals and communities throughout its life cycle
> - meets market criteria for performance and cost
> - is sourced, manufactured, transported, and recycled using renewable energy
> - maximises the use of renewable or recycled source materials
> - is manufactured using clean production technologies and best practices
> - is made from materials healthy in all probable end of life scenarios
> - is physically designed to optimise materials and energy
> - is effectively recovered and utilised in biological and/or industrial cradle-to-cradle cycles.
>
> *Source*: Sustainable Packaging Coalition [10, p. 1]

Now a Global Approach

These developments prompted The Consumer Goods Forum, an association representing over 650 global retailers, manufacturers, service providers and other stakeholders across 70 businesses, to initiate the Global Packaging Project (GPP) in November 2008.

The GPP aims to provide 'a common language to enable intelligent and informed discussion between our businesses on sustainable packaging' [34, p. 2]. The framework includes principles of sustainable development (environmental, social and economic aspects and a life cycle approach) and proposals for sustainable packaging that is:
- designed holistically with the product to optimise environmental performance
- made from responsibly sourced materials
- able to meet market criteria for performance and cost
- manufactured using clean production technologies
- efficiently recoverable after use
- sourced, manufactured, transported and recycled using renewable energy.

The principles are supported by detailed indicators and metrics [35], which are designed to 'facilitate understanding and communication about the relative sustainability of packaging' (p. 6).

Another global initiative is the development of international standards on packaging sustainability by the International Standards Organisation (ISO), which are intended to 'facilitate global trade and a harmonised approach to environmental protection' [36]. The draft standards were circulated for feedback in late 2010 and are expected to be finalised in 2012.

1 Developing the Strategy

A guide for corporate decision-makers has also been developed by EUROPEN (the European Organization for Packaging and the Environment) and ECR Europe ('Efficient Consumer Response') [37]. It provides an overview of packaging sustainability issues and practical steps to optimise a packaging sustainability strategy.

1.5 Developing the Business Case

A business case to invest in packaging for sustainability is necessary to clarify goals and targets and obtain the organisational commitment and resources required to achieve them. The business case will reflect the phase of evolution of the business and the business's products, services and corporate model. However, there are some common drivers.

1.5.1 Corporate and Brand Positioning

Stakeholder Assessment of Sustainability Performance is Increasing

As businesses evolve to integrate sustainable development, new ways of assessing and reporting their long term potential and performance are also evolving. Public reporting of sustainability performance (voluntary and mandatory) is increasing, and third-parties are reviewing, assessing and reporting on corporate performance for a range of purposes including investment decisions.

An example of external reporting is the Dow Jones Sustainability Indexes [38]. The indexes are based on the principle that corporate sustainability is an investable concept (see Case Study 1.7). They provide benchmarks on sustainability

> **Case Study 1.7 Benchmarking Sustainability Performance: The Dow Jones Sustainability Indexes**
>
> The Dow Jones Sustainability Indexes follow a best-in-class approach and include sustainability leaders from each industry on a global and regional level respectively. The annual review of the Dow Jones family is based on a thorough analysis of corporate economic, environmental and social performance, assessing issues such as corporate governance, risk management, branding, climate change mitigation, supply chain standards and labour practices. It accounts for general as well as industry-specific sustainability criteria for each of 57 industries defined according to the Industry Classification Benchmark. *Source*: Dow Jones Sustainability Indexes [9]

performance and are published as global and regional reports (European, Eurozone, North American, United States, Asia/Pacific).

An important feature of sustainability reporting and indexes such as the Dow Jones is that they allow comparisons not only with a business's past performance but also with businesses in the same region or sector. In the future, sustainability performance is likely to be assessed more frequently and intensely in relation to competitors.

Sustainability Performance Must be Communicated

Sustainability performance is becoming an important element of the marketing strategy for a business and its products and services. Performance is often communicated through sustainability reports that include strategies, targets, achievements and challenges. These reports should provide a balanced and reasonable representation of the sustainability performance—including both positive and negative contributions [39].

⇒ Read more about marketing and communicating sustainability in Chap. 3.

1.5.2 Supply Chain Requirements

Retailers are Driving Sustainable Development

In recognition of their unique position to influence suppliers and consumers, retailers across the globe are driving changes in packaging to support sustainability (see Table 1.6). Many are also involved in national or regional initiatives, such as 'The Courtauld Commitment' in the United Kingdom and the 'Retail Forum for Sustainability' in the European Union.

Sustainability is Becoming a Pre-requisite for Doing Business

Brand owners are also driving changes in packaging to support their sustainable development goals. Nike, for example, requires its suppliers to comply with its packaging restricted substances list and design requirements (see Case Study 1.8).

Environmental guidelines such as those used by Nike are often supported by a supplier questionnaire or 'scorecard' to measure performance. See the examples of Walmart (see Sect. 7.3.2) and McDonald's (Case Study 1.9).

'Green procurement' is also being adopted by government agencies and other large organisations to reduce the environmental impacts of their supply chains, and this often includes packaging. High profile public events are starting to specify 'closed loop' waste management systems that require food service providers to recover packaging and recycling bins to be available for all food and packaging waste. The Sydney Olympics achieved a 70% recovery rate by ensuring that all packaging was either recyclable or biodegradable [46].

Table 1.6 Examples of packaging initiatives driven by retailers

Organisation or Company	Packaging initiatives
European retailers	The retail forum for sustainability is a voluntary initiative established under the auspices of the European Commission. It aims to exchange best practices and identify opportunities or barriers that might further or hinder progress towards more sustainable production and consumption. Members commit to a series of environmental actions under the retailers' environmental action programme. Twenty retailers and seven retail associations have joined. [40]
UK retailers	Forty-six retailers, brandowners and suppliers have signed the second stage of the Courtauld Commitment in the United Kingdom. Signatories agree to support the Waste and Resources Action Programme (WRAP) to: • reduce the carbon impact of grocery packaging by 10% through reduced packaging weight, increased recycling and increased recycled content • reduce household food and drink wastes by 4% • reduce grocery product and packaging waste in the grocery supply chain by 5% (solid and liquid wastes). [41]
Tesco (UK)	In 2007, Tesco announced a plan to cut packaging on own-label and branded products by 25% within three years. The company also pledged to put 'carbon labels' on all their products to provide information on their carbon footprint from production through to consumption. [42]
Sainsbury (UK)	Sainsbury has a target of reducing own-brand packaging weight, relative to sales, by 33% by 2015 against a 2009 baseline. It will also be working with all stakeholders to make more of its packaging recyclable. [43]
Marks & Spencer (UK)	In January 2007, Marks & Spencer announced as part of their 'Plan A' environmental program that it would cut its use of non-glass packaging by 25% by 2012. Other initiatives included plans to increase use of more sustainable raw materials for packaging such as recycled materials and Forest Stewardship Council certified boards. [14]
Walmart (US)	In September 2006, Walmart announced a 5% packaging reduction target to be achieved by 2013. An environmental scorecard is used to evaluate the sustainability performance of all suppliers and to encourage continuous improvement. [7]
Carrefour (France and International)	Carrefour's commitments include deploying 'best practices' for the design of their packaging to ensure a reduction of waste at source, and raising awareness of own-brand products to the importance of environmental considerations in the design of products and packaging. Over 10 years to 2009, the company saved 13,000 tonnes of material per year by optimising the design of own-label product packaging. [44]
The Warehouse (NZ)	The Warehouse's Packaging Guide states that: 'The Warehouse requires that packaging conforms to our environmental principles of reducing unnecessary packaging, facilitating the re-use or recycling of packaging materials and restricting or eliminating particular types of packaging materials'. It includes an environmental design checklist based on the New Zealand Packaging Accord. [45]

Case Study 1.8 Packaging Restricted Substances List and Design Requirements: Nike

Nike's packaging specifications support the company's commitment to sustainable development and, in some cases, include bans or restrictions on the use of materials and constituents that Nike believes may be a risk to human health or the environment.

The specifications go well beyond compliance. The company has a target to reduce point-of-purchase packaging by 30%, and to help achieve this goal it imposes limits on the amount of empty space and the number of layers in consumer packaging. Minimum levels of post-consumer recycled content are specified for different material types, and suppliers must certify that their packaging is recoverable through one of three routes: material recovery, energy recovery or organic recovery.

Suppliers must provide all of the relevant information and reference documents to demonstrate compliance with the packaging requirements on a regular basis through the routine technical file completion process.

Source: Nike [11]

Case Study 1.9 Packaging Sustainability Performance Expectations: McDonald's

McDonald's is implementing a global packaging scorecard to better inform decisions about packaging. It focuses on six priorities:
- minimising weight
- maximising use of recycled materials
- preference for raw materials from third-party certified sources
- minimising the quantity of harmful chemicals used in production
- reducing CO_2 and other greenhouse gas emissions
- maximising end-of-life options such as recycling.

Implementation of the scorecard will involve close collaboration with and support from suppliers. McDonald's website states:

> 'Our suppliers will be held accountable for achieving mutually established waste reduction goals, as well as continuously pursuing sound production practices that minimise environmental impact. Compliance with these policies will receive consideration with other business criteria in evaluating both current and potential McDonald's suppliers'.

Source: McDonald's [15, 16]

1.5.3 Solid Waste Reduction

It is essential that businesses operating within the packaging supply chain, whether they are users of packaging or suppliers of materials or packaging products, integrate packaging waste reduction within their broader sustainable development goals. Businesses need to be aware of the waste and recycling targets of their relevant jurisdictions and how these will impact their operations. Packaging decisions can no longer be conducted without an understanding of their impact on waste generation and the associated strategies to reduce impacts including increased efficiency, reuse, recycling and use of recycled materials. A business case can show that a focus on waste reduction will improve efficiencies, reduce current costs and avoid future costs.

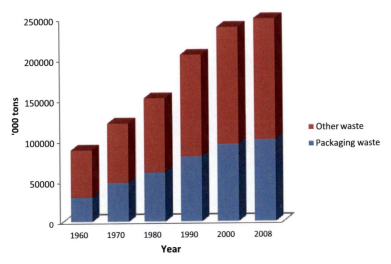

Fig. 1.9 Packaging waste generation as a proportion of municipal solid waste (MSW), United States, 1960–2008. *Source*: Based on United States EPA [50, p. 19]. Municipal waste generation is before materials recovery or combustion. It excludes construction and demolition waste and certain other wastes

Packaging Use Must be Decoupled from Packaging Waste

Consumption of packaging continues to increase, due to a range of social, demographic and economic trends. These include increasing populations and incomes, particularly in developing countries, as well as lifestyle trends such as the increasing consumption of convenience and takeaway meals. In its 2010 assessment report of progress under the European Packaging Waste Directive, the European Environment Agency reported that between 1998 and 2006 packaging consumption and waste generation continued to increase, from 160 to 179 kg per capita. There was, however, a slight decoupling of packaging waste generation (15.5% growth) from gross domestic product (almost 20% real growth) [47].

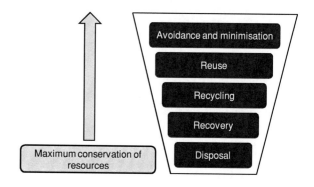

Fig. 1.10 Waste hierarchy

Globally, there will be continued and increasing pressure to reduce all forms of waste and identify and implement strategies that decouple waste generation from economic activity [48, 49].

Packaging Waste is Increasing

The amount of solid waste generated in developed countries has increased greatly over the past 50 years. Waste generation increased at a steady rate in the United States from 1960 but started to stabilise in the 2000s. Waste generation solely from packaging has increased in absolute terms and as a percentage of total municipal solid waste (MSW). Between 1960 and 2008, it increased from 27 million tons to 77 million tons, and from 34% of MSW in 1960 to 40% in 2008 [50, p. 19] (Fig. 1.9).

Waste reduction strategies adopted in different regions, countries or states are generally based on the waste reduction hierarchy (an example is shown in Fig. 1.10), although specific regional issues and current performance lead to variations in principles and priorities (see Chap. 4). Despite differences in approaches, the underlying sustainable development goals are to:
- decrease the use of packaging
- divert packaging from landfill
- increase recycling
- establish reliable data collection protocols that allow performance to be measured and reported.

Packaging Recovery Rates are Improving—Slowly

The overall recovery rate for MSW in the United States increased from 6.4% in 1960 to 33.2% in 2008 but this has not had a significant impact on the amount of waste requiring disposal (see Fig. 1.11). As recovery rates for packaging materials have improved, policy makers and waste management companies have shifted their focus to the organic fraction of the waste stream, which can be recovered through technologies such as composting and anaerobic digestion. Almost 13% of all MSW in the United States is from food scraps. (See Fig. 1.12).

1 Developing the Strategy

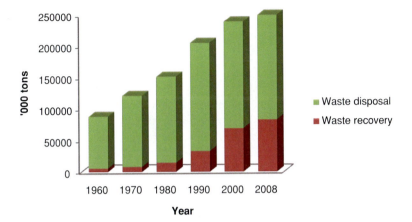

Fig. 1.11 Waste disposal and recovery, United States, 1960–2008. *Source*: Based on United States EPA [50, pp. 2–3]. Municipal waste generation is before materials recovery or combustion. It excludes construction and demolition waste and certain other wastes

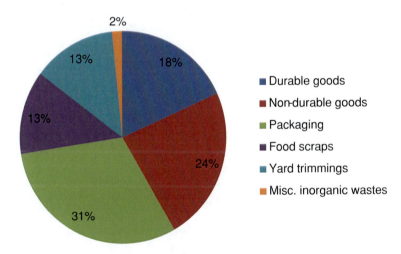

Fig. 1.12 Categories of products generated in municipal solid waste, United States, 2008. *Source*: Based on United States EPA [50, p. 10]. Municipal waste generation is before materials recovery or combustion. It excludes construction and demolition waste and certain other wastes

Up to 50% of Packaging is Not Recycled

Recycling rates for packaging in developed countries have improved significantly since the 1960s, but in many countries between one-third and one-half of all packaging consumed is not recycled (see Table 2.14 in Chap. 2). In countries such as the United States, Canada, the United Kingdom and Australia, packaging that is not recycled is disposed to landfill, while in many European and Asian countries, waste is incinerated.

Packaging is Lost in Litter

Packaging materials are 'lost' through disposal as litter (Photo 1.2). Litter remains problematic in developed countries despite decades of community education and improved infrastructure for waste collection. It is an even bigger problem in developing countries, where waste collection facilities are inadequate to cope with increasing amounts of non-biodegradable packaging.

Photo 1.2 Packaging litter in Yarra River, Melbourne (Photo: Karli Verghese)

Packaging can Reduce Product Waste

Waste and landfill reduction have played a significant role in shaping environmental policies for packaging. With increasing use of life cycle thinking, the role of packaging to reduce *product* waste will also become important, but reducing *packaging* waste is still necessary.

1 Developing the Strategy

1.5.4 Resource Efficiency

For many businesses, the business case for investing in packaging for sustainability rests firmly on its potential to reduce packaging and supply chain costs (see Table 1.7).

Resource Efficiency Reduces Costs

Eco-efficiency strategies reduce the environmental impacts as well as costs of packaging. Examples include eliminating unnecessary layers, reducing the thickness of packaging (down-gauging), reducing void space, switching to bulk or reusable packaging, and redesigning the product. These initiatives not only reduce materials consumption and waste from packaging manufacture; they generally also cut energy and water consumption, waste and emissions in the packaging supply

Table 1.7 Cost of packaging waste

Cost item	Description	Who bears the cost
Packaging material	The cost of packaging may be a small percentage of the product cost (e.g., expensive electronic equipment) or a relatively high percentage (e.g., bottled water)	Product manufacturer
Handling and labour costs	These include the labour and equipment costs associated with packaging the product (e.g., loading and wrapping pallets), warehousing and unpacking the product	Product manufacturer, retailer/distributor or customer
Shipping (freight) costs	Cost-efficient transport requires optimum use of available space in transport vehicles	Product manufacturer, retailer/distributor customer
Storage costs	These include the cost of storing empty packaging before use, the impacts of packaging design on the efficiency of storage in warehouses, and the costs of storing used packaging (e.g., baled cardboard or film ready for collection)	Product manufacturer retailer/distributor customer
Disposal or recycling costs	These include handling and labour costs of managing waste packaging (e.g., removing waste, baling cardboard) as well as collection and disposal charges	Retailer/distributor or customer
Reuse costs	The costs associated with reusable packaging include labour, freight, cleaning and repairing containers	Product manufacturer
Damaged products	Packaging is used to protect products in transit. If it fails, the cost of the lost product can be many times the cost of the packaging itself	Product manufacturer

Source: Verghese and Lewis [52, p. 4386]

chain. Using less material reduces all of the impacts associated with raw materials. More efficient designs, such as smaller pack sizes and improved use of pallets, also reduce transport impacts by increasing the number of sales units that can be transported in a single truck or container.

There are many examples of cost savings. Walmart's collaborative program with suppliers, which includes a target to reduce packaging by 5% by 2013, is already achieving results (see Case Study 1.10). Businesses that have signed the Singapore Packaging Agreement have reaped savings of US$8.34 million and eliminated 4,520 tonnes of packaging waste [51].

> **Case Study 1.10 Economic and Environmental Benefits: Packaging Waste Reduction at Walmart**
>
> Walmart aims to reduce its packaging by 5% by 2013. Conservative estimates of the annual savings that this will achieve include:
> - 667,000 metric tons of CO_2 not emitted into the atmosphere
> - 213,000 trucks off the road annually
> - 66.7 million gallons of diesel fuel saved.
>
> For example, when it partnered with private label suppliers to improve the packaging of its Kid Connection toy line, it was able to use 497 fewer shipping containers and generated savings of more than US$2.4 million per year.
> *Source*: Wal-Mart [7]

1.5.5 Climate Change

One of the most significant issues for a business to address in its sustainability strategy is climate change. Serious and irreversible risks of climate change associated with global emissions of greenhouse gases such as carbon dioxide (CO_2) have become a major concern to governments around the world. This can be attributed to several events, including publication of the 'Stern Review' on the economics of climate change [53], the most recent reports by the Intergovernmental Panel on Climate Change [54] and the release of Al Gore's film *An Inconvenient Truth*.

It is not possible in this book to give anything but a brief summary of climate change science and policies. With that in mind, it is true to say that current scientific evidence indicates that climate change is occurring, and that greenhouse gas emissions from human activities are considered to be the main cause [55]. Average air temperatures rose by 0.7°C over the 100 years to 2009, and models predict a long term warming of about 3°C (within an uncertainty range of 2–4.5°C) [56].

Emissions of carbon dioxide, methane and nitrous oxide began to rise 200–300 years ago, around the time of industrialisation, and have accelerated in more recent years. Global greenhouse gas emissions due to human activities increased 70% between 1970 and 2004 [55, p. 5].

Efforts to develop a co-ordinated global response to climate change have been patchy. The Kyoto Protocol, an international agreement linked to the United Nations Framework Convention on Climate Change, commits 37 industrialised countries and the European Community to meet an average 5% reduction target for greenhouse gas reductions by 2012 on 1990 levels. The current agreement ends in 2012, and negotiators have so far failed to agree on new targets. However, individual countries are well-advanced in the design or implementation of policies to promote energy efficiency, alternative energy sources and reduced emissions from non-energy sources [57, pp. 217–219]. For example, China is aiming to achieve a 40–45% reduction in the emissions intensity of its gross domestic product by 2020 compared to 2005 and gain 15% of its energy from renewable sources. The European Union has a target of 20% reduction in energy consumption by 2020 compared to 'business as usual' projections. Many developed countries have introduced an emissions trading scheme or carbon tax to promote energy efficiency and emissions reduction or are planning to do so.

Take Action Now to Avoid Future Costs

The packaging supply chain, like all sectors of the economy, needs to take action to minimise future costs arising from climate change regulations and initiatives. Individual businesses can achieve this by improving energy efficiency in operations and transport, switching to alternative fuels for transport, investing in renewable energy and 'carbon offsets' and by redesigning products to reduce emissions over their life cycle.

These actions can contribute to the business case for packaging sustainability, for example by:
- improving corporate and brand positioning if a business decides to position itself or its products as 'carbon neutral'
- reducing costs of manufacturing and transport
- meeting customer expectations for reduced emissions in the supply chain.

UK supermarket chain Tesco announced in 2007 that it intended to put 'carbon labels' on all of its products to provide information on its carbon footprint, from production through to consumption [42]. Many businesses are aiming to become 'carbon neutral' by generating sufficient renewable energy for their needs or buying carbon offsets (see Sect. 3.5.8). For example, Taylors Wines in Australia has measured and offset all of the carbon emissions associated with its Eighty Acres range [58]. The data was collected through an LCA and independently verified. The initiative was recognised by industry peers when the company received an award for best 'Green Brand Launch' by industry magazine *Drinks Business* in London in 2010 [59].

1.6 Conclusion

This chapter has outlined issues to be considered in the business case for investing in packaging for sustainability.

A well-constructed corporate strategy will clearly identify the role of packaging in sustainability based on:

- understanding the environmental life cycle of products and their packaging
- the impact and requirements of any packaging-specific environmental regulations
- the impact and requirements of current and future environmental regulations throughout the supply chain.

However, achieving sustainable development goals through packaging requires new knowledge, skills and tools as well as changes to existing business processes and work flows. The business needs to be capable of addressing each of the elements featured in Fig. 1.13, most of which are outlined in specific chapters of this book: regulations and life cycle thinking (Chaps. 4 and 5); packaging design and materials selection (Chaps. 2 and 6); and marketing and communication (Chap. 3). Chapter 8 outlines business units and processes involved in packaging for sustainability and strategies for creating or accelerating packaging's contribution to sustainable development.

There are now many decision-support tools available to embed sustainable development into packaging decision-making; for example, product positioning, packaging design and procurement (see Chap. 7). Throughout this book, we outline and demonstrate some of them together with case studies.

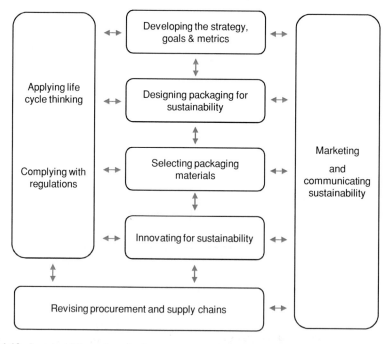

Fig. 1.13 Sustainability and packaging

References

1. White P (2009) Building a sustainability strategy into the business. Corp Gov 9(4):386–394
2. World Business Council for Sustainable Development (2006) WBCSD learning tool helps companies to adopt, implement and integrate eco-efficiency. http://www.wbcsd.org/plugins/DocSearch/details.asp?type=DocDet&ObjectId=MTgwMjc (cited 13 December 2010)
3. McDonough W, Braungart M (2002) Cradle to cradle: remaking the way we make things. North Point Press, New York
4. Dow Jones (2010) Sustainability Index. Corporate sustainability. http://www.sustainability-index.com
5. Busser S, Jungbluth N (2009) The role of flexible packaging in the life cycle of coffee and butter. Int J LCA 14(Suppl 1):S80–S91
6. United Nations Department of Economic and Social Affairs (UNDESA) and the United Nations Environment Programme (UNEP) (2010) Ten year framework of programmes on sustainable consumption and production, revised draft 7 April 2010. Marakesh Process Secretariat—United Nations Department of Economic and Social Affairs and United Nations Environment Programme
7. Walmart (2008) Sustainable packaging fact sheet: Walmart is taking the lead on sustainable packaging. http://nbis.org/nbisresources/packaging/walmart_packaging_factsheet.pdf (cited 7 August 2011)
8. Lewis H, Sonneveld K, Fitzpatrick L, Nicol R (2002) Towards sustainable packaging, Discussion paper. http://www.sustainablepack.org/database/files/filestorage/Towards%20Sustainable%20Packaging.pdf (cited 14 May 2009)
9. DJSI (2008) SAM, Dow Jones Indexes and Stoxx Ltd. announce results of Dow Jones sustainability indexes review. http://www.sustainability-index.com/djsi_pdf/news/PressReleases/SAM_PressReleases_080904_Review08.pdf (cited 20 September 2010)
10. Sustainable Packaging Coalition (2005) Definition of Sustainable Packaging, Version 1. http://www.sustainablepackaging.org/pdf/Definition%20First%20Page.pdf (cited 2 April 2009)
11. Nike (2010) Packaging restricted substances list and design requirements. http://www.nikebiz.com/responsibility/considered_design/documents/Nike_PRSL.pdf (cited 7 August 2011)
12. The Consumer Goods Forum (2009) A global language for packaging and sustainability. A framework and a measurement system for our industry. The Consumer Goods Forum, Paris
13. Grant T, Beer T (2008) Life cycle assessment of greenhouse gas emissions from irrigated maize and their significance in the value chain. Aust J Exp Agric 48(3):375–381
14. Marks and Spencer (2010) Your M&S: how we do business 2010. London
15. McDonald's (2011) McDonald's global environmental commitment—effectively managing solid waste. http://www.aboutmcdonalds.com/mcd/csr/about/sustainable_supply/resource_conservation/global_environmental.html (cited 9 March 2011)
16. McDonald's (2011) Green packaging design—EcoFilter. http://www.aboutmcdonalds.com/mcd/csr/about/sustainable_supply/resource_conservation/sustainable_packaging.html?DCSext.destination=http://www.aboutmcdonalds.com/mcd/csr/about/sustainable_supply/resource_conservation/sustainable_packaging.html (cited 9 March 2011)
17. Gazulla C, Raugei M, Fullana-I-Palmer P (2010) Taking a life cycle look at crianza wine production in Spain: where are the bottlenecks? Int J Life Cycle Assess 15(4):330–337
18. P&G (2010) Now & for Generations to Come. 2010 Sustainability Overview. Procter and Gamble
19. Mohan AM (2010) P&G elucidates new sustainability strategy. 19 October 2010]; http://www.greenerpackage.com/corporate_strategy/pg_elucidates_new_sustainability_vision
20. P&G (2010) P&G Supplier Environmental Sustainability Scorecard. http://www.youtube.com/watch?v=5rSrnT2E1dI&feature=related and http://www.pgsupplier.com/sites/default/files/Rick_Hughes_Scorecard_P_G_V1.wmv
21. Helen Lewis Research (2009) Quickstart 3. Plastics and Chemical Industry Association and Sustainability Victoria, Melbourne

22. Commission of the European Communities, (1994) European Parliament and Council directive 94/62/EC of 20 December 1994 on packaging and packaging waste
23. United Nations (1987) Report of the World Commission on Environment and Development: Our Common Future, Transmitted to the General Assembly as an Annex to document A/42/427—Development and international cooperation: environment. UN WCED, Geneva
24. Elkington J (1997) Cannibals with forks: The Triple Bottom line of 21st century business. Capstone Publishing Limited, UK
25. Commission of the European Communities (2008) Communication from the Commission on sustainable consumption and production and sustainable industrial policy action plan. Brussels
26. Verfaillie H, Bidwell R (2000) Measuring eco-efficiency: a guide to reporting company performance. World Business Council for Sustainable Development (WBCSD)
27. Alston K (2010) Cradle to cradle design: Positive sustainability agenda for products and packaging; the limitations of eco-efficiency. Address to Packaging for Tomorrow, 7 December 2010. 3 Pillars Network, Melbourne
28. Dunphy D, Griffiths A, Benn S (2003) Organisational change for corporate sustainability. Routledge, London
29. Willard B (2010) The 5-stage Sustainability Journey. http://sustainabilityadvantage.com/2010/07/27/the-5-stage-sustainability-journey/ (cited 20 September 2010)
30. Nidumolu R, Prahalad CK, Rangaswami MR (2009) Why sustainability is now the key driver of innovation. Harvard Business Review. 2009 (September): pp 56–64
31. Balkau F, Sonnemann G (2010) Managing sustainability performance through the value-chain. Corp Gov 10(1):46–58
32. Sustainable Packaging Coalition (2005) Definition of sustainable packaging, version 1.0. http://www.sustainablepackaging.org/about_sustainable_packaging.asp (cited 19 October 2007)
33. Sustainable Packaging Coalition (2009) Sustainable packaging indicators and metrics. Sustainable Packaging Coalition (SPC), Charlottesville, VA
34. The Consumer Goods Forum (2010) A global language for packaging and sustainability. Paris
35. The Consumer Goods Forum (2011) Global protocol on packaging sustainability 2.0: draft for consultation
36. Linde A (2010) Packaging and sustainability—developing international standards on packaging and the environment. Address to Packaging for Tomorrow, 7 December 2010. Melbourne
37. EUROPEN and ECR Europe (2009) Packaging in the sustainability agenda: a guide for corporate decision makers. The European Organisation for Packaging and the Environment (EUROPEN) and ECR Europe, Brussels
38. Dow Jones Sustainability Indexes (2010) Dow Jones sustainability indexes. http://www.sustainability-index.com/ (cited 9 March 2011)
39. GRI (2006) Sustainability reporting guidelines. http://www.globalreporting.org/NR/rdonlyres/ED9E9B36-AB54-4DE1-BFF2-5F735235CA44/0/G3_GuidelinesENU.pdf (cited 27 March 2009)
40. Commission of the European Communities (2010) The retail forum. http://ec.europa.eu/environment/industry/retail/about.htm (cited 19 December 2010)
41. WRAP (2010) The Courtauld Commitment. Waste & Resources Action Program (WRAP): Banbury, Oxon, UK
42. Leahy T (2007) Tesco, carbon and the consumer: Address to a Joint Forum for the Future and Tesco event. http://www.tesco.com/climatechange/speech.asp (cited 29 October 2007)
43. J Sainsbury plc (2010) Sainsbury's Corporate Responsibility Report, London, UK
44. Carrefour Group (2009) At the heart of life: 2008 sustainability report. Levallois-Perret Cedex, France
45. The Warehouse Limited (2002) Packaging Guide. http://www.thewarehouse.co.nz/is-bin/intershop.static/WFS/TWL-Site/TWL-B2C/en_NZ/content/Suppliers/Terms%20of%20Trade/Packaging_Guide_-_v1-02.pdf (cited 7 August 2011)

46. Closed Loop Recycling (2010) Sydney 2000 Olympic games case study. http://www.closedloop.com.au/casestudy.php?id=8 (cited 14 December 2010)
47. European Environment Agency (2010) Generation and recycling of packaging waste. http://www.eea.europa.eu/data-and-maps/indicators/generation-and-recycling-of-packaging-waste/generation-and-recycling-of-packaging-1 (cited 19 December 2010)
48. Commission of the European Communities (2008) Directive 2008/98/EC of the European Parliament and of the Council of 19 November 2008 on waste and repealing certain other Directives, Brussels
49. Environment Protection and Heritage Council (2009) National waste policy: less waste, more resources. Environment Protection and Heritage Council (EPHC), Adelaide
50. United States Environment Protection Agency (2009) Municipal solid waste generation, recycling and disposal in the United States: facts and figures, Washington
51. Shafawi M, Loh D (2010) Singapore Packaging Agreement helps signatories save S$8.34m. Singapore News, 25 October. http://www.channelnewsasia.com/stories/singaporelocalnews/view/1089154/1/.html (cited 7 August 2011)
52. Verghese K, Lewis H (2007) Environmental innovation in industrial packaging: a supply chain approach. Int J Prod Res 45(18/19):4381–4401
53. Stern N (2006) Stern Review Report on the Economics of Climate Change. http://www.hm-treasury.gov.uk/independent_reviews/stern_review_economics_climate_change/stern_review_report.cfm (cited 29 October 2007)
54. Intergovernmental Panel on Climate Change (2007) Climate change 2007: impacts, adaptation and vulnerability. Working Group II contribution to the Intergovernmental Panel on Climate Change fourth assessment report. Summary for policy makers. http://www.ipcc.ch/SPM6avr07.pdf (cited 29 October 2007)
55. Pachauri RK, Reisinger A (eds.) (2007) Climate change 2007: synthesis report. Intergovernmental Panel on Climate Change (IPCC), Geneva
56. Australian Academy of Science (2010) The science of climate change: questions and answers, Canberra
57. Australian Government: Prime Minister's Task Group on energy efficiency (2010) Report of the Prime Minister's Task Group on energy efficiency, Canberra
58. Taylors Wines (2009) Taylors Eighty Acres 100% carbon neutral based on an ISO 14044 compliant life cycle assessment model—a world first. http://www.taylorswines.com.au/winenews/94 (cited 15 February 2010)
59. Taylors Wines (2010) Taylors Wines declared world's best "Green Brand Launch" at the Drinks Business Green Awards 2010. http://www.taylorswines.com.au/winenews/100 (cited 15 February 2010)

Chapter 2
Designing for Sustainability

Helen Lewis

Abstract Packaging for sustainability initiatives can be identified through strategic and operational planning processes, but they are primarily delivered through design. Packaging design is an already complex process that considers many aspects of marketing, packaging function and cost. Designing for sustainability adds further complexity to this process. To integrate these new requirements into the packaging design process efficiently and effectively we propose the use of a packaging sustainability framework. A framework is presented that applies a triple bottom line approach to packaging design based on four design principles: effectiveness (fit for purpose), efficiency (efficient use of materials, energy and water), cyclic material flows (renewable/recyclable materials and minimal waste) and safety (non-polluting and non-toxic). Each principle is outlined, and practical design strategies and case studies are provided to demonstrate their application.

Contents

2.1	Introduction	43
2.2	Designing Packaging for Sustainability	44
2.3	Packaging Sustainability Framework	47
	2.3.1 Effective Packaging	48
	2.3.2 Efficient Packaging	51
	2.3.3 Cyclic Packaging	53
	2.3.4 Safe Packaging	56
2.4	Applying the Packaging Sustainability Framework	59
	2.4.1 Designing for Effectiveness	61
	2.4.2 Designing for Efficiency	64
	2.4.3 Designing Cyclic Packaging	70
	2.4.4 Designing for Safety	86

H. Lewis (✉)
Centre for Design, RMIT University, GPO Box 2476, Melbourne, VIC 3001, Australia
e-mail: lewis.helen@bigpond.com

K. Verghese et al. (eds.), *Packaging for Sustainability*,
DOI: 10.1007/978-0-85729-988-8_2, © Springer-Verlag London Limited 2012

2.5 Selecting Materials 95
2.6 Conclusion 102
References... 102

Figures

Figure 2.1 The design approach 43
Figure 2.2 The four sustainable packaging principles 47
Figure 2.3 Energy for one person s weekly consumption of food 67
Figure 2.4 Comparing the impacts of over-packaging and under-packaging 68
Figure 2.5 The decision-making process for degradable packaging 85

Tables

Table 2.1   Functions of packaging 45
Table 2.2   Examples of 'win-win' packaging sustainability strategies 46
Table 2.3   Potential triple bottom line benefits of effective packaging 48
Table 2.4   Potential triple bottom line benefits of efficient packaging 51
Table 2.5   Potential triple bottom line benefits of cyclic packaging 53
Table 2.6   Recycling options for common packaging materials 54
Table 2.7   Potential triple bottom line benefits of safe packaging 56
Table 2.8   Design for accessibility strategies 63
Table 2.9   Strategies to improve materials efficiency 67
Table 2.10 Energy efficiency strategies 70
Table 2.11 Suitability of products for returnable packaging 75
Table 2.12 Strategies for reusable secondary or tertiary packaging 75
Table 2.13 Energy savings from the use of recycled rather than virgin material 76
Table 2.14 Packaging material recycling rates by geographical region (%) 80
Table 2.15 Design strategies to improve recycling 82
Table 2.16 Common phthalate plasticisers used in PVC 90
Table 2.17 Examples of heavy metals in packaging 91
Table 2.18 Strategies to prevent the incidence or impact of litter 95
Table 2.19 Evaluating packaging materials against the four principles of packaging sustainability ... 96
Table 2.20 Evaluating thermoplastic polymers against the four principles
            of packaging sustainability 99

Photo

Photo 2.1   The BPA-free water bottle 89

Case Studies

Case Study 2.1 The Keep Cup 49
Case Study 2.2 Design for accessibility (Duracell) 50

2 Designing for Sustainability

Case Study 2.3 Efficient, cyclic and safe: Dell's goals for computer packaging 52
Case Study 2.4 Compostable packaging - Gingerbreak Folk 55
Case Study 2.5 Forest Stewardship Council -certified cartons: Tetra Pak 57
Case Study 2.6 Cadbury Dairy Milk chocolate bars.. 60
Case Study 2.7 Rethinking technical requirements... 62
Case Study 2.8 Eliminating packaging at Sainsbury's 64
Case Study 2.9 Materials efficiency: Bunnings 65
Case Study 2.10 Redesigning the product: laundry detergent.. 69
Case Study 2.11 Energy efficiency: the square milk bottle 71
Case Study 2.12 CHEP reusable crates 73
Case Study 2.13 Reusable kitchen worktop packaging 73
Case Study 2.14 Recycled PET (rPET) in Marks and Spencer's 'Food to Go' range 77
Case Study 2.15 Source reduction versus recycled content paperboard............................ 79
Case Study 2.16 Amazon's 'Certified Frustration Free Packaging'.................................... 81
Case Study 2.17 Food waste 84
Case Study 2.18 BPA Heinz baby food... 88

2.1 Introduction

Most businesses have a process for product development that encompasses packaging optimisation or redesign and new packaging design. The process is generally led by a packaging specialist, either within or external to the business. It involves extensive consultation with a range of stakeholders to develop the best solution that meets many potentially conflicting objectives: cost, function, consumer acceptability, transport efficiency, shelf presence, promotion and now sustainable development.

Build Sustainable Development into the Design Process

Designing for sustainability involves considering sustainability objectives as early as possible in, and regularly throughout, the design process. This provides the greatest potential to influence the design to achieve the best sustainability outcomes with least cost (see Fig. 2.1).

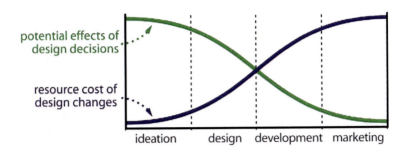

Fig. 2.1 The design approach. *Source*: McDonough Braungart Design Chemistry [96]

Sustainability Changes the Design Process

Sustainability in packaging design requires new information to be considered including:

- environmental life cycle of the product and its packaging
- the role of packaging in achieving sustainable development goals
- packaging environmental regulatory requirements
- systems in place for the recovery, use and disposal of packaging at end-of-life.

⇒ See Chaps. 4 and 5 to learn more about regulations and life cycle thinking respectively.

Increasingly, decision-support and eco-design tools are used to accelerate the integration and consistency of application of sustainable development into the design process. They also help educate the design team and relevant stakeholders on the impact of decisions on sustainability issues.

⇒ See Chap. 7 for more on applying tools to packaging design for sustainability.

2.2 Designing Packaging for Sustainability

Packaging is used for the containment, protection, handling, delivery, presentation, promotion and use of products. A number of packaging components make up the 'packaging system' (see Table 1.1) with each component selected for a particular purpose (see Table 2.1).

Aim to Create Economic, Social AND Environmental Value

Strategies to improve sustainable development, such as increased efficiency, recyclability and elimination of toxic components, must be balanced against all relevant performance criteria during production, distribution, storage and use [1]. The idea of 'balancing', however, implies trade-offs. While this is often necessary, the aim should be to design and manufacture packaging that simultaneously delivers economic, social and environmental value. This may require a departure from 'business as usual' to find 'win–win' solutions and new innovative ways of achieving the required objectives. Some examples of potential win–win solutions are provided in Table 2.2.

Commit to Innovation

Packaging development has a long history of technical innovation and enabling product innovation [2]. Examples include ready-to-eat fresh meals, re-sealable packs and longer-life packaging. These have taken advantage of advances in materials and food processing technologies and have sought to meet changing consumer tastes and lifestyle choices.

2 Designing for Sustainability

Table 2.1 Functions of packaging

Function	Features
Protection	Prevent breakage (mechanical protection)
	Prevent spoilage (barrier to moisture, gases, light, flavours and aromas)
	Prevent contamination, tampering and theft
	Increase shelf life
Promotion	Description of product
	List of ingredients
	Product features and benefits
	Promotional messages and branding
Information	Product identification
	Product preparation and usage
	Nutritional and storage data
	Safety warnings
	Contact information
	Opening instructions
	End-of-life management
Convenience	Product preparation and serving
	Product storage
	Portioning
Utilisation	Provision of consumer units
	Provision of retail and transport units
Handling	Transport from producer to retailer
	Point-of-sale display
Waste reduction	Enables centralisation processing and re-use of by-products
	Facilitates portioning and storage
	Increases shelf life
	Reduces transport energy

Source: EUROPEN and ECR Europe [91, p. 7]

Designing for sustainability requires a commitment to rethink the design of the product-packaging system. There are potential trade-offs between objectives, for example:

- material efficiency of a plastic pouch vs. the recyclability of a plastic bottle
- environmental benefits of enhanced recyclability vs. the cost of changing the packaging
- elimination of heavy metal-based inks or pigments vs. the marketing advantage of vibrant and durable colours.

However, approaches to the design process should begin by rethinking the problem in a more open and creative way. For example, can material efficiency AND recyclability be achieved by concentrating the product and selling it in a much smaller container? Does the recyclable package offer additional commercial or marketing benefits that justify the additional expense of the new design?

Table 2.2 Examples of 'win–win' packaging sustainability strategies

Design strategy	Cost	Functionality	Environment	Risks/potential impacts
Lightweighting	Reduced material costs Reduced transport costs	Easier for consumer to carry (i.e. less weight) Potential to improve openability/usability by simplifying the design (e.g. by removing unnecessary components)	Benefits throughout the life cycle due to avoided material production and waste Improved transport fuel efficiency (less material being distributed) and therefore less air emissions generated	Too much lightweighting can increase product damage and waste in the supply chain, which normally has higher impacts than the avoided packaging
Returnable transport packaging	Avoided costs of single-use packaging (The up-front costs of returnable packaging are higher than single-use packaging but a high return rate will generate cost savings) Avoided costs of baling and recycling single-use corrugated packaging	Potential to improve product protection (stronger containers)	Benefits throughout the life cycle due to avoided production and waste of single-use packaging (There will be some new impacts associated with cleaning and transport, and these need to be weighed against environmental benefits and minimised as much as possible)	Some new impacts associated with cleaning and transport that need to be weighed against environmental benefits and minimised as much as possible Up-front costs of returnable packaging are higher than single-use packaging but a high return rate will generate cost savings
Design for recycling	Simplification of the design may reduce packaging components and inventory costs (e.g. by reducing ink coverage or by eliminating labels)	Rethinking the design provides an opportunity to build in additional functionality.	Benefits throughout the life cycle due to avoided production (i.e. if virgin material is replaced with recycled material) Packaging material diverted from landfill	There may be increased costs associated with some recyclable materials

Use a Packaging Sustainability Framework

To integrate the new approaches required to address sustainability efficiently and effectively in packaging design, a packaging sustainability framework is useful as a decision-support tool. In this chapter we present and demonstrate the use of a framework that brings together traditional packaging design considerations with triple bottom line sustainability considerations. The framework is the latest evolution of work commenced by the Sustainable Packaging Alliance in 2002 [3]. It has been refined for this book in line with the evolution in thinking about packaging's role in sustainable development as outlined in Sect. 1.4.

⇒ See Sect. 1.2 in Chap. 1 to learn more about the triple bottom line.

The framework uses four principles to guide decisions about design, manufacturing, transport, use and recovery of packaging. Examples of strategies that can help to achieve these principles are provided together with a detailed case study demonstrating how to use the framework (see Appendix 1).

2.3 Packaging Sustainability Framework

In order to contribute to sustainable development, packaging needs to be (Fig. 2.2):
- *effective* in delivering the functional requirements of the packaging
- *efficient* in its use of materials, energy and water throughout its life cycle
- *cyclic* in its use of renewable materials and recoverability at end-of-life
- *safe* for people and the natural environment.

Fig. 2.2 The four sustainable packaging principles

Each principle is outlined in Sects. 2.3.1–2.3.4 respectively, and practical design strategies and case studies demonstrating their application are provided in Sect. 2. Section 8.2.8 also proposes metrics for aligning the design strategy with corporate sustainable development goals.

2.3.1 Effective Packaging

Effective packaging is fit for purpose and achieves its functional purpose with minimal environmental and social impact.

> **Effective ★**
> Fit for purpose

According to The Consumer Goods Forum [4, p. 11], well-designed packaging meets all its functional requirements while minimising the economic, environmental and social impacts of the product and its packaging. This reflects the concept of the triple bottom line and is a good definition of 'effective' packaging. Examples of the triple bottom line benefits of effective packaging are provided in Table 2.3.

Table 2.3 Potential triple bottom line benefits of effective packaging

Economic benefits	Reduced product damage
	Increased product sales
	Compliance (labelling)
Social benefits	Consumer convenience
	Accessible packaging (e.g. easy to open for older consumers)
Environmental benefits	Reduced production waste
	Reduced product damage in the supply chain

Demonstrate the Triple Bottom Line Benefits

The effectiveness principle requires designers to:
- demonstrate how the packaging design is 'fit for purpose'
- identify the economic, social and environmental value provided by the packaging
- re-examine conventional design objectives such as technical performance, convenience, cost and so on from a sustainability perspective.

> ⇒ The Sustainable Packaging Coalition guidelines [5] provide suggestions on rethinking conventional design criteria such as cost, technical performance, asset protection, etc.

Packaging must be Essential

Effective packaging fulfils a number of essential functions, such as [1, p. 10]:
- ensuring the contents are delivered to the consumer in good condition
- protecting the contents from hazards such as vibration, heat, odour, light penetration, micro-organisms and pest infestation
- being easy to open (but difficult to open accidentally) and pilfer-resistant
- allowing liquids to pour without spillage
- enabling all of the product to be dispensed
- being as easy as possible to carry
- for consumer goods, being attractive enough to buy
- providing information about the product, the business that bears responsibility for it, and instructions for handling or use.

The specific functional benefits of each component of the packaging system and the structure of the packaging system as a whole should be challenged and validated throughout the design process.

Explore New Opportunities

Businesses have always focused on the functional aspects of packaging design, but a focus on sustainability can open up new opportunities or a reassessment of the role of packaging. For example [5]:

- Are there opportunities to prevent theft in retail stores without relying on the packaging; for example, by modifying fixtures and displays?
- Is the package fit-for purpose but not over-engineered?
- Can the cost of the package be reduced through a more efficient design or by using materials that attract lower recycling fees (some countries have differential recycling fees—see Chap. 4 for more information)?

Applying the effectiveness principle should identify new opportunities for innovation including the creation of new product concepts that reduce the need for packaging (see The Keep Cup Case Study 2.1).

Case Study 2.1 The Keep Cup

The reusable 'KeepCup' for takeaway coffee demonstrates a new way of thinking for out-of-home packaging. Promoted as the 'first barista standard reusable coffee cup', the KeepCup has a similar shape to conventional coffee cups and is easily filled by a cafe espresso machine. Since its launch in 2009, over 1,000,000 of the cups have been sold in Australia, the United States and Europe. The product also highlights for consumers the production of waste associated with their purchase and consumption of takeaway beverages.

A streamlined life cycle assessment was used to compare the KeepCup with a conventional paper coffee cup. If both cups are used daily for 12 months, the KeepCup achieves a 97% reduction in global warming potential, a 98% reduction in water use and a 96% reduction in waste to landfill [6].

Photo: KeepCup

Design for Accessibility

'Design for accessibility' is becoming an essential design requirement for social sustainability. One of the most important access issues is ease of opening. Stringent requirements for packaging functions such as product protection, tamper evidence, and prevention of theft are often pursued at the expense of openability. Another accessibility issue is the ability of consumers with poor eyesight to read labels.

Design for accessibility has many implications for consumer health and safety including:

- packaging related injuries: many of these occur when people resort to a knife or scissors to open packages [7]
- inability to open packaging and thereby access products: consumers with functional disabilities associated with diseases such as arthritis sometimes cannot open packaging, and this problem is increasing as a result of the aging population in Western countries. Companies such as Duracell are redesigning packaging to address the needs of people with restricted strength or movement in their hands (see the Duracell Case Study 2.2)
- risk of product misuse: the poor readability of small text on labels is a problem—also arising from the aging population—and means that important information such as directions for use, safety warnings and disposal guidelines are sometimes not read.

Case Study 2.2 Design for Accessibility (Duracell)

In 2001, Duracell announced a new form of packaging to simplify the often difficult task of replacing hearing aid batteries. Batteries in the newer, more compact hearing aids are tiny, and wearers need to change their hearing aid batteries up to 50 times each year. The problem is magnified by poor eyesight and arthritis, which are common to many users. Duracell's EASYTAB^{TM} packaging makes it easier to remove the batteries and insert them into the hearing aid.

Historically, hearing aid batteries were packaged in a circular case that required consumers to 'dial' a battery into an opening for removal. This allowed more than one battery to fall from the opening when the user shook it from the opening. If the batteries fell to the floor, particularly onto thick carpet, they were difficult to find. The batteries were packaged with a small tab that adhered to the battery to prevent exposure to oxygen before use. Removing this tab to activate the battery was very difficult for users with limited dexterity.

The new EASYTAB^{TM} design features a brightly coloured tab attached to each battery, which is used as a tool to remove the battery and insert it in the hearing aid. If the battery is accidentally dropped, the tab is clearly visible. Once the battery is inside the hearing aid, the tab is removed to activate the battery.

Sources: Business Wire [8], The Center for Universal Design [9]

See Sect. 2.4.1 for a discussion and case study examples of design strategies for effective packaging.

2.3.2 Efficient Packaging

Efficient $
Materials, energy, water

'Efficient' packaging is designed to minimise resource consumption (materials, energy and water), wastes and emissions throughout its life cycle.

A common theme in the sustainable development literature is the need to go beyond incremental improvements and look for 'step changes' or significant improvements in eco-efficiency. For example, the authors of *Natural Capitalism* [10] have argued for 'radical' improvements in resource productivity to reduce depletion of resources and pollution and to lower costs'. Some researchers have estimated that for the world's resource use to be sustainable we need a 75–90% improvement in resource efficiency [11–13]. Examples of the triple bottom line benefits of efficient packaging are provided in Table 2.4.

Table 2.4 Potential triple bottom line benefits of efficient packaging

Economic benefits	Reduced resource costs—materials, energy, water
	Increased supply chain efficiency
	Cost savings passed on to consumers
Social benefits	More affordable products
	Reduced weight or volume
Environmental benefits	Reduced consumption of resources—materials, energy, water
	Reduced waste and emissions from production of virgin materials
	Reduced energy consumption and emissions from transport
	Reduced product waste

Apply Life Cycle Thinking

Life cycle assessment (LCA) studies show that minimising packaging and maximising supply chain efficiency are two of the three most important actions that reduce the environmental impacts of packaging [14]. (The other is use of renewable energy.)

⇒ See Chap. 5 to learn more about life cycle thinking.

As a general guide, reducing the weight of packaging by 20% will reduce environmental impacts of the packaging by about 20%. In contrast, recycling, while still desirable for many reasons, such as resource conservation, consumes energy and generates waste and emissions during transport and reprocessing [14].

Economic and Environment Win–Win

The benefits of more efficient packaging include:

- cost savings in the supply chain, which can be captured by the business or passed on to suppliers, customers and consumers
- less demand for materials, energy and water, which in some cases are being extracted from the natural environment at an unsustainable rate [15]
- less pollution and waste that must be absorbed by the natural environment by creating more efficient supply chains.

⇒ See Sect. 2 for strategies for designing for efficiency.

A number of businesses have adopted efficiency goals for packaging. Walmart, for example, plans to reduce packaging by 5% by 2013 compared to 2008 to reduce carbon dioxide emissions by 667,000 metric tons annually. This will also create US$10.98 billion in savings, including a US$3.4 billion saving to Walmart [16]. In 2008, Dell announced plans to reduce its packaging by 8.7 million pounds (3,946 tonnes), and by 2010 the company had already made significant progress (see Case Study 2.3).

Case Study 2.3 Efficient, Cyclic and Safe: Dell's Goals for Computer Packaging

In late 2008 Dell announced plans to eliminate 20 million pounds (9,072 tonnes) of packaging for its desktop and laptop computers by 2012 and to make the remainder of its packaging 'greener'. In the process, the company hoped to save an estimated $US8.1 million.

Dell's goals for packaging improvement are called the '3 Cs':
- cube—reduce the size
- content—use recycled or sustainable materials
- curb—ensure that it is easily recyclable.

These correspond to several of the sustainable packaging principles highlighted in this book, i.e. efficient, cyclic and safe packaging. In 2010 Dell reported a number of achievements, including:
- introduction of bamboo packaging for cushioning its Inspiron Mini 10 and 10v inside an outer box made from 25% post consumer materials
- elimination of 8.7 million pounds of packaging
- increased recyclability of packaging, with a shift to moulded pulp, high density polyethylene (HDPE) cushion, expanded polyethylene (EPE), bamboo and corrugate
- introduction of a 'multipack' for large orders that combines multiple orders into one box.

Sources: Greener Design [17], Dell [18]

See Sect. 2.4.2 for a discussion and case study examples of design strategies for efficient packaging.

2.3.3 Cyclic Packaging

Cyclic ()
Renewable and/or recyclable

'Cyclic' packaging is designed to maximise the recovery of materials, energy and water throughout its life cycle.

Match Materials with the Metabolic Cycle

As McDonough and Braungart state in their book *Cradle to cradle*, there is no waste in nature [19]. To minimise waste, packaging materials should be designed to become 'nutrients' for another process. Natural and renewable materials such as paper and wood should become nutrients for the biological metabolism; for example, in organic processes such as composting. Manufactured materials such as glass and plastics should become nutrients for the technical metabolism; for example, in industrial processes such as mechanical (material) recycling [19].

⇒ See Sect. 1.2 to learn about biological and technical metabolisms.

Examples of the triple bottom line benefits of cyclic packaging are provided in Table 2.5.

Table 2.5 Potential triple bottom line benefits of cyclic packaging

Economic benefits	Reduced material costs (recycled materials)
	Cost savings passed on to consumers
Social benefits	Reduced aesthetic impacts of litter
	Extension of life for existing landfills
Environmental benefits	Reduced consumption of resources—materials, energy, water
	Reduced waste and emissions from production of virgin materials
	Reduced packaging waste requiring disposal/recovery

Aim for Closed Loop Recycling

> **Closed loop recycling** involves reprocessing materials back into the same application, e.g. packaging to packaging.
> **Down-cycling** occurs when a material is reprocessed into an alternative, lower value application that often prevents further recycling, e.g. packaging into garden mulch.

It is generally more sustainable to recycle a material back into the same application (closed loop recycling) than down-cycle. A good example is a glass bottle, which can be re-melted in the glass furnace and manufactured back into a new bottle or jar.

Other materials are more difficult to reprocess back into the same application and may need to be 'down-cycled' into a lower value applications. For example, recycled plastic might not be suitable for the manufacture of new packaging because it does not meet food contact regulations, or it may not be able to compete with virgin resin because of the higher costs of processing. Therefore, it can only be down-cycled into products such as garden furniture and plant pots.

⇒ Refer to Table 2.14 for recycling rates of common packaging materials globally.

Design for recyclability aims to remove barriers to closed loop recycling to ensure that recovered materials can be reprocessed into high value applications. Some examples of closed loop material recycling and down-cycling (as well as barriers to them) are given in Table 2.6.

An emerging technology for the recovery of biodegradable plastic packaging is composting: a form of 'organic recycling'. These materials can potentially be

Table 2.6 Recycling options for common packaging materials

Material	Closed loop opportunities	Barriers to closed loop recycling	Examples of down-cycling opportunities
Polyethylene terephthalate (PET)	Jars or bottles, up to 100%	Quality (suitability for food contact), contamination with PVC	Fibre for clothing and other textile products
High density polyethylene (HDPE)	Bottles or tubs, up to 100%	The wide range of HDPE resins on the market, which may result in an inconsistent product, colour contamination	Crates, bollards, outdoor furniture, lumber
Polyvinyl chloride (PVC)	Bottles or tubs, up to 100%	Low cost of virgin resin, small quantity of post-consumer material	Pipe fittings, footwear, flooring
Glass	Jars or bottles, up to 100%	High cost of transport, contamination with ceramics/other glass, mixing of different coloured glass	Road base, asphalt, filtration media, blasting abrasive
Aluminium	Cans	Minimal	Car and truck components, doors, windows, siding
Steel	Cans	Minimal	Reinforcing rod, pipe, wire, appliances
Paper/cardboard	Boxes, cartons, bags	Quality of the recycled fibre (fibre length, colour, contamination)	Animal litter, insulation, mulch

collected in a source-separated organic stream (garden and/or food waste) for processing into organic products such as soil conditioner or mulch. Some of these materials may also be suitable for home composting.

⇒ See Chap. 6 to learn more about biodegradable materials.

Avoid Cross-Contamination Between Metabolisms

In Chap. 1 we introduced the work of McDonough and Braungart, who described two recovery mechanisms for products: the biological metabolism such as composting and technical metabolism such as an industrial recycling process. They argue that products should be designed for one of these metabolisms, and with care to ensure that a product designed for one system does not contaminate the other. Contamination could occur, for example, if a biodegradable plastic shopping bag, designed for composting, ends up in a conventional plastics recycling system, or if a polyethylene plastic bag ends up in a composting system.

One company that has carefully considered all of these issues is biscuit manufacturer Gingerbread Folk, which uses a biodegradable material certified to an international standard and advises consumers on appropriate disposal (see Case Study 2.4).

Case Study 2.4 Compostable Packaging: Gingerbreak Folk

Gingerbread Folk pack their biscuits in NatureFlex resin from Innovia Films. The raw material for the film is cellulose extracted from wood fibre. The label advises consumers that:

> 'We care about the planet, that's why the wrapper is compostable. When finished, please place this wrapper in your home compost—really, it is OK to do this'.

Photo: Gingerbread Folk

More Economic and Environmental Win–Wins

The benefits of recycling packaging often include significant environmental savings when recycled materials replace virgin materials in production. For example, it has been estimated that recycled aluminium requires only 7% of the energy required for virgin aluminium, and recycled high density polyethylene (HDPE) only requires 21% of the energy required for virgin HDPE [20, p. xi].

See Sect. 2.4.3 for discussion and case study examples of design strategies for cyclic packaging.

2.3.4 Safe Packaging

'Safe' packaging is designed to minimise health and safety risks to humans and ecosystems throughout its life cycle.

Designing for sustainability considers a broader range of potential impacts on the health of humans and ecosystems than traditional packaging design, such as:

- ecological impacts of growing natural raw materials, particularly from land degradation and biodiversity loss
- ecological and health impacts of pollution from manufacturing processes
- risks associated with migration of hazardous substances into food and beverages
- occupational health and safety risks in the supply chain
- impacts of packaging litter on wildlife, particularly in marine environments.

⇒ See Sect. 2.4.4 for strategies for designing for safety.

There are triple bottom line benefits of considering these impacts, as shown in Table 2.7.

Table 2.7 Potential triple bottom line benefits of safe packaging

Safe	
Economic benefits	Reduced costs of disposal (hazardous or toxic waste)
	Reduced risk of product recalls
	Carbon credits or reduced cost of carbon emissions
Social benefits	Reduced health and safety risks for consumers and neighbours
Environmental benefits	Reduced eco-toxicity impacts
	Reduced contribution to global warming

Take Responsibility for Sustainability Impacts of Raw Materials

> **Ecological and environmental stewardship** are terms given to programs that aim to reduce the social and environmental impacts of farming, forestry or fishing practices.

Designing for safety must consider the environmental and social impacts of raw materials, particularly those derived from forestry or farming activities. This is often referred to as 'ecological stewardship'. Timber, fibre-based packaging materials and biopolymers from agricultural products can impact on biodiversity and the sustainability of natural ecosystems. Forestry operations, for example, may reduce or damage old growth forests. The procurement of 'renewable' materials needs to minimise any potential impacts; for example, by only using paper or cardboard from sustainably managed forests. Food security issues also need to be addressed; for example, by investigating the impact of diverting food crops such as corn to manufacture packaging. The Forest Stewardship Council (FSC) certifies materials according to ecological stewardship criteria, and businesses may specify only certified materials, as illustrated in the TetraPak Case Study 2.5.

> **Case Study 2.5 Forest Stewardship Council-Certified Cartons: Tetra Pak**
>
> Tetra Pak has been a member of the Swedish Forest Stewardship Council (FSC) since 2006, and their long term goal is to use FSC-certified fibre for all of their liquid food cartons. In September 2009, the company announced that beverage cartons with the FSC logo would be available to customers in Sweden, Denmark and Belgium. The cartons were already available in China, France, the United Kingdom and Germany.
>
> FSC is an independent non-government organisation that promotes responsible management of the world's forests (more detail is provided in Sect. 2.4.4).
>
> *Source*: Tetra Pak [21]

Implement and Support Cleaner Production Technologies

> **Cleaner production** aims to reduce waste and emissions in manufacturing by changing management practices, processes and product design, rather than treating waste and emissions before disposal (the traditional 'end-of-pipe' solution).

Pollution from manufacturing processes in the packaging industry have a range of environmental and health impacts. Emissions of volatile organic compounds (VOCs) from printing processes contribute to ground-level ozone pollution, and the wastewater from chlorine bleaching of paper during the manufacturing process contains organochlorine compounds such as dioxins.

⇒ See Sect. 2.4.4 for more on chlorine bleaching of paper.

Designing for safety requires:
- understanding the processes used in manufacturing and printing packaging
- changing design specifications to shift to less polluting processes where available.

Validate Safety of Packaging

Food packaging systems must protect the integrity of the product so that consumer health is not compromised. Some constituents in packaging, such as Bisphenol A (BPA) and phthalates, can migrate in small amounts into food products. While there is scientific uncertainty about their health effects, there is mounting evidence that they are potentially toxic and should be avoided where possible [22]. A risk management approach to packaging safety requires:

⇒ See Sect. 2.4.4 for more on BPA and phthalates.

- understanding in detail the materials and constituents used in the packaging
- obtaining Materials Safety Data Sheets or other documentation from suppliers
- monitoring the latest published research on migration of substances into food and other consumer products
- consulting with suppliers, researchers and safety authorities if there are any concerns
- as a precautionary measure, taking steps to replace any materials or constituents that may pose a health risk.

Design for Safe Handling

The implications of packaging design for occupational health and safety in the packaging supply chain also need to be considered. For example, attention must be paid to any risks associated with storage and handling in the supply chain. Any packaging that requires a knife to open is a potential hazard to workers or consumers. Packaging should be designed for easy opening without the use of sharp instruments. The weight of packed products is also an issue, particularly for work that involves shifting or dispensing products. Weight is generally not an issue at the consumer level, although the larger capacity of reusable shopping bags often results in overloading, making the bags heavier and more difficult to handle by cashiers [23].

Design for Litter Reduction

Packaging litter has many sustainability impacts, including:
- injury or death of wildlife. It is estimated that 6.4 million tonnes of litter enter the oceans every year [24, p. 101]. While the impact of packaging is relatively small, a number of reports have highlighted wildlife impacts associated with packaging [25]
- damage to nautical equipment
- aesthetic impacts in waterways, along beaches and in other public places
- injuries to people; for example, cuts from broken glass
- costs of litter clean-ups.

The packaging design team can help to minimise the incidence or impact of litter; for example, by minimising the number of separable components or by communicating an anti-litter message. Litter statistics published by industry associations and/or non-government organisations can be used to better understand the products, packaging and brands that are littered most frequently. This information can then be used to see if any of the business's packaging portfolio falls within the most littered items.

See Sect. 2.4.4 for a discussion and case study examples of design strategies for safe packaging.

2.4 Applying the Packaging Sustainability Framework

In this section, design strategies and case studies are presented to illustrate how each of the four packaging for sustainability principles can be addressed:
- designing for effectiveness (see Sect. 2.4.1)
- designing for efficiency (see Sect. 2.4.2)
- designing for cyclic packaging (see Sect. 2.4.3)
- designing for safety (see Sect. 2.4.4).

⇒ See Chap. 8 for more on integrating sustainability in the product development process.

The packaging sustainability framework is a systematic approach to design that can be applied by assessing each of the four principles and the way they work together. It should be used particularly at the initial ideas stage of the product development process, where there is the most freedom to explore alternative strategies.

The design process should optimise the choice of projects in line with the business's sustainable development goals and metrics. In practice, the final design decision may require trade-offs to address competing goals and metrics. This is illustrated in the case study about Cadbury (Case Study 2.6) in which *more* material has been used to improve functionality and recyclability.

Case Study 2.6 Cadbury Dairy Milk Chocolate Bars

Cadbury Australia redesigned the packaging of its Dairy Milk chocolate bars in 2008, primarily to improve recyclability. The original packaging was made from a non-recyclable metallised paper. The new packaging consists of two recyclable components—an aluminium foil enclosed in a lightweight carton. While the overall weight of the primary packaging increased, market research found that consumers would be more likely to recycle a carton than the lighter weight alternative, a paper wrap. The redesign also provided an opportunity to introduce an innovative feature that allows the product to be resealed after opening, which is less messy and maintains the freshness of the product.

The new Dairy Milk packaging

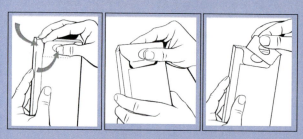

Opening/reclose feature (patent pending)

Source: Chessell [26]
Images supplied by Cadbury Australia

2.4.1 Designing for Effectiveness

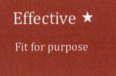

- Meeting consumer needs
- Functionality – technical performance, convenience, accessibility
- Opportunities for innovation

Effective ★
Fit for purpose

By focusing on the effectiveness principle the design team confirms:
- the role of each packaging component and the packaging system as a whole
- how the packaging protects the product and creates consumer value.

It may also help to generate ideas for new product-packaging concepts with the potential to deliver better value to the business and the consumer with less environmental impact.

Is the Packaging Necessary?

The first challenge is to enquire whether the package is necessary. In the process of answering this question, the design team gains a better understanding of the basic needs met by each package component and the packaging system as a whole [27, p. 21]. In some situations it may be possible to eliminate the package or a component of the packaging system that adds little or no value to the protection of the product.

Market research can be used to understand how and where a product is consumed and whether certain features of the packaging are actually required or used by customers and consumers. For example, fresh salad packaging often includes disposable cutlery because it is intended to be consumed away from home. It is therefore important to know where the salads are actually consumed and the extent to which the cutlery is used. If most salads are consumed at home or in a workplace, with ready access to durable cutlery, then this feature could potentially be removed, saving cost and environmental impact.

Optimise Function of all Components AND the System

Product containment and protection are the primary role of packaging. Depending on the product and its supply chain, the packaging system may need to protect its contents from:
- climatic influences, such as light, humidity and temperature
- mechanical hazards, such as impacts, accelerations, abrasions and vibrations
- gas and odour exchange
- contamination by micro-organisms or pest-infestation.

Secondary and tertiary packaging facilitates distribution by bundling products together for transport and handling. Secondary and tertiary packaging choices are inter-dependant with the primary packaging, and the complete system must be optimised.

From a sustainability perspective, it is important to ensure that all functional requirements are met without over-engineering the packaging system. Rethinking all of the technical requirements may open up new opportunities to reduce material or energy consumption or to improve productivity in the supply chain (see Case Study 2.7 on a hypothetical product).

> **Case Study 2.7 Rethinking Technical Requirements**
>
> A hypothetical food product has a 12 month shelf life, uses a film as the primary packaging and is packed in a display carton and then a corrugated case. The film provides a sufficient barrier to enable an 18-month shelf life without oxidation. The display carton is used for promotion by providing a shelf display. The corrugated case ensures the product survives national distribution.
>
> Decide how many of the current (or new) product and distribution assumptions can be broken or challenged (e.g., shelf life, distribution modes). This is the hardest part, and sometimes requires looking at some different commercial environments. If your product is a typical supermarket product, look in a hardware store or a pharmacy for clues on how other products might be working.
>
> Provided the product tastes as expected, and is relatively undamaged, the consumer is not particularly concerned about the display carton or corrugated case. Can these therefore be avoided? Could the shelf life be managed if it was shorter, say 10 months? This might allow a lower material gauge or less complex barrier film. The flow wrapper used to form the primary packaging might use a seal and end crimps. The consumer places no value on the size of these, so can they be removed to reduce the surface area of the primary packaging?
>
> Is the display carton necessary, or is it only a method of 'bundling' a number of units? Could this be a bundling film wrap, or is that layer of the system necessary at all? Could the outer case count be reduced, removing the necessity for the display carton altogether?
>
> With the corrugated case, could the distribution packaging be reduced? Does the business specify airbag suspension trucks? Are the maximum static and dynamic stack heights allowed really necessary? Truck height will typically allow a 2.4 m high stack, but do you produce 2 × 1,200 mm high, and line haul on rails? Do you fully utilise this height in the truck? Doing so may significantly reduce the impacts associated with trucking. Also, count

(continued)

2 Designing for Sustainability

Case Study 2.7 (continued)

> only the surface area of the carton that you need to protect and contain the product. For example, a regular slotted carton generally has large areas of overlap in the closure flaps. This area does not add value to the protection and containment function of the case. It is there to allow manufacture of the box. Remove any surface area overlap that is not necessary from the calculations.
>
> So, a possible packaging system for this product could consist of a lighter gauge film with no lost seal area, no display carton and a smaller count shipping case with a better pallet and truck space utilisation.
>
> *Source*: Bryce Hedditch, SustainPak [28]

Design for Accessibility

Designing for 'accessibility' requires making packaging easy to open by the 'average' consumer as well as the elderly and consumers of any age with a disability or arthritis.

The openability of packaging can be promoted to consumers as a market differentiator. The Arthritis Foundation in the United States, for example, has developed an accreditation and labelling scheme for 'ease-of-use: user-friendly products and packaging' [29]. Like all design for sustainability strategies, openability can be easily addressed by integrating these requirements into the design process as early as possible (Table 2.8). Readability of labels by all consumers, including those with poor eyesight, also needs to be considered.

Table 2.8 Design for accessibility strategies

Cans with pull-tabs can be improved by deepening the pre-cut around the edge to make it easier to pull the lid up
Packages using a tear notch should indicate clearly and accurately where the notch actually is
Jars with rounded plastic lids and no serration should flatten the lids to a sharp edge and incorporate serration for grip
Foil lids should incorporate an opening tab that is big enough to grip
Screw-tops need to balance vacuum suction with how easy it is to open the product
Child safety and anti-tampering is of paramount importance but can be maintained by using intelligent opening systems such as lining up dots or arrows instead of 'squeeze in, push down and twist'
Reading instructions are imperative for safety reasons or efficacy, and design can be improved with these simple guidelines: • simple sans serif typefaces such as Arial or Helvetica are recommended for maximum readability • good contrast contributes to legibility. The text should be printed with the highest possible contrast • lower case text is easier to read, and using text consisting entirely of capital letters should be avoided.
Source: Judith Nguyen from Arthritis Australia, cited in Packaging News [92]

2.4.2 Designing for Efficiency

Efficient $
Materials, energy, water

- Material efficiency
- Minimising product waste
- Energy efficiency

By focusing on the efficiency principle the design team confirms:
- the amounts of packaging used and required
- the environmental benefits provided by the packaging through product protection
- the life cycle environmental impacts of the packaging components and system arising from energy consumption.

> **Right-sizing** is reducing the size or weight of the package but not to the point at which the product becomes vulnerable to breakage or spoilage [27, p. 36].

Is the Packaging Necessary?

The first step in efficient design is to identify any components of the packaging system that are not necessary and could be eliminated (see Case Study 2.8 on Sainsbury's). This step should be taken when first considering design for effectiveness. A proper assessment of efficiency considers the interaction between all components of the packaging system throughout the distribution chain and looks for any that can be eliminated, keeping in mind that a reduction in the weight of a primary pack may require stronger secondary packaging or result in more product damage. This is why packaging needs to be *optimised* rather than minimised.

> **Case Study 2.8 Eliminating Packaging at Sainsbury's**
>
> Sainsbury's in the United Kingdom has announced that its 'basics' range of cereals will be stocked in plastic bags rather than a bag inside a carton. When fully implemented across the product range, this will result in 165 tonnes less packaging per year.
>
> *Source*: Ditching Cereal Boxes [30]

Opportunities to Minimise Material Use

Packaging should be manufactured with the minimum amount of material required to be effective. There is significant room to reduce material use in packaging: a European evaluation of packaging efficiency for 468 common products found that on average, the product contributed 80% of the weight of the packed pallet but only 50% of the volume [31, p. 7]. A Dutch study concluded that the most significant environmental gains for packaging can be made by choosing smaller-sized packaging and/or a more easily stackable shape [32]. Both strategies allow more products to be packed in a container or truck, reducing the cost and environmental impact of transport. Metrics used to measure changes in material use include packaging weight, packaging-product ratio and cube utilisation (a volumetric measurement of packaging design efficiency). See Case Study 2.9 where Bunnings made improvements to a hardware product.

Case Study 2.9 Materials Efficiency: Bunnings

Bunnings is Australia and New Zealand's leading retailer of home improvement and outdoor living products and a major supplier of building materials with 239 stores and more than 30,000 employees. As part of its wider commitment to environmental sustainability, the company is implementing a range of energy and water efficiency and waste minimisation initiatives. In terms of packaging these have included the elimination of single-use plastic bags (action in Australia commenced in 2003) and the introduction of recycling programs for packaging received in store. During 2008/2009 Bunnings' recycling rates doubled for the second consecutive year from 25 to 50%.

In 2008 the company engaged consulting group Net Balance to undertake an audit of product packaging to review its environmental performance. The audit found many examples of efficient or recyclable packaging but it also identified numerous examples of 'over-packaging'. For example, some electrical extension leads were packed individually in plastic bags while others were sold with only a sales tag providing essential information and a bar code. Many products sold in hardware stores require very little protection and in these cases there is an opportunity to eliminate or reduce packaging. In addition to its impact on resource consumption and waste to landfill, unnecessary packaging adds costs to the business. These include:
- the hidden costs of packaging in the product
- opportunity costs—it takes up additional shelf space and reduces the ability of the business to keep stock on hand
- staff unpacking and re-packing time
- disposal costs.

The recommendations of the audit have been implemented through ongoing work with suppliers to reduce unnecessary packaging. A Working

(continued)

Case Study 2.9 (continued)

Group was established with the company's 10 largest local suppliers, and as a result sustainable packaging principles were integrated into packaging specifications for imported products. Bunnings continues to work toward reducing unnecessary packaging in keeping with its long term strategy to reduce waste to landfill.

An early example of a packaging improvement is shown below.

Old packaging   New packaging

These wrenches used to be individually wrapped in plastic film and then unitised in a flexible PVC bag. They now have minimal packaging—a product cable to hold the wrenches together and a swing tag.

Sources: Bransgrove [33], Bunnings Group Limited [34]

Photos: Bunnings

Reduce Packaging Weight

The next step is to identify opportunities to reduce the size or weight of all packaging components (Table 2.9).

Optimise the Product-Packaging System: Avoid Under-Packaging

Understanding the product environmental lifecycle and the role of packaging allows an assessment of whether a product is 'under-packaged'. This is particularly important, as the environmental impact of products may be many times that of the packaging (see Sect. 1.1).

⇒ See Chap. 5 for more on LCA.

It has been estimated that the energy required to make food packaging, for example, is approximately 10% of the energy used to produce, protect, distribute, store and prepare the food it contains (Fig. 2.3) [35, p. 4].[1]

---

[1] The percentage is higher for some products, e.g. 16% for cereals, 23% for fresh fruit, 20% for fruit produce, 28% for alcohol, 23% for snack foods and 46% for soft drinks [35].

2 Designing for Sustainability

Table 2.9 Strategies to improve materials efficiency

Down-gauge (in thickness and weight) as much as possible
Eliminate unnecessary void space, layers and components
Eliminate labels by printing directly onto the packaging
Optimise the quantity of product in the consumer package to meet the needs of the consumer while also, wherever possible, reducing the packaging-product ratio
Consider using a larger volume pack, although it is important to ensure that this does not result in more product waste
Increase the volume density by concentrating products such as juice, soups and detergents
Design lightweight refill packs
Strengthen or weaken certain components to reduce overall material use
Minimise use of inks where this will not compromise the consumer appeal of consumer units
Ensure primary packs fit snugly into secondary units
Optimise secondary packaging dimensions to ensure good pallet optimisation
Use point-of-sale displays to convey messages and image rather than increasing the packaging on every item
Investigate whether plastic slip-sheets can be used instead of pallets
Investigate the potential to replace secondary packaging with a bulk reusable transit packaging system
Review competitors' products and international best practice to identify new design or lightweighting options

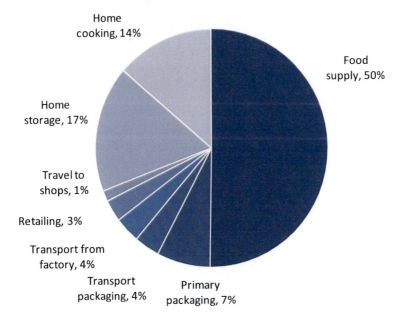

Fig. 2.3 Energy for one person's weekly consumption of food. *Source*: Based on INCPEN [35, p. 4]

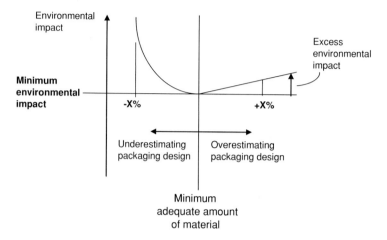

Fig. 2.4 Comparing the impacts of over-packaging and under-packaging. *Source*: Based on Erlov et al. [36, p. 4]

A model developed by Packforsk (Fig. 2.4) compares the environmental consequences of underestimating and overestimating the amount of packaging required for a product [36]. The growth in environmental impact that results from over-packaging is linear. However, the growth in environmental impact that results from under-packaging is exponential because it is linked to the impact of the packaging as well as the lost product. Over-packaging by 10% means that 10% of the resources needed to produce and transport the packaging are unnecessary and therefore wasted. Under-packaging may result in packaging failure, which usually leads to 100% waste of the resources used to produce and distribute both the product and its packaging [1, p. 11].

Sometimes less packaging can reduce rather than increase the amount of product waste. This has occurred with the redesign of distribution packaging for appliances and electrical equipment from corrugated boxes to clear film. While film is not as strong as a box, material handling workers tend to be more careful because the product is visible and damage cannot be concealed. The result is less product damage and waste in the supply chain [37].

Redesign the Product

There may be an opportunity to redesign the product to reduce packaging consumption and transport impacts. Examples include concentrated detergents (Case Study 2.10) and 'flat packaged' furniture.

2 Designing for Sustainability

> **Case Study 2.10 Redesigning the Product: Laundry Detergent**
>
> Unilever has shifted to more concentrated liquid detergents that require less packaging and are more efficient to transport. In Australia, for example, the introduction of concentrated Omo and Surf 'Small & Mighty' detergents, and the associated switch from 1.4 L to 475 mL bottles, resulted in:
> - a reduction of 82 tonnes of plastics per year
> - 32 tonnes less material in landfill
> - environmental savings from materials use, manufacturing, transport, and recycling.
>
> In 2008 it won the Packaging Council of Australia (PCA) Sustainability Award.
>
>
>
> Photo: Helen Lewis
>
> *Sources*: Unilever [38], Packaging Council of Australia [39]

Optimise the Product-Packaging System: Dispense ALL the Product

Efficient packaging design ensures that the product-packaging system is designed to allow complete dispensing of the product.

Any product residue left behind in the packaging when it is disposed in a recycling or rubbish bin represents an environmental and financial cost associated with poor packaging design. The resources consumed and environmental impacts arising from the production of inputs and the product are often higher than those associated with the packaging itself (see Sect. 1.1). The lost product is a financial cost to the consumer, who has paid for a product that cannot be fully consumed.

Strategies to help ensure that packaging can be fully emptied include:
- designing bottles with a wide neck
- using perforations that allow cartons to be opened all the way across the top
- selecting appropriate materials
- modifying the rheological properties (flow) of the product
- using packs that can be stored inverted—with the opening at the bottom.

Reduce Energy Consumption

Efficient design aims to reduce energy consumption at every stage of the product environmental life cycle to help conserve fossil fuels and reduce greenhouse gas emissions.

There are many strategies that can be used to minimise energy consumption throughout the supply chain, including optimising the design of all packaging components for transport efficiency. The selection of materials is also important, because some materials have lower 'embodied energy' than others; that is, they use less energy in raw materials extraction and manufacturing processes. For each material type, a lighter weight pack will also use less energy than a heavier one to manufacture and transport.

Energy efficiency strategies are presented in Table 2.10 and in the Superior Dairy Case Study 2.11, energy benefits of the square milk bottle.

Table 2.10 Energy efficiency strategies

During production
• Minimise the amount of material used, for example through lightweighting (see Table 2.9)
• Select materials that are more energy efficient; i.e. that have relatively low 'embodied energy'
• Maximise the amount of recycled content (recycled material uses significantly less energy—see Table 2.13)
• Purchase materials from suppliers with an effective energy efficiency program (e.g. ask for data on energy consumed to generate a unit of product)
During transport
• Reduce the size of packaging; for example, by concentrating the product or reducing void space, to increase pallet utilisation and therefore reduce the number of truck movements
• Switch to bulk distribution of raw materials and components to increase the amount of product being carried on each truck and therefore reduce the number of truck movements
• Reduce the weight of packaging to reduce fuel consumption
During consumption
• Provide clear and prominent information for consumers on whether or not refrigeration is required (some consumers refrigerate products unnecessarily; e.g. some spreads and sauces)
• Provide clear and prominent information on energy-efficiency; e.g. labelling on laundry detergents should promote minimum doses (to reduce overall use of detergents) and cold-water washing (to reduce energy consumption in appliances)
• Use long life packaging for products to eliminate the need for refrigeration in transport; e.g. aseptic packaging

2.4.3 Designing Cyclic Packaging

Cyclic ()
Renewable and/or recyclable

- Renewable materials and energy
- Design for reuse
- Design for recycling
- Recycled materials

Case Study 2.11 Energy Efficiency: The Square Milk Bottle

Photo: Helen Lewis

In 1998 Superior Dairy in Canton, Ohio, redesigned the conventional gallon milk bottle to eliminate the need for milk crates. Milk bottles are normally transported in plastic milk crates that can be stacked for transport. The crates require a lot of material to manufacture and need to be loaded, unloaded, collected, shipped back to the dairy empty, washed, stored and replaced when lost or stolen—a common problem for the highly functional milk crate.

Through their sister company, Creative Edge, the dairy undertook a major redesign of the milk bottle to make it self-stacking. The bottle has a square shape, slightly thicker walls and a recessed spout. These design features allow the bottles to be stacked six-high on a pallet.

Walmart picked up the innovative bottle design in 2008 for their own brand 'Member's Mark' milk, which is sold in their discount store Sam's Club. When the product was launched, Sam's Club reported that the trucks used for shipping can accommodate approximately 9% more milk: 4,704 gallons per truck or approximately 384 more bottles. This results in a significant saving in energy and greenhouse gas emissions associated with transport from the dairy to the retailer, and a cost saving for Walmart. A percentage of the cost saving is being passed on to consumers.

Sources: Sam's Club square case-less milk jug packaging [40], Mans [41]

By focusing on the cyclic principle the design team confirms how to:
- reduce consumption of virgin materials
- reduce reliance on non-renewable resources
- maximise the recovery of packaging materials.

> **A renewable resource is** a natural resource that is depleted at a rate slower than the rate at which it regenerates. Packaging materials that are theoretically 'renewable' include wood, paper and some biodegradable polymers (those made from natural products such as corn or cellulose).

Reassess Reusable/Refillable Consumer Packaging

Most packaging was originally reusable or refillable, particularly for beverages. However, in most developed countries reusable glass bottles have been replaced by single-use containers. There are a number of reasons for this shift:
- the introduction of self-service supermarkets and the decline of home delivery services
- industry consolidation to achieve economies of scale and the increasing size of distribution networks, particularly international networks, which add to transport costs for the return of empty bottles
- an increase in the proportion of beverages consumed away from home
- a decline in return rates for refillable bottles, which reduced their financial viability and environmental benefits
- the opposition of brand owners and retailers to reusable packaging for a range of commercial, health and safety reasons.

This shift has been less pronounced in countries with specific regulatory measures in place to encourage reusable packaging. In Germany, for example, there is an industry cooperative that supplies refillable glass and PET bottles to over 230 mineral water bottlers [42, p. 212]. The users of this system tend to be small businesses, and the bottled water is generally only transported a few kilometres. In Norway, refillable soft drink containers have a market share of approximately 98%, and their market share for beer is around 44% [42, p. 213].

Self-dispensing systems are common in specialty organic or health stores, where customers are encouraged to bring their own packaging to the store for filling. Recent developments in the United Kingdom indicate that retailers and manufacturers may be willing to introduce refill systems for mainstream products. UK-based organisation WRAP (Waste & Resources Action Programme) has undertaken research on the potential for these to be introduced for beverages in retail stores [43], and in 2009 and 2010 funded a self-dispensing trial for liquid laundry products [44]. This research will be important in determining whether refillable packaging should be reconsidered for some mainstream consumer applications.

Identify Supply Chain Packaging Reuse Strategies

In contrast to consumer packaging, the reuse of secondary and industrial packaging has increased over the past decade [42]. Reusable systems include plastic trays and crates, intermediate bulk containers, wooden or plastic pallets,

2 Designing for Sustainability

beer kegs, roll cages and moulded plastic containers for specialty products (see Case Study 2.12 and 2.13).

Case Study 2.12 CHEP Reusable Crates

An LCA compared a returnable plastic crate with single-use corrugated packaging (100% recycled content) for transporting fresh produce from farm to retail store. The study concluded that the crate generated 70% less greenhouse gas emissions, 95% less solid waste and used 85% less water. According to CHEP their returnable crate system also delivers increased functionality and financial value in the supply chain.

Photo: CHEP

Sources: CHEP [45], Crates Offer Produce a Green Premium [46]

Case Study 2.13 Reusable Kitchen Worktop Packaging

WRAP and home improvement retailer B&Q assessed the feasibility of a reusable packaging system for kitchen worktops. Worktops require a significant amount of packaging to avoid transit damage and scuffing of worktop corners, edges and presentation surfaces. Any damage can result in the product being rejected by the customer and then scrapped.

Single-use cardboard packaging was replaced in a trial with a purpose-designed, reusable plastic 'Carrierpac' (45% recycled content and recyclable at end-of-life). The Carrierpac was found to be quicker to pack and unpack,

(continued)

Case Study 2.13 (continued)

and there were no reports of product damage (eliminated product losses and increased customer satisfaction).

The Carrierpac was adopted by B&Q, reducing annual packaging use by 1,100 tonnes and damage rates from 6% to less than 1% (saving 900 tonnes of worktops from landfill per year), and saving B&Q £1m per annum. Since the launch, some Carrierpacs have reached 80 reuses, with average reuse now running at over 40 trips. If adopted by other leading manufacturers and retailers it could reduce waste by over 5,000 tonnes per year.

Photo: WRAP and B&Q

Source: WRAP [48]

If used appropriately, reusable transport packaging may generate a range of sustainability benefits [48, p. 2]:
- *cost savings*—reduced packaging and waste disposal costs, reduced product damage and reduced cost of returns and rejects
- *consumer benefits*—increased satisfaction, no bulky waste to recycle or dispose of and improved product presentation
- *company and employee benefits*—reduced risk of personal injury to packing and delivery teams, improved customer service, improved company image, and marketing opportunities
- *environmental benefits*—less packaging waste and reduced product damage.

PIRA International has undertaken a detailed analysis of the costs, benefits and feasibility of multi-trip boxes and crates used to transport products between the packer and retailer, which are becoming widely used in some sectors [49]. They concluded that reuse systems are not always appropriate as an alternative to the conventional shrink wrap and corrugated packaging systems, depending on a range of factors including distribution costs, size and shape of the primary pack, branding, susceptibility to damage, product turnaround, supply chain, the level of

2 Designing for Sustainability

Table 2.11 Suitability of products for returnable packaging

Returnable transport packaging	Non-returnable packaging
Loose product	Highly branded product
Certain bagged product	Products with high distribution costs
Easily damaged product	Imported product
Manually packed product	Large items
Fast turnaround, closed loop product	Products produced and packed on high speed packaging lines

Sources: PIRA [49], Saphire [50]

Table 2.12 Strategies for reusable secondary or tertiary packaging

To optimise its environmental performance, reusable packaging should:
• be durable and designed for easy maintenance including cleaning
• be collapsible and/or nest-able to optimise return transport capabilities
• be as lightweight as possible
• incorporate recycled content where possible
• have facilities available for cleaning, repair or reconditioning
• be recyclable at end-of-life

automation and transport distances. Accordingly, some sectors appear to benefit most from reusable transport packaging systems (Table 2.11).

Saphire [50] found that reusable transport packaging is more likely to be feasible under the following circumstances:

- short distribution distances
- frequent deliveries
- a small number of parties
- company-owned vehicles.

These conditions are most likely to exist in closed loop distribution systems where the container always goes back to the same point of origin.

Many companies use standard off-the-shelf pallets, crates, boxes, drums and intermediate bulk containers to transport products, components and raw materials. In other cases, reusable packaging needs to be designed to meet specific needs and to minimise environmental impacts (see Table 2.12).

Use of Recycled Materials

Every attempt should be made to maximise the use of materials with recycled content as they:

- generally consume less energy to manufacture (see Table 2.13)
- reduce consumption of virgin material and reliance on non-renewable resources
- often generate less pollution and greenhouse gas emissions because they avoid the manufacture of virgin materials.

In some cases recycled material may also offer a cost advantage.

Table 2.13 Energy savings from the use of recycled rather than virgin material

Material	Energy saving from use of recycled material (%)
Corrugated board—unbleached	22
Steel	79
Aluminium	93
HDPE	79
PET	76
PVC	80
Glass	57

Source: Grant et al. [20, p. xi]

The use of materials with recycled content may be limited by:
- the function of the packaging or packaged product
- supply constraints
- health and safety standards.

⇒ Food contact regulations are discussed in Chap. 4.

Recycled polymers can only be used in direct food contact applications if they meet stringent safety standards. The exception is non-processed fruit and vegetables. In other applications the recycled polymer needs to be certified by the appropriate food safety authority. The test standard that is often applied is the US Food and Drug Administration's (FDA) standard for recycled materials in direct food contact [51]. Recycled resins that meet the FDA standard have either undergone a feedstock (chemical) recycling process or a 'super clean' mechanical process involving several cleaning and decontamination stages. Multilayer co-injection techniques, which provide a functional barrier between the recycled resin and the contents of the container, are more expensive and have largely been replaced by monolayer processes.

Recycled PET (rPET) is generally blended with virgin resin at rates of up to 50% in order to meet the strict technical and aesthetic requirements for food grade packaging. It is increasingly being used for primary packaging; for example:
- A percentage of rPET is used in Coca Cola bottles in the United States, Netherlands, Belgium, Switzerland, Germany, Sweden, Australia, Japan and Mexico, and an extensive trial in the United Kingdom established that it can be combined with virgin resin at rates of up to 50% [52]
- In 2007 McDonald's Australia replaced its virgin polystyrene dessert cups with PET cups containing 35% recycled content [53]
- Direct Pack Inc. in collaboration with Global PET began production of its 100% rPET takeaway food containers ('The Bottle Box') in California in 2009 [54].

A large scale trial of recycled rPET in retail packaging was undertaken by WRAP in the United Kingdom between 2004 and 2006 [55]. As part of this trial, a percentage of recycled resin was incorporated in a selection of Marks & Spencer's takeaway salad bowls (50%) and juice bottles (30%), and Boots' toiletry bottles (30%). The trial demonstrated that rPET could be successfully incorporated within the containers, and both companies expressed their willingness to continue rolling out the use of rPET across additional lines. More detail is provided in the case study on Marks & Spencer's 'food to go' range below (Case Study 2.14).

Case Study 2.14 Recycled PET (rPET) in Marks and Spencer's 'Food to Go' Range

In June 2005 Marks & Spencer announced the launch of its rPET range of packaging to coincide with the re-launch of their 'Food to Go' range of food and beverage products. Market research into customers' perceptions about Marks & Spencer's packaging highlighted some concerns about rigid plastics and polystyrene. In response, the company developed a range of sustainable packaging initiatives including the use of rPET in a range of 'to go' packaging lines. The initial trial was undertaken with the support of WRAP and Closed Loop London between August 2004 and February 2006.

The target percentage of recycled resin was based on the level that was considered technically and aesthetically feasible as well as the need to include a 'meaningful' percentage rather than a 'tokenistic attempt to appear to be offering a greener packaging solution' [55, p. 54]. As a result of these considerations, the decision was taken to incorporate 50% rPET in the thermo-formed sheet used to make salad bowls and 30% rPET for blow moulded juice and 'smoothie' bottles. The containers were labelled with a 'closed loop' recycling symbol and the words '50% recycled content' and '100% recyclable'. Collection bins were provided in some stores to collect packaging labelled with the 'closed loop' symbol, including the PET containers and paperboard sandwich packs.

Results were as follows:
- The recycled product was safe, meeting regulatory requirements for plastics in contact with food
- There were no problems with material clarity and colour
- The rPET was able to be processed on existing equipment with only minor changes to the equipment used to manufacture the bottles. There was no impact on production efficiency
- Raw material costs were comparable or slightly better than for virgin PET
- There was continuity of supply for the recycled material
- Customer feedback was very positive.

An important finding was the need to closely specify material standards for the rPET to ensure that high quality standards were achieved for the packaging.

(continued)

Case Study 2.14 (continued)

> The next stage of the roll out was expected to incorporate rPET in additional lines, including more juice and smoothie bottles, flavoured milk bottles, pre-prepared fruit salad trays, dessert pots and prepared vegetable trays and boxes.
>
> *Sources*: Churchwood et al. [55], Marks & Spencer [56]

Recycled PET is also used to make bottles for detergents and other household products, although in these applications PET competes with PVC and HDPE. For this reason the market is highly price sensitive.

Recycled HDPE is widely used to manufacture bottles for non-food products, such as detergents, bleach and other household chemicals. It is very rarely used for food and beverage packaging, but Marks & Spencer's organic milk bottles are now manufactured from a blend of recycled HDPE resin certified to the FDA standard (10%) and virgin HDPE [57]. Post-consumer HDPE is also blended with LDPE or LLDPE to produce films for carry bags and rubbish bags.

Recycled paper and cartonboard can be used for some food-contact packaging as long as the sources of the fibre are known (some sources are not acceptable, such as paper from mixed waste) and the recycled material has been processed and cleaned to a level that meets all food safety requirements. Swiss researchers have found traces of mineral oils in recycled cartons at unacceptably high levels [58]. The oils are from printing inks in newsprint, which cannot be removed completely during the recycling process. One solution is to pack foodstuffs that are especially susceptible to mineral oil migration in an inner liner bag [58]. Another is to improve the efficiency of the recycling process to improve the removal of mineral oils. Recycled fibre can be used in secondary packaging or in the inside liners of multi-wall corrugated paperboard (for example, a double or triple wall). These components can be manufactured from 100% recycled fibre without seriously affecting performance.

> ⇒ Detailed guidelines on the use of recycled content in plastics packaging are available from the Sustainable Packaging Coalition [59].

A high level of recycled content may require an increase in board weight (see Case Study 2.15). These types of trade-offs need to be considered by the design team in the context of the business's overall corporate and sustainable development goals and metrics (see Chap. 8).

To optimise the use of recycled material in packaging, it's necessary to [5]:
- determine whether the technical requirements of the packaging can be met using recycled material, and if so how much can be used
- find suppliers with dependable sources of recycled materials that meet the business's packaging requirements
- set internal goals for the use of recycled material.

> **Case Study 2.15 Source Reduction Versus Recycled Content Paperboard**
>
> As a general rule, source reduction (e.g. through lightweighting) is preferable to recycling. Recycled materials have a lower environmental impact against most indicators than the equivalent virgin material (see Table 2.13), but using less material in the first place has an even lower impact.
>
> For folding cartons, using 100% recycled paperboard may require a slight increase in the weight of the packaging compared to solid bleached sulphate (SBS) or coated unbleached kraft (CUK). This example provides an exception to the rule that source reduction is always preferable because the environmental impact for recycled paperboard is substantially less than the environmental impact for virgin board on a weight-for-weight basis. The recycled board is environmentally preferable even if the carton is 10–20% heavier.
>
> A smaller percentage of recycled fibre (20–30%) can be added to SBS or CUK board without increasing its weight.
>
> *Source*: The Paper Task Force [60, pp. 100–101]

Design for Mechanical Recycling

> **Recyclable** means: 'a characteristic of a product, packaging or associated component that can be diverted from the waste stream through available processes and programs and can be collected, processed and returned to use in the form of raw materials or products' [61, p. 13].

The recyclability of a packaging material depends on two things:
- its technical recyclability: the ease with which it can be reprocessed and used to manufacture new products
- the availability of collection, sorting and reprocessing facilities for the material.

> ⇒ Consult with material recovery facility (MRF) operators and recyclers on the recyclability of any new packaging system, particularly if it uses a combination of materials.

A systems approach is therefore required; one that considers both the design of the package and the availability of a recovery system. A material may be *technically* recyclable, but if material recovery facilities (MRFs) and recyclers do not have the technology to separate and reprocess it, or if there is no viable end-market for the material, then it is effectively non-recyclable (see Fig. 6.2 for a description of MRFs).

> ⇒ More detail on the recyclability of individual materials is provided in Chap. 6.

Technical recyclability depends on the characteristics of the material itself as well as recycling technologies. As new technologies are developed and become commercially available, more materials are likely to be considered 'recyclable'.

Recyclability also depends on the availability of recovery and recycling services, which vary by geographic region (Table 2.14). The packaging materials most widely collected through kerbside and 'drop-off' services are glass

Table 2.14 Packaging material recycling rates by geographical region (%)

Country	Glass	Paper and paperboard	Plastics	Steel	Aluminium	Total metals	Wood	Total
European Union	64	77	28	NA	NA	67	41	59
Austria	86	84	33	NA	NA	67	19	67
Belgium	100	92	38	NA	NA	91	72	80
Bulgaria	71	98	20	NA	NA	0	0	55
Cyprus	10	39	14	NA	NA	70	22	26
Czech Republic	65	94	46	61	31	56	37	66
Denmark	128	61	22	NA	NA	87	33	57
Germany	84	80	43	91	74	90	30	67
Estonia	62	57	38	NA	NA	18	39	50
Finland	81	88	18	NA	NA	70	10	52
France	62	89	21	57	40	64	21	57
Greece	18	80	14	54	34	51	75	48
Hungary	21	87	17	NA	NA	65	20	46
Italy	60	70	28	59	54	67	54	57
Ireland	76	77	22	NA	NA	68	76	61
Latvia	35	58	23	NA	NA	50	24	40
Liechtenstein	63	77	3	100	100	100	0	88
Lithuania	36	68	29	NA	NA	57	32	43
Luxembourg	92	71	39	NA	NA	80	31	63
Netherlands	81	74	26	NA	NA	84	32	61
Norway	99	82	30	66	NA	66	NA	68
Poland	40	69	28	21	82	30	48	48
Portugal	46	82	15	NA	NA	63	71	57
Romania	17	61	15	NA	NA	55	9	31
Slovakia	55	86	42	NA	NA	74	5	61
Spain	56	61	23	NA	NA	63	61	52
Sweden	95	74	42	77	69	74	17	59
United Kingdom	55	79	23	56	31	52	77	59
Australia	46	65	31	38	70	49	NA	56
Japan	NA	61	NA	88	91	NA	NA	NA
Korea	72	69	NA	NA	NA	NA	NA	NA
New Zealand	62	78	23	47	88	NA	NA	60
United States	28	62	12	65	39	NA	15	43

Notes NA not available. Rates are calculated as material recycling as a percentage of material consumption, although the methodology for data collection varies between countries. *Sources* Europe (2007) [93]; Australia (2007) [94, p. 6]; US (2007) [95, p. 7]; New Zealand (2007) [97]; Japan (2006) [98, pp. 73, 74, 79]; Korea [99, pp. 25–26] (2004)

containers, aluminium and steel cans, selected plastic containers (often PET; sometimes other plastics), paper bags, cartons and corrugated boxes. Milk and juice cartons (liquid paperboard) are also collected in some areas.

Recyclable packaging is generally collected as a mixed (commingled) stream and then sorted at a MRF, although sometimes individual materials, particularly paper and paperboard, are collected separately. At the MRF, individual materials are sorted and compressed or baled for transport to reprocessors, who then use these materials to manufacture new raw materials or products.

⇒ Detailed guidelines for specific materials have been published by the Packaging Resources Action Group [62] and Recoup [63].

The use of multiple materials can inhibit recycling or cause problems in the recycling process. For example, plastic 'windows' on pasta boxes, plastic film on tissue boxes and the moulded plastic on blister packs are separated in the paper recycling process but will end up in the waste stream. Plastic or wax coatings on paper also reduce the amount of fibre that can be reclaimed. An example of a business working to improve the recyclability of its packaging is Amazon (see Case Study 2.16). Its 'frustration free packaging' is marketed as easier to open but has a range of other benefits including recyclability.

If more than one material is used (for example, plastics and paperboard), consumers should be advised to separate the two materials before recycling (see Sect. 3.5.2). The use of adhesives to attach different materials, such as foam cushions to corrugated board, should also be avoided. Specific strategies for individual materials are provided in Table 2.15.

Case Study 2.16 Amazon's 'Certified Frustration Free Packaging'

Amazon is working with its vendors to supply products in 'frustration free packaging', which means it is:
- easy to open
- recyclable
- ships in its own package without an additional shipping box.

Certified packaging can be opened without the use of a knife or box cutter. It is recyclable because it does not include any additional components such as plastic clamshell casings, plastic bindings and wire ties.

Amazon has also developed software to determine the 'right sized' box for each product based on dimension and weight. This helps to avoid over-packaging.

Source: Amazon [64]

Table 2.15 Design strategies to improve recycling

Material	Recycling strategies
Plastic packaging	Specify a plastic that is recyclable in the intended markets for the product (i.e. a recovery system is available to most consumers)
	Try to use only one material, or material combinations that are compatible in the recycling process (see Table 6.7)
	Avoid multi-layer containers
	Try to ensure that polymers used for auxiliary components such as labels, closures, liners and cap seals match that of the container
	If auxiliary components are manufactured from a different material to the container, ensure that the different materials can be easily separated during the washing process
	Consult with recyclers to establish if any components will be problematic in the recovery process or end-product
	Un-pigmented polymers are more valuable as recyclate than pigmented. If colour is required, try to limit it to labels
	Avoid fillers that change the density of the plastic or minimise their use, as they lower the quality of the recycled material
	PVC and PET are incompatible in the recycling process. Avoid PVC labels, closures or tamper-proof seals on consumer packs made from PET
	Avoid wet-strength paper labels on plastic packaging, as they do not disintegrate into pulp during the wash phase and will contaminate the polymer
	Avoid metallic labels and aluminium closures and seals, as they can severely impact the viability of polymer recycling
	Avoid pressure-sensitive adhesives that cover the entire back of the label, as they are difficult to remove and contaminate the recycled polymer
	Incorporate recycled content where possible (subject to food contact requirements)
	Label rigid packaging with the relevant identification code and recycling symbol (see Sect. 3.5.2)
Paper and boxboard	Avoid the use of wax or aluminium coatings, which reduce the yield of recycled fibre
	Avoid plastic/aluminium laminates
	Check with recyclers to ensure that polymer coatings and varnishes, if required, are compatible with the recycling process
	Minimise or avoid the use of non-paper components (e.g., foam pads, plastic film windows, metal tear strips, plastic handles, etc.)
	Minimise the use of inks
	Do not use inks, dyes and coatings that contain heavy metals
	Minimise the use of adhesives; e.g. by using mechanical fastenings such as interlocking tabs
	Check with local recyclers to find out whether they would prefer a water-soluble adhesive or one that can be easily separated out in the pulping process (e.g. some hot melts)
	Avoid highly wet-strength paper (e.g. labels) or cartonboard. These generally don't break down, causing blockages in the pulping process

(continued)

2 Designing for Sustainability

Table 2.15 (continued)

Material	Recycling strategies
	If extra labels are required, they should be made of a paper-based material
	Incorporate recycled content where possible (subject to food contact requirements)
	Label recyclable packaging with the relevant recycling symbol (see Sect. 3.5.2)
Glass	Avoid dark green, dark blue or black glass. These may contaminate recyclable glass (which is sorted into clear (flint), amber and green glass with strict specifications)
	Avoid components that are problematic in the glass recycling process, such as cobalt blue pigment, metal tamper-evident rings and metal-based inks for on-glass printing [5]
	Label with the relevant recycling symbol (see Sect. 3.5.2)
Steel and aluminium	Avoid inks and coatings that might be a contaminant or result in problematic emissions at the refinery (e.g. lead based inks and chlorinated plastics) [5]
	Avoid features made from other metals; e.g. aluminium foil on steel cans [5]
	Use appropriate labelling to encourage consumers to recycle them after use (see Sect. 3.5.2)

Assess the Role of Biodegradable Polymers

Biodegradable polymers are increasingly used because of their potential benefits at end-of-life; for example, as a raw material for composting processes. Many are also made from renewable materials and could, if widely used, reduce reliance on oil for manufacturing plastics.

⇒ Read more about the advantages and disadvantages of biodegradable polymers in Chap. 6.

Biodegradable polymers may be a good option for:
- short life products that are insensitive to moisture and oxygen, that do not require heating in-pack, and are non-carbonated [1]
- packaging that is currently *not* recyclable through existing material recycling systems, such as film and bags
- packaging that tends to contaminate food recovery systems (see Case Study 2.17).

When using biodegradable polymers the design process needs to ensure that the materials will *actually* be recovered by assessing that:
- the material has been certified to a recognised international standard for biodegradability or composting
- the infrastructure exists to collect and reprocess the material.

> **Case Study 2.17 Food Waste**
>
> The waste from fast food outlets comprises fresh produce scraps and packaging from back-of-house, and food and packaging from the restaurant. If the food service packaging is made entirely from paper, cartonboard or compostable polymers (plates, cups, straws, napkins, etc.) then the waste does not need to be separated. It can all be sent to a composting or anaerobic digestion facility.
>
> The waste from supermarkets also includes fresh produce past its prime and any associated packaging, such as plastic bags, film, trays and clamshells. The use of biodegradable polymers for this type of packaging facilitates the recovery of food waste by reducing contamination from plastic packaging (a strategy being pursued by retailers such as Sainsbury's).
>
> Bags made from biodegradable polymers can be used for the collection of compostable food waste and yard waste (weeds and fallen leaves) from households. Once again, this means that the packaging does not need to be separated before the organic material is disposed of in a home composting bin or delivered to a commercial composting facility.

Where biodegradable polymers are used, the packaging should be designed so that it does not compromise the degradation process. For example, coatings and pigments may interfere with degradability and compost certification. Toxic heavy metals in pigments and printing inks (for instance, lead, cadmium, mercury or chromium) could also have eco-toxicity impacts. Potential risks should be assessed to minimise any ecological or health effects during manufacture, use and disposal [65].

Packaging designed for organic recovery, whether through a commercial or home composting system, needs to be labelled appropriately. Consumers should be advised that the material is biodegradable and be given information on how they can dispose of it correctly (see Sect. 3.5.3).

Key steps in the decision-making process for biodegradable packaging are shown in Fig. 2.5. When selecting paper and biodegradable polymers, the designer needs to confirm that they are based on sustainably harvested feedstocks. LCA should also be used to understand and validate the environmental benefits. All materials, whether sourced from renewable or non-renewable materials, have impacts upon the environment, and performing an LCA ensures that all necessary life cycle stages are assessed; that is, that materials are compared on a level playing field.

2 Designing for Sustainability

```
┌─────────────────────────────────────────────────────────────────┐
│ Will a biodegradable polymer improve the sustainability of the  │
│ packaging?                                                      │
└─────────────────────────────────────────────────────────────────┘
                                  ↓
┌─────────────────────────────────────────────────────────────────┐
│ Identify the environments where the packaging is likely to end  │
│ up:                                                             │
│   • terrestrial, e.g., composting, anaerobic digestion, litter  │
│     in soil, litter above ground                                │
│   • aquatic, e.g., fresh water, sea water                       │
│   • wastewater, e.g., domestic, commercial or trade waste.      │
│ Will it contaminate any recycling streams? Discuss concerns     │
│ with recyclers and consider labeling.                           │
└─────────────────────────────────────────────────────────────────┘
                                  ↓
┌─────────────────────────────────────────────────────────────────┐
│ Select a shortlist of materials that will facilitate degradation│
│ in the right environment, e.g., compostable to EN 13432 or      │
│ water-soluble.                                                  │
└─────────────────────────────────────────────────────────────────┘
                                  ↓
┌─────────────────────────────────────────────────────────────────┐
│ Determine the functional requirements of the material, e.g.,    │
│ processing requirements, shelf life.                            │
└─────────────────────────────────────────────────────────────────┘
                                  ↓
┌─────────────────────────────────────────────────────────────────┐
│ Select the most appropriate biodegradable material for the      │
│ application.                                                    │
└─────────────────────────────────────────────────────────────────┘
                                  ↓
┌─────────────────────────────────────────────────────────────────┐
│ Design for degradability:                                       │
│   • Minimise wall thickness, pigments and coatings              │
│   • Avoid pigments and inks containing heavy metals or other    │
│     substances that might contaminate the end product           │
│     (particularly for compost).                                 │
└─────────────────────────────────────────────────────────────────┘
                                  ↓
┌─────────────────────────────────────────────────────────────────┐
│ Labelling:                                                      │
│   • Ensure that any degradability claims comply with Trade      │
│     Practices legislation and ISO 14021                         │
│   • Consumers should be advised about appropriate disposal or   │
│     recovery.                                                   │
└─────────────────────────────────────────────────────────────────┘
```

Fig. 2.5 The decision-making process for degradable packaging. *Source*: Based on PACIA [67, p. 12]

Coca-Cola, for example, has developed a bottle made from 30% plant derivatives (sugar cane), which it claims can be recycled through conventional recycling systems without contaminating the recycled polymer [66].

Support Renewable Energy Growth

Renewable energy is generated from sources such as water (hydro power), wind, biomass (e.g. incineration or anaerobic digestion) and solar. Renewable energy is beneficial because it reduces depletion of non-renewable resources and greenhouse gas emissions.

Strategies to promote the use of renewable energy include:
- generating power on-site; for example, through the installation of solar panels on roofs
- using renewable transport fuels; for example, biofuels, where these do not conflict with food security and are found to have the lowest impact
- purchasing 'renewable energy credits' to match the electricity used by the company
- purchasing 'carbon credits', which offset the greenhouse gas emissions of a product or business. The money paid to organisations for carbon credits is used to fund projects such as infrastructure for renewable energy or tree planting to absorb greenhouse gas emissions.

Frito-Lay has purchased renewable energy credits to offset the electricity consumed at all of their US-based manufacturing facilities and has installed solar panels at its manufacturing facility in Modesta, California [68]. These initiatives have allowed the company to use the 'Green-e' label on their SunChips brand of potato chips.

2.4.4 Designing for Safety

Safe +
Non-polluting & non-toxic

- Avoiding hazardous substances
- Cleaner production
- Ecological stewardship
- Litter reduction

By focusing on the safe principle the design team will:
- understand the complete life cycle of their packaging component
- identify and avoid the use of hazardous substances in their products
- identify and avoid the production of hazardous substances (including greenhouse gases) throughout the life cycle of the packaging components they use
- identify strategies to reduce litter and the impacts of litter in relevant ecosystems.

Identify and Avoid Hazardous Substances

Conventional risk management principles involving risk identification and hazard risk analysis should be applied to the selection of materials, inks, pigments, coatings, plasticisers and other substances used to produce or use the packaging. A risk management approach involves the following steps [69]:

1. Define the review mechanism.
2. Identify opportunities, risks and barriers.
3. Assess the factors that are within the control of the organisation.
4. Ensure that those within the control of the organisation are acted on.
5. Report on the process.

The design team needs to fully understand the production and manufacturing processes for their packaging and products. A risk assessment should identify any substances used or emitted at any stage of the life cycle of packaging components and their use, including recovery, reuse and reprocessing, that might be toxic to workers, consumers or ecosystems. Information should then be sourced to appropriately assess the safety risk and ensure that packaging is designed to avoid the substances or, as a minimum, that known public safety standards are met. Information, including acceptable limits where applicable, should be included in life cycle maps and packaging specifications.

⇒ The European Printing Industry Association has developed an 'exclusion list' for printing inks and related substances based on health and safety concerns. The list includes substances classified as carcinogenic, mutagenic or toxic in relevant European directives and pigment colourants based on antimony, arsenic, cadmium, chromium (VI), lead, mercury and selenium [70].

Bisphenol A (BPA) and phthalates (see below) are two examples of substances that were considered safe but are now the subject of further research and development to overcome potential risks associated with their widespread use.

A document published by Ciba Speciality Chemicals (now BASF) provides information on the compliance of specific pigments and dyes with the European Packaging Directive, US and other international regulations [71].

Bisphenol A

BPA is a chemical used to make polycarbonate and epoxy resins. Polycarbonate is used in the manufacture of reusable baby bottles and reusable outdoor drink bottles, while epoxy resins line most metal food and beverage cans. BPA prevents packaging materials imparting any taste to the product, and it is highly stain-resistant [72].

⇒ Proposals to restrict the use of BPA are discussed in Sect. 4.2.3.

> ⇒ For a review of the literature on health and environmental risks linked to BPA, see reports by Environment California [74] and Green Century Capital Management [72].

BPA leaches out of both plastics and has been found at very low concentrations in food and beverages packaged in these materials. For example, a test of canned foods by Consumer Reports in the United States found BPA in almost all of the 19 products tested [73]. It is absorbed by the human body: a study cited by Environment California [74] found that BPA is present in the urine of 95% of Americans. Many peer-reviewed studies have linked these low dosages to a wide range of developmental and health problems, including prostate effects, breast cancer, heart disease, obesity, attention deficit, altered immune system and early puberty. Pregnant women, infants and children have been found to be most at risk. Environment Canada also noted that BPA enters the environment through waste water, washing residues and leachate from landfills, and has potential to build up in waterways and harm fish and other organisms [75].

As a result, some national, state and local governments have moved to regulate the use of BPA in packaging for infants and young children, particularly in baby bottles and infant formula packaging. Walmart, Toys 'R' Us and Wholefoods have voluntarily stopped selling baby bottles made with BPA, and some food and packaging manufacturers are investigating alternatives to BPA in their packaging [72] (see the Heinz Case Study 2.18).

Case Study 2.18 BPA: Heinz Baby Food

Heinz is considered an industry leader in phasing out the use of BPA because the company has eliminated it from the epoxy resin lining in its baby food cans and has started to remove BPA from baby food jar lids in the United Kingdom [72]. This move is largely in response to consumer concerns about its potential health impact rather than any explicit acknowledgement of a health risk:

'Heinz … is pleased to be recognized for our leadership in moving to alternative materials that are Bisphenol A (BPA) free. Heinz has been a leader in food safety ever since our founder started this company in 1869. Although scientific bodies worldwide have concluded that minute levels of BPA are safe, Heinz is proactively exploring alternatives to BPA in response to consumer opinion' [78].

Sources: Green Century Capital Management and As You Sow [72], Heinz [78]

Alternative polymers to polycarbonate include polyamide for baby bottles, and tritan copolyester for reusable drink bottles. There are also alternatives to epoxy coatings on metal cans, including polyester coatings and natural oils and resins, but these tend to cost more and are less effective for highly acidic foods such as tomatoes [72]. Japanese businesses voluntarily reduced their use of BPA between 1998 and 2003 after BPA was detected in canned drinks. According to the Environmental Working Group in the United States [76], companies switched to either a PET lining or an epoxy resin with much lower BPA migration. Another option is polypropylene-lined cans [77].

One of the challenges for manufacturers and regulators is the need to ensure that alternatives to BPA are also thoroughly tested and found to be safe.

Many food and beverage containers, such as the water bottle shown in Photo 2.1, are now BPA-free.

Photo: Helen Lewis

Photo 2.1 BPA-free water bottle

Phthalates

Phthalates are a group of chemicals widely used in personal care products (shampoos, lotions, liquid soaps and so on) and some packaging. They look like clear vegetable oil and are used as 'plasticisers'; for example, to make PVC more flexible. Phthalates can comprise 10–50% of flexible PVC by weight [79].

Like BPA, phthalates can be absorbed in the body through migration into food (in the case of packaging) or through other forms of contact. They appear to act as endocrine disrupters in the human body, and research studies have linked phthalate exposure to health problems including reduced male fertility and rising rates of testicular and prostate cancer. While it is certain that everyone is exposed to low levels of endocrine-disrupting chemicals (including phthalates), there is still a considerable amount of scientific uncertainty about their health impacts [79, p. 14]. A particular concern to health advocates and regulators is the exposure of small children to phthalates in toys because they are more likely to put toys in their mouth. Children are more at risk from ingested or inhaled pollutants because they have less well-developed detoxification mechanisms.

Phthalates are also a common environmental pollutant, as they have been used in a wide range of products since the 1940s. However, the toxicity risks are limited because they readily biodegrade in aerobic environments and their concentrations are generally below levels likely to have toxicity or reproductive impacts on living organisms [80].

While restrictions on the use of phthalates have targeted children's products rather than PVC in general (see Sect. 4.2.3), the risks of using PVC for food and beverage packaging need to be carefully assessed. Based on a comprehensive review of the available data, one academic noted that while 'there is a lack of scientific evidence showing that phthalates have an adverse effect on humans at levels likely to be encountered either environmentally or during normal use of phthalate containing products…the possibility that such a link will be established in future should not be discounted' [79, p. 17].

A common application of PVC in food packaging is the ring of rubbery material, or gasket, which forms the seal inside the metal lid of a screw-topped jar. Products packed in glass jars were tested by the Australian Consumers Association for the presence of phthalates. Of the 25 products tested, 12 contained phthalates at levels above the maximum limits permitted in the European Union [81].

There are many different phthalates used in PVC (see Table 2.16), but the most common is DEHP. (See Table 2.16 for the full scientific name of the phthalate and others mentioned in this paragraph.) This is also the most dominant plasticiser found in the environment. In Europe, DEHP is mainly being replaced by DIDP and DINP, which have been given a lower risk rating by the European Union [80, p. 26]. DEHP, DBP and BBP are classified in the European Union as reproductive toxicants [82]. There are three types of non-phthalate plasticiser suggested as replacements for problematic phthalates: adipates, citrates and cyclohexyl-based plasticisers, although these tend to be more costly and are yet to undergo risk assessments in the European Union [80, p. 26]. A recent innovation is the development by Danish company Danisco of a biodegradable plasticiser to replace phthalates in PVC. The plasticiser is manufactured from castor oil and acetic acid and has been approved for food contact in Europe [83].

Table 2.16 Common phthalate plasticisers used in PVC

Chemical name	Abbreviation
Dimethyl phthalate	DMP
Diethyl phthalate	DEP
Dibutyl phthalate	DBP
Disobutyl phthalate	DIBP
Di-n-hexyl phthalate	DHP
Benzyl butyl phthalate	BBP
Diethylhexyl phthalate	DEHP
Dioctyl tere-phthalate	DOTP or DEHT
Diisooctyl phthalate	DIOP
Diisononyl phthalate	DINP
Diisodecyl phthalate	DIDP

Source: [80, p. 21]

Table 2.17 Examples of heavy metals in packaging

Packaging component	Heavy metal	Source
Glass packaging	Lead	Recycled glass (e.g. lead crystal, automobile glass, mirrors, TV screens)
Plastic crates and pallets	Lead, cadmium and chromium[a]	Black, brown, green, dark blue, orange, red and yellow pigments. Some heavy metals (no longer used in virgin polymers) made from recycled material
Coloured plastic nets	Lead and chromium[a]	Red, yellow and orange pigments
Plastic caps	Cadmium	Yellow, orange, red and green pigments
Plastic shopping bags	Lead and chromium[a]	Gold, yellow, orange, red and green pigments
Plastic non-food bottles	Lead, cadmium and chromium[a]	Yellow, orange and green pigments
Plastic foils coated with aluminium	Lead and chromium[a]	Red, gold and silver coatings

[a] Not all of the chromium was chromium VI. This tends to be associated with red and orange pigments
Source: Based on PIRA International and ECOLAS [42]

Identify and Avoid Heavy Metals

The European Packaging and Packaging Waste Directive specifies that the combined weight of heavy metals (lead, cadmium, mercury and hexavalent chromium) in packaging or packaging components should not exceed a concentration of 100 ppm. 'Toxics in packaging' laws in the United States have the same limit but are stricter than the European Directive because they also prohibit the 'intentional' introduction of any amount of the four restricted metals. Some recycled materials contain heavy metals, but this is acceptable under the European Directive and similar state laws in the United States.

Testing in Europe and the United States has found continuing use of heavy metal based pigments, inks and stabilisers for packaging (see examples in Table 2.17). US tests have also found high levels of heavy metals in shopping bags, particularly lead, mercury and chromium [84], arising from the use of solvent-based inks. A high percentage of flexible PVC bags have also failed tests, including 'zipper bags' used to package bedding and other home furnishings and pouches for pet toys and chews. Almost all of these were imported from Asia.

Support or Use Cleaner Production Initiatives

A full understanding of manufacturing and printing processes may highlight opportunities to reduce the environmental impacts of packaging with cleaner production technologies. Two common pollutants that can be minimised by changing specifications at the design or procurement stage are discussed below: volatile organic compounds (VOCs) and organochlorine compounds.

Emissions of Volatile Organic Compounds

VOCs are natural or synthetic organic substances that have a tendency to vaporise during handling or use, and emissions can be harmful or toxic if inhaled. They can also combine with sunlight and nitrous oxides to generate low-level ozone [85]. Sources of VOC emissions in the packaging industry include solvent-based inks and adhesives (including laminates), as well as cleaners used in printing processes.

Alternatives to solvent-based inks include water-based, ultra-violet curable and litho inks, although these tend to require more energy and may not be suitable for all applications [85, pp. 68–69].

According to Envirowise [85], water-based adhesives or hot melts can be used in some applications instead of solvent-based adhesives to reduce VOC emissions. Hot melt adhesives can cause problems in the paper recycling process, however, because they break up. Because of their similar density to water and fibre, they are difficult to remove. Care should be taken to specify adhesives with a higher or lower density, which are therefore easier to remove from the pulp (such as newer ethylene–vinyl acetate (EVA) hot melts and fast drying polyurethane rubber adhesives). Water-based adhesives do not generate any VOCs but may require more energy for drying and are not suitable for all applications [85].

Henkel has developed a solvent-free lamination adhesive (polyurethane) for food packaging, which according to the company reduces emissions, energy costs and cure times [86].

Chlorine Bleaching Processes for Paper

Elemental chlorine has traditionally been used as the bleaching agent in pulp mills to produce white paper. The wastewater from these mills contains organochlorine compounds such as dioxins that are toxic in the natural environment. Chlorine dioxide is less polluting than chlorine gas and is increasingly used by paper mills. Chlorine combines with lignin (the 'glue' that holds the wood fibre together) to create organochlorine compounds that end up in wastewater, whereas chlorine dioxide breaks apart the lignin and creates organic compounds that are water-soluble and similar to those occurring in the natural environment [87]. Processes that have replaced all of the elemental chlorine with chlorine dioxide are referred to as elemental chlorine-free (ECF). While a significant improvement, ECF processes still generate chlorinated compounds, which make the wastewater too corrosive to recycle. The result is that effluent is treated and discharged to receiving waters [60].

There are alternatives to traditional chlorine bleaching:
- replacing chlorine compounds with oxygen-based compounds in the first stage of the bleaching process, which allows the waste water from this stage to be reused
- replacing all chlorine compounds in the bleaching process with oxygen-based chemicals such as ozone or hydrogen peroxide, potentially allowing all the wastewater to be reused. (In reality most mills moving to a totally chlorine free process still discharge wastewater to the receiving environment [60].)

Processes that have eliminated all chlorinated bleaching agents are referred to as totally chlorine-free (TCF). The Chlorine Free Products Association in the United States has introduced an eco-labelling scheme for TCF and processed chlorine-free (PCF) products [88]. The PCF logo can be used for recycled papers that meet minimum recycled content standards and are bleached without any chlorine compounds (see Sect. 3.5.6).

> ⇒ Chapter 3 provides information on the use of logos and labels.

To reduce the environmental impact of bleaching processes for paper and paperboard packaging, it's necessary to:

> ⇒ Chapter 6 describes the environmental impacts associated with paper recycling.

- use unbleached fibre where feasible, or
- if white paper or paperboard is required, specify TCF or PCF fibre.

Greenhouse Gas Emissions

Greenhouse gas emissions are generated at every stage of the packaging life cycle: during material extraction or harvesting, manufacturing, filling, transport, use and disposal. Most of these emissions, particularly carbon dioxide, are associated with energy consumption, but methane is also generated when organic materials break down in landfill.

Many of the strategies to reduce energy consumption and associated greenhouse gas emissions have already been discussed, including reducing the size or weight of packaging and using recycled rather than virgin materials. Emissions can also be reduced in other aspects of the business; for example by:

- undertaking an energy audit, which will identify opportunities to reduce energy consumption in manufacturing, administration and distribution processes
- purchasing renewable energy or 'carbon offsets'.

Some businesses are using 'carbon labels' to inform consumers about the greenhouse gas emissions associated with the production of food and packaging (see Sect. 3.5.8). The aim of these labels is twofold: to drive efficiencies in the supply chain and to encourage consumers to purchase lower carbon products.

Ecological Stewardship

It is important to know the source of raw materials, particularly for timber products (pallets and crates) and the fibre used to manufacture paper bags, paperboard packaging and corrugated boxes. Timber and paper products from sustainably managed forests should be specified, with preference for those certified by a third party organisation such as the Forest Stewardship Council (FSC) (see Sect. 3.5.7). A number of other national schemes have been assessed and approved by the Program for the Endorsement of Forest Certification, a non-government organisation which has its own labelling scheme for certified products. Demand from pulp and paper manufacturers for woodchips certified by the FSC is starting to drive change in forestry operations. For example, Australian suppliers of wood and woodchips faced a downturn in demand in 2009, particularly from Japanese customers who didn't want to buy woodchips from native forests [89]. As a result, the Tasmanian state government has asked Forestry Tasmania and the largest woodchip exporter, Gunns, to seek FSC certification.

There are no certification schemes for sustainable sourcing of other packaging materials, but similar issues need to be considered during the design and procurement process:
- How and where was the material extracted/harvested?
- How are these impacts managed?
- Do suppliers comply with all relevant legislation?

Similarly, it is important to understand the raw materials and processes used to manufacture biopolymers. Is the raw material grown using sustainable agriculture principles? Are biopolymers competing for food supplies and helping to drive up prices?

Litter Reduction

Design for litter reduction is important for products likely to be consumed away from home, such as single-serve beverages, sweets, snacks and salads. Structural design can assist by minimising the number of parts that break away from the main pack and are likely to end up as litter. For example, the 'ring-pull tabs' on aluminium drink cans used to completely detach from the can after opening. These were sharp and caused cuts when people accidently stood on the tabs. The tab was redesigned so that after lifting it is levered beneath the opening and stays attached to the can [90].

For packaging such as takeaway food packs and straws that often end up in the litter stream, the use of a biodegradable material such as paper or cartonboard is preferable. Biodegradable polymers certified to a relevant standard may reduce the impacts of litter, but there is insufficient public information available on how fast and to what the extent they break down in open environments, such as soil or the ocean, instead of a controlled composting environment. Messages on the label can also be used to encourage consumers to dispose of the packaging appropriately, in a litter or recycling bin (see Sect. 3.5.5). Table 2.18 includes strategies to prevent the incidence or impact of litter.

Table 2.18 Strategies to prevent the incidence or impact of litter

Minimise the number of separable components that can be littered (e.g. straws, tamper evident seals, trays, spoons and forks)
Provide information to the consumer to encourage responsible disposal
Work with recyclers and local/state governments either directly or through industry associations and non-government organisations to implement public place recycling bins
Where appropriate consider the use of a biodegradable material certified to a relevant standard

2.5 Selecting Materials

> 'There is no such thing as a fundamentally good or bad packaging material: all materials have properties that may present advantages or disadvantages depending on the context within which they are used' [91, p. 8].

The choice of packaging materials has a significant impact on sustainability, but it is not possible to say that a particular material should always be avoided or favoured. The impacts and benefits of a material are highly dependent on how and where it is sourced, manufactured, used and recovered.

Tables 2.19 and 2.20 show how the sustainable packaging framework can be used to evaluate the advantages and disadvantages of materials for a particular application. These are generic examples only—the specific benefits will depend on the product, its packaging requirements, the supply chain, the availability of recycling facilities and so on.

A more detailed description of the life cycle impacts of common packaging materials is provided in Chap. 6.

Table 2.19 Evaluating packaging materials against the four principles of packaging sustainability

Sustainable packaging design principles		Material			
		Aluminium	Steel	Glass	Paper and board
Effective	+ve	Broad range of barrier properties (inert)	Impervious to gases, aromas, light and microorganisms	Impervious to liquids, gases, aromas, and microorganisms	Versatile (e.g. rigid or flexible, can be formulated to give a range of properties)
		Less dense than steel or glass	Opaque (good for food preservation)	Clarity allows the consumer to see the contents	Can have a good printing surface
		Can be pressed and drawn into many different shapes	Can be pressed and drawn into many different shapes	Can be hot-filled	Impermeable to liquid and gas if laminated
		Opaque (good for food preservation)	High impact resistance	Odourless	Transfers heat well
			No refrigeration required		Relatively inexpensive
	−ve	Consumer cannot see the contents	Corrodes in contact with water unless coated	Low impact resistance, leading to high wall thickness requirements	Non-laminated and non-waxed grades are not waterproof, leading to property degradation
			Consumer cannot see the contents	Higher density than plastic or board, which may lead to high environmental impacts from transport	Not a barrier against microorganisms or animals
					Low impact resistance
					Consumer cannot see the contents
Efficient	+ve	Lightweight	No refrigeration required	Can be made with high recycled content	Low density
	−ve	Energy-intensive (virgin material)	Energy-intensive (virgin material)	Energy-intensive (virgin material)	Energy- and water-intensive (virgin material)
Specific design strategies		Downgauge as much as possible	Downgauge as much as possible	Downgauge as much as possible	Downgauge as much as possible
		Maximise recycled content	Maximise recycled content	Maximise cullet content	Maximise recycled content

(continued)

Table 2.19 (continued)

Sustainable packaging design principles		Material			
		Aluminium	Steel	Glass	Paper and board
Cyclic	+ve	Can be recycled an infinite number of times with minimal loss of technical properties	Can be recycled an infinite number of times with minimal loss of technical properties	Can be recycled an infinite number of times with minimal loss of technical properties	Renewable raw material Can be recycled many times but fibre length and strength reduces following subsequent reprocessing
	-ve	Non-renewable resources Many foils are not recyclable due to food contamination, e.g. confectionary wrappers	Non-renewable resources	Non-renewable resources (but abundant) Recovery is limited by mixing of colours in the recycling stream and inability to utilise all fine glass particles	Each cycle reduces cellulose fibre length, reduces properties. This limits the number of cycling loops possible Wax-coated paper and board are not recyclable and liquid paperboard is difficult to recycle
	Specific design strategies	Maximise recycled content Add recycling logo to label Minimise contamination for recycling stream	Maximise recycled content Add recycling logo to label	Maximise recycled content Some colours may be less recyclable in some regions Add recycling logo to label	Recycled content If possible, avoid wax coatings Avoid combining with other materials, e.g. plastics Add recycling logo

(continued)

Table 2.19 (continued)

Sustainable packaging design principles		Material			
		Aluminium	Steel	Glass	Paper and board
Safe	+ve	Minimal direct wildlife impacts in litter	Minimal direct wildlife impacts in litter	Inert—no migration into food products Minimal direct wildlife impacts in litter	Biodegrades in the natural environment – no direct hazard to wildlife
	−ve	Does not degrade	Does not readily degrade	Durable in litter and broken glass is a hazard	Relatively high greenhouse gas emissions from manufacturing/recycling Greenhouse gas emissions from landfill Fibre may be harvested from old growth or unsustainably managed forests Bleaching generates waterborne pollutants, e.g. dioxins
	Specific design strategies	Advise consumers not to litter	Advise consumers not to litter Avoid Bisphenol A (BPA) liners in packaging for children or infants	Advise consumers not to litter	Source from certified forests Avoid bleached packaging Otherwise, use totally chlorine-free (TCF) or process chlorine-free (PCF) bleaching Advise consumers not to litter

2 Designing for Sustainability

Table 2.20 Evaluating thermoplastic polymers against the four principles of packaging sustainability

Sustainable packaging design principles		Thermoplastic polymer material			
		Non-renewable		Renewable	
		Non-degradable[a]	Degradable[b]	Non-degradable[c]	Degradable[d]
Effective	+ve	Wide range of mechanical properties Versatile, can be processed into complex shapes Translucent to opaque, coloured High impact resistance Resistance to most chemicals Lower density than metals and glass			
	−ve	Limited thermal stability Properties may be affected by light		Limited thermal stability Properties may be affected by light Some may not be as water resistant as synthetic polymers. Density may be higher than non-renewable polymers	
Efficient	+ve	Lightweight Can be processed at low temperatures (less energy) Most manufacturing processes generate little waste (scrap can be recycled in-house)			
	−ve	Some processes generate non-recyclable scrap during manufacturing (e.g. laminated film)		Due to higher density and lower strength, may require more material than fossil-based thermoplastics in order to fulfill function Some polymers must be modified to avoid moisture uptake, which adversely affects processing	
	Specific design strategies	Reduce material weight, e.g. downgauge			
Cyclic	+ve	Technically recyclable – depends on packaging format and availability of recycling services	Compostable if certified to a recognised international standard, e.g. for commercial or home composting	Renewable raw materials	Renewable raw materials Compostable through an organic recycling process if certified to a recognised international standard

(continued)

Table 2.20 (continued)

Sustainable packaging design principles		Thermoplastic polymer material			
		Non-renewable		Renewable	
		Non-degradable[a]	Degradable[b]	Non-degradable[c]	Degradable[d]
	−ve	Degradation can occur during recycling (e.g. colour contamination, mixed grades)	Non-renewable Oxodegradable polymers are not compostable and may contaminate recycling of other thermoplastics	Not compostable or currently commercially recyclable. May contaminate recycling of non-renewable thermoplastics	Appropriate composting services may not be widely available. May contaminate recycling of non-renewable thermoplastics
Specific design strategies		Confirm compatibility with recycling services Design for recycling, e.g. specify a polymer that is widely recycled, use only one material	Design for degradation, e.g. minimal inks Include specific disposal guidance for consumers	Include specific disposal guidance for consumers	Confirm compliance with composting standards Design for composting, e.g. minimal inks Include a composting message for consumers
Safe	+ve	Lower greenhouse emissions than metals or glass	Lower greenhouse emissions than metals or glass	May have a higher specific gravity than non-renewable polymers, minimising hazard to marine wildlife	May degrade May have a higher specific gravity than non-renewable polymers, minimising hazard to marine wildlife
	−ve	Some polymers contain BPA or phthalates Some pigments may contain heavy metals Polymers with a specific gravity of less than 1 can pose a hazard to marine wildlife	Some pigments may contain heavy metals Impacts in litter are not well understood	Some pigments may contain heavy metals May be a hazard to wildlife in litter	Some pigments may contain heavy metals May generate higher greenhouse emissions in manufacturing and in landfill Impacts in litter are not well understood

(continued)

2 Designing for Sustainability

Table 2.20 (continued)

Sustainable packaging design principles	Thermoplastic polymer material			
	Non-renewable		Renewable	
	Non-degradable[a]	Degradable[b]	Non-degradable[c]	Degradable[d]
Specific design strategies	Minimise or avoid heavy metals Validate safety for food applications e.g. leaching of monomers, additives Avoid BPA and phthalates in packaging for children or infants Advise consumers not to litter	Validate safety of degradation additives in the natural environment Avoid pigments with heavy metals Advise consumers not to litter	Avoid pigments with heavy metals Advise consumers not to litter	Avoid pigments with heavy metals Advise consumers not to litter

[a] Includes all conventional commodity polymers, e.g. PET, HDPE, PVC, LDPE, PP
[b] Includes biodegradable polymers made from crude oil or natural gas, e.g. aliphatic aromatic copolyesters as well as oxodegradable polymers made by combining a conventional oil/gas derived polymer with a prodegradant additive
[c] Includes renewable polymers made from modified starch specifically designed not to degrade, including those blended with non-renewable thermoplastics
[d] Includes some biodegradable polymers derived from corn starch, wood or other renewable material

2.6 Conclusion

Design is critical to the achievement of packaging sustainability goals. Most of the decisions that impact on sustainable development, including the choice of materials and processing methods, are made at the design stage. For this reason, life cycle thinking must be embedded in the product-packaging development and review processes to achieve better outcomes.

A framework for embedding sustainable development principles into the packaging design process has been presented in this chapter. However, implementing this framework requires a good understanding of:

- the function of packaging components
- the values and expectations of consumers (see Chap. 3)
- the corporate, brand and product sustainability positioning (see Chap. 3)
- global packaging regulations and emerging policy trends (see Chap. 4)
- the environmental life cycle impacts of products, packaging and materials (Chaps. 5 and 6).

The selection and use of appropriate decision-making tools (Chap. 7) to embed sustainable development in product and packaging design processes should also be considered as part of the packaging for sustainability strategy.

References

1. Envirowise (2008) Packguide: a guide to packaging eco-design. Envirowise, Didcot, Oxfordshire
2. Packaging Council of Australia (2010). The Donald Richardson Packaging Collection—That was then, This is Now. http://www.pca.org.au/site/index.php/articles/article/1/546/The%20Donald%20Richardson%20Packaging%20Collection (cited 29 April 2011)
3. Lewis, H., Sonneveld, K., Fitzpatrick, L. and Nicol, R (2002) Towards Sustainable Packaging. Discussion Paper. http://www.sustainablepack.org/database/files/filestorage/Towards%20Sustainable%20Packaging.pdf (cited 14 May 2009)
4. The Consumer Goods Forum (2010) A global language for packaging and sustainability, Paris
5. SPC (2006) Design guidelines for sustainable packaging, version 1.0. Sustainable Packaging Coalition Charlottesville, Virginia
6. KeepCup (2011) FAQs. http://www.keepcup.com/the-keepcup/faq (cited 14 March 2011)
7. Britten N (2003) 60,000 are injured by opening packaging, The Telegraph, 11 February 2003.http://www.telegraph.co.uk/news/uknews/1421698/60000-are-injured-by-openingpackaging.html#. Accessed 25 Oct 2009
8. Business Wire (2001) Duracell launches its most powerful hearing aid battery ever with breakthrough consumer benefits, 16 April 2001. http://www.thefreelibrary.com/Duracell+Launches+Its+Most+Powerful+Hearing+Aid+Battery+Ever+With...-a073238213. Accessed 7 Nov 2010
9. The Center for Universal Design (2001) Duracell listens to its customers. Case studies on universal design. http://www.ncsu.edu/ncsu/design/cud/projserv_ps/projects/case_studies/Duracell.htm (cited 10 November 2010)

10. Hawken P, Lovins A, Lovins LH (1999) Natural capitalism. Little Brown and Company, Boston
11. von Weizsacker E, Lovins A, Lovins H (1997) Factor 4: doubling wealth—halving resource use. Allen & Unwin, Sydney
12. Schmidt-Bleek B (2000). Factor 10 manifesto. January 2000. Downloadable paper. http://www.factor10-institute.org/files/F10_Manifesto_e.pdf (cited 8 August 2008)
13. Weaver P, Jansen L, van Grootveld G, van Spiegel E, Vergragt P (2000) Sustainable technology development. Greenleaf Publishing, Sheffield
14. Parker G (2008) Measuring the environmental performance of food packaging: life cycle assessment. In: Chiellini E (ed) Environmentally compatible food packaging. Woodhead Publishing Limited, Cambridge, pp 211–237
15. United Nations Environment Program (2005) Ecosystems and well-being: synthesis. A report of the Millennium Ecosystem Assessment. http://www.millenniumassessment.org/documents/document.356.aspx.pdf (cited 1 October 2007)
16. Wal-Mart (2006) Wal-Mart launches 5-year plan to reduce packaging. http://walmartstores.com/pressroom/news/5951.aspx (cited 17 September 2010)
17. Greener Design (2008) Dell's packaging plan: cut 20 million pounds, save $8 million, 16 December 2008. http://www.greenbiz.com/news/2008/12/16/dells-packagingplan-cut-20-million-pounds-save-8-million. Accessed 31 Dec 2008
18. Dell (2008) Corporate responsibility summary report.
19. McDonough W, Braungart M (2002) Cradle to cradle: remaking the way we make things. North Point Press, New York
20. Grant T, James K, Lundie S, Sonneveld K (2001) Stage 2 report for life cycle assessment for paper and packaging waste management scenarios for Victoria. EcoRecycle Victoria, Melbourne
21. Tetra Pak (2010) Tetra Pak extends FSC^{TM}-labelled beverage cartons to customers in Sweden, Denmark and Belgium. Media release, 24 September
22. Smith R, Lourie B (2009) Slow death by rubber duck. University of Queensland Press, St Lucia
23. KPMG (2008) Trial of a government and industry charge on plastic bags. http://www.ephc.gov.au/sites/default/files/PS_PBags_Rpt_KPMG_Final_Report_on_the_Trial_of_a_Charge_on_Plastic_Bags_20081030.pdf (cited 23 January 2009)
24. Osborne D (2010) Dumped in the surf. Sustainability, 2:101–103
25. EPHC (2008) Decision Regulatory Impact Statement (RIS): Investigation of options to reduce the impacts of plastic bags. http://www.ephc.gov.au/pdf/Plastic_Bags/200805__Plastic_Bags__Decision_RIS__Options_to_Reduce_Impacts__incl_AppendicesCD.pdf (cited 26 September 2008)
26. Chessell K (2009) Brand owners responsibility when designing new packaging, Presentation to Australian Packaging Summit, 25 August. Melbourne
27. Boylston S (2009) Designing sustainable packaging. Laurence King Publishing, London
28. Hedditch B (2009) Director, SustainPak, H. Lewis, Editor. Melbourne
29. Arthritis Foundation (2010) Easy to use products. http://www.arthritis.org/ease-of-use-new.php (cited 20 October 2009)
30. Ditching Cereal Boxes (2011) Whats New in Food Manufacturing & Technology. (January/February): p 22
31. EUROPEN and STFI-Packforst (2009) The European shopping baskets: packaging trends for fast moving consumer goods in selected European countries. EUROPEN and STFI-Packforst, Brussels
32. Thinking about the box (2009). What's New in Food Technology & Manufacturing (Nov/Dec): p 12–13
33. Bransgrove G (2009) Putting over-packaging under the hammer, Presentation to Australian Packaging Summit 25 August. Melbourne
34. Bunnings Group Limited (2009) Bunnings and sustainability. http://www.bunnings.com.au/contact-us_bunnings-sustainability.aspx (cited 12 April 2009)

35. INCPEN (2009) Table for one: the energy cost to feed one person. Report http://www.incpen.org/docs/TableForOne.pdf (cited 11 November 2009)
36. Erlov L, Lofgren C, Soras A (2000) Packaging—a tool for the prevention of environmental impact. Packforsk, Kista
37. Twede D (1995) Less waste on the loading dock: competitive strategy and the reduction of logistical packaging waste. Yale School of Forestry and Environmental Studies, New Haven
38. Unilever (2001) Unit dose: a sustainability step for fabrics liquids. Report. http://www.unilever.com/images/capsules_tcm3-4586_tcm13-40198.pdf. Accessed 18 March 2011
39. Packaging Council of Australia (2008) Carter Holt Harvey Sustainability Award. http://pca.org.au/results08/apa/page.php?page=cartersustain (cited 15 March 2011)
40. Sam's Club square case-less milk jug packaging (2008) Sustainable is Good 1 July. http://www.sustainableisgood.com/blog/2008/07/sams-clubsquar.html (Accessed 17 Dec 2009)
41. Mans J (2007) Computer-controlled filler runs all fat levels of milk. Packaging Digest
42. PIRA International Ltd and ECOLAS N.V (2005) Study on the implementation of Directive 94/62/EC on Packaging and Packaging Waste and options to strengthen prevention and re-use of packaging, final report. 2005: Surrey, UK
43. WRAP (2009) Beverages: self-dispensing. Waste & Resources Action Program (WRAP). Bunbury, UK
44. WRAP (2009) WRAP retail trials could signal a 'reusables revolution' in packaging. Media release. http://www.wrap.org.uk/downloads/Reusables_-_Supporting_Infomation.1f2bdded.7589.pdf. Accessed 8 Nov 2010
45. CHEP, What is black and blue and green all over? Comparative lifecycle assessment in the fresh supply chain. 2010: Melbourne
46. Crates offer produce a green premium (2010) in PKN Packaging News, Yaffa Publishing, Sydney. p 34
47. WRAP (undated) Reusable 'Carrierpac' protects product and cuts down on packaging. Case study. Waste & Resources Action Program (WRAP). http://www.wrap.org.uk/downloads/15203-07_BQ_CS_LoRes.77db7a77.3742.pdf. Accessed 24 Dec 2009
48. WRAP (undated) Reusable transit packaging reduces costs, reduces packaging, improves efficiency and builds brand value. Case study. http://www.wrap.org.uk/downloads/15203-05_Overall_CS_LoRes.064901c4.3740.pdfh. Accessed 8 Nov 2010
49. PIRA International (2000) Strategic futures: multi-trip versus single-trip secondary packaging. Pira International, Leatherhead
50. Saphire D (1994) Delivering the goods: benefits of reusable shipping containers. INFORM, Inc., New York
51. FDA (2006) Use of recycled plastics in food packaging: chemistry considerations. Second edition: http://www.fda.gov/food/guidancecomplianceregulatoryinformation/GuidanceDocuments/FoodIngredientsandPackaging/ucm120762.htm (cited 9 January 2007)
52. Rodgers M (2006) Large scale demonstration of the viability of recycled PET (rPET) in retail packaging. Coca-Cola Enterprises Ltd, Banbury
53. Recycling on the menu: McDonald's Australia becoming more sustainable. Waste Streams, 2006 (Dec 2006/Jan 2007): p 17
54. First ever true bottle-to-box recycled product, the ECO-CHIC 100% recycled PET Bottle Box (2009) Press release. http://www.earthtimes.org/articles/show/first-ever-true-bottle-to-box-recycled,1025766.shtml# (cited 29 December 2009)
55. Churchwood, G., A. Ebel, E. Kosior, O. Tait, A. Jenkins, and S. Owen (2006) Large-scale demonstration of viability of recycled PET (rPET) in retail packaging: Closed Loop London (M&S, Boots). WRAP: Banbury, UK
56. Marks & Spencer (2005) Marks & Spencer closes the recycling loop. Media release. http://corporate.marksandspencer.com/investors/press_releases/company/08062005_MarksSpencerClosesTheRecyclingLoop (cited 30 December 2009)
57. Marks & Spencer (2007) Marks & Spencer launches the world's first plastic milk bottle using recycled material. Press release 2007. http://corporate.marksandspencer.com/investors/press_

releases/company/30032007_MarksSpencerLaunchesTheWorldsFirstPlasticMilkBottleUsing RecycledMaterial (cited 30 Dec 2009)
58. From packaging to food: mineral oil migration. Food Processing, 2010 (July/August): p. 63–66
59. Sustainable Packaging Coalition (2010) Guidelines for increasing recycled content in plastic packaging. Charlottesville, Virginia
60. The Paper Task Force (1995) Paper Task Force recommendations for purchasing and using environmentally preferable paper. Environmental Defence Fund: New York
61. ISO 14021:1999, Environmental labels and declarations - Type II environmental labelling - Principles and procedures. International Standards Organization (ISO): Geneva
62. Packaging Resources Action Group (2009) An introduction to packaging and recyclability. http://www.wrap.org.uk/retail_supply_chain/research_tools/tools/packaging_and.html. Accessed 5 June 2010
63. Recoup (2009) Plastics packaging: recyclability by design. Woodston, UK
64. Amazon (2010) Amazon certified frustration free packaging FAQs. http://www.amazon.com/gp/help/customer/display.html?nodeId=200285450 (cited 2 January 2010)
65. Lewis H (2009) Quickstart 4: Degradable polymers in product design. Sustainability Victoria and Plastics and Chemicals Industries Association. Melbourne
66. Bardelline J (2009) Coca-Cola to Test Dasani Bottles Made with Sugar Cane, Molasses. Greenbiz.com, 15 May 2009. http://www.greenbiz.com/print/25190. Accessed 20 Feb 2010
67. PACIA (2007) Using Degradable Plastics in Australia. A Product Stewardship Guide Commitment. Australian Government. Department of the Environment and Water Resources, Canberra
68. SunChips adds Green-e label to package (2007) Environmental Leader, 14 September 2007. http://www.environmentalleader.com/2007/09/14/sunchipspackaging-includes-green-e-label/. Accessed 15 Feb 2010
69. Australian Packaging Covenant: a commitment by governments and industry to the sustainable design, use and recovery of packaging (2010) APC Secretariat: Sydney
70. European Printing Industry Association (2007) Exclusion list for printing inks and related products. http://www.cepe.org/EPUB/easnet.dll/GetDoc?APPL=1&DAT_IM=020A3B&TYPE=PDF (cited 31 December 2009)
71. Ciba Specialty Chemicals (2004) Colorants for use in food packaging, toys and consumer goods
72. Green Century Capital Management and As You Sow (2009) Seeking safer packaging: ranking packaged food companies on BPA
73. Consumers Union (2009) Concern over canned foods. http://www.consumerreports.org/cro/magazine-archive/december-2009/food/bpa/overview/bisphenol-a-ov.htm (cited 16 February 2010)
74. Environment California. Bispenol A overview. Undated. http://www.environmentcalifornia.org/environmental-health/stop-toxic-toys/bisphenol-a-overview%20 (cited 31 December 2009)
75. Health Canada (2008) Government of Canada protects families with Bisphenol A regulations. Press release. http://www.hc-sc.gc.ca/ahc-asc/media/nr-cp/_2008/2008_167-eng.php (cited 31 December 2009)
76. Environmental Working Group (2007) Bisphenol A: toxic plastics chemical in canned food: companies reduced BPA exposures in Japan. http://www.ewg.org/node/20938 (cited 26 January 2010)
77. Voith M (2009) Can conundrum. Chem Eng News 87(29):28
78. Heinz (2009) CSR Report 2009. http://www.heinz.com/CSR2009/social/business/food_safety.aspx (cited 1 January 2010)
79. Coghlan P (2001) A discussion of some of the scientific issues concerning the use of PVC. CSIRO Molecular Science and Australian National University. A report to the Vinyl Council of Australia, Canberra
80. Scheirs J (2003) End-of-life environmental issues with PVC in Australia. Report to Department of Environment, Water, Heritage and the Arts, Excelplas Polymer Technology, Melbourne

81. Australian Consumers Association (2008) What's lurking under the lid? Choice, August, p 24–25
82. Phthalates Information Centre Europe (2010) Classification and labelling. http://www.hthalates.com/index.asp?page=56 (cited 6 January 2010)
83. Biodegradable plasticiser developed as phthalate replacement (2006) Food Production Daily, 6 January 2006. http://www.foodproductiondaily.com/Packaging/Biodegradable-plasticiser-developedas-phthalate-replacement. Accessed 5 June 2008
84. Toxics in Packaging Clearinghouse (2007) An assessment of heavy metals in packaging. Report to the US Environment Protection Agency
85. Envirowise (2008) Packaging design for the environment: reducing costs, quantities (Report # GG360). Envirowise, Harwel International Business Centre, Didcot
86. Towards sustainability with solvent-free adhesives (2009) In: Packaging News, Australia, p 30
87. AET Elemental chlorine-free: what the experts say. undated, The Alliance for Environmental Technology (AET). http://www.aet.org/reports/communication_resources/pamphlets/experts_final.pdf. Accessed 7 Jan 2010
88. CFPA (2010) Chlorine Free Products Association (CFPA). Undated. http://www.chlorinefreeproducts.org/ (cited 7 January 2010)
89. Darby A (2010) Buyers force Tasmanian woodchip mills to use only plantation timber. In: The Sydney Morning Herald, Sydney, p 9 (5 February 2010)
90. Wikipedia (2010) Beverage can, 2 January 2011. http://en.wikipedia.org/wiki/Beverage_can
91. EUROPEN and ECR Europe (2009) Packaging in the sustainability agenda: a guide for corporate decision makers. The European Organisation for Packagign and the Environment (EUROPEN) and ECR Europe: Brussels
92. Designing for society (2009) Packaging News, June, p 23
93. Commission of the European Communities (2009) Quantities of packaging waste generated in the Member State and recovered or incinerated at waste incineration plants with energy recovery within or outside the Member State, Brussels 2007
94. NPCC (2008) Mid-term performance review data report, 5 December 2008. http://www.packagingcovenant.org.au/documents/File/MTR%20Data%20Report%20Final(1).pdf
95. US EPA (2008) Municipal solid waste generation, recycling, and disposal in the United States: facts and figures for 2007, 4 May 2009. http://www.epa.gov/epawaste/nonhaz/municipal/pubs/msw07-fs.pdf
96. Alston K (2010) Cradle to cradle design: Positive sustainability agenda for products and packaging; the limitations of eco-efficiency, Presentation to Packaging for tomorrow. 7 December 2010, 3 Pillars Network, Melbourne
97. Packaging Council New Zealand (2008) Mass balance—consumption and collection, 4 May 2009. http://www.packaging.org.nz/packaging_info/packaging_consum.php
98. Ministry of Economy, Trade and Industry (2008) Towards a 3R-oriented sustainable society: legislation and trends, Tokyo
99. OECD (2007) OECD environmental data: compendium 2006–2008. Organisation for Economic Cooperation and Development (OECD)

Chapter 3
Marketing and Communicating Sustainability

Helen Lewis and Helaine Stanley

Abstract Sustainable development creates new challenges for marketing and communication strategies. This is particularly so for packaging because consumers tend not to be aware of the potential environmental benefits of packaging other than those relating to disposal (recyclability, reuse, biodegradability and over-packaging). Optimal life cycle-based solutions may be counter-intuitive to consumer perceptions. However, in addition to products and brands, individuals and businesses 'buy' corporate philosophies and policies. While developing its strategy, a business should therefore determine how to market and position itself, its brands and its products from the perspective of sustainability. Packaging has a major role in this positioning, which affects both packaging and label design. This chapter provides an overview of research conducted on consumer attitudes and purchasing behaviour relating to 'green' products in general and packaging in particular. On balance, consumers have negative rather than positive associations with packaging. Different approaches can be taken to packaging design and communication within a business's broader marketing strategy. These are provided together with an introduction to the use of environmental claims and labels often associated with packaging.

Contents

3.1	Introduction	109
3.2	Understanding the Citizen Consumer	112
	3.2.1 Responsiveness to 'Green' Marketing	112
	3.2.2 Perceptions of Packaging	115

H. Lewis (✉) · H. Stanley
Centre for Design, RMIT University, GPO Box 2476, Melbourne, VIC 3001, Australia
e-mail: lewis.helen@bigpond.com

H. Stanley
e-mail: Helaine.Stanley@rmit.edu.au

3.3	Sustainability Marketing Principles	118
	3.3.1 Reflect the Organisation's Sustainable Development Values and Goals	118
	3.3.2 Tell the Truth: Honesty in Marketing	119
	3.3.3 Comply With Regulations	124
3.4	The Sustainability Marketing Strategy	125
	3.4.1 Developing the Strategy	126
	3.4.2 Understand and Engage With Stakeholders	126
	3.4.3 Understand and Engage with Consumers	130
	3.4.4 Identify Messages and Improvement Strategies	132
	3.4.5 Communicating the Messages	134
	3.4.6 Confirm the Role of Packaging	136
3.5	Environmental Claims and Labels	137
	3.5.1 Types of Claims and Labels	137
	3.5.2 Material Recycling	138
	3.5.3 Organic Recycling	142
	3.5.4 Recycled Content	145
	3.5.5 Litter Reduction	145
	3.5.6 Chlorine Free	146
	3.5.7 Forest Stewardship	147
	3.5.8 Carbon Reduction	148
	3.5.9 Renewable Energy Use	148
3.6	Conclusion	150
References		150

Figures

Figure 3.1	A multi-level approach to sustainable development marketing	111
Figure 3.2	The lost opportunity on the "path to purchase"	114
Figure 3.3	Green claims on non-food supermarket products, Australia, 2008	123
Figure 3.4	Sustainability purchase matrix	132
Figure 3.5	Resin identification codes for plastics	142
Figure 3.6	Checklist for determining the sustainability marketing and communication approach	149

Tables

Table 3.1	Consumer behaviour, United Kingdom, 2000 and 2008	113
Table 3.2	Breakdown of shoppers by green purchasing development level, United States, 2009	113
Table 3.3	How consumers regard the importance of different green claims, Australia, 2005	115
Table 3.4	Concerns about environmental and ethical issues, United Kingdom, 2008	116
Table 3.5	Specific products that are perceived to be 'over-packaged', United Kingdom, 2008	117
Table 3.6	Consumer priorities for packaging redesign	117
Table 3.7	Best practice standards for environmental claims	122
Table 3.8	The six sins of greenwash	124
Table 3.9	Corporate stakeholders and communication strategies	128
Table 3.10	Sustainability impact matrix for roasted coffee	133
Table 3.11	Types of environmental labels, per ISO 14024, 14021 and 14025	138

3 Marketing and Communicating Sustainability 109

Table 3.12	Packaging sustainability principles (cyclic and safe) and relevant labels	139
Table 3.13	Correct use of the SPI resin identification codes	142
Table 3.14	Standards for degradable plastics	146

Photo

| Photo 3.1 | Example of self-declared claims about degradability | 143 |

Case Studies

Case Study 3.1	Hewlett Packard's global citizenship policies	119
Case Study 3.2	Orange Power learning from experience	121
Case Study 3.3	Mobil's degradable bags	125
Case Study 3.4	Cascade Green	127
Case Study 3.5	McDonalds and the "great clamshell debate"	129
Case Study 3.6	Nude Food Movers	131
Case Study 3.7	Visy marketing the company	135
Case Study 3.8	Patagonia "the good and the bad" of products and packaging	135
Case Study 3.9	On-Pack Recycling Label	141
Case Study 3.10	Misleading degradability claims on packaging (US)	144

3.1 Introduction

A business should determine through its strategic planning process how it will market and position itself together with its brands and products from a sustainability perspective. Through its business processes, it should then align all reporting, marketing and communication activities. Packaging design is affected by these decisions because:

⇒ See Chap. 1 for more information on developing the sustainability strategy

- packaging plays an important role in brand and product promotion
- how to use and dispose of the product and its packaging should be communicated
- packaging and label design requirements change as strategic choices about corporate positioning, product mix and product design occur.

Continually Review Marketing Objectives

Marketing strategies should continually re-evaluate how to:
- achieve organisational objectives including sustainable development
- meet consumer needs and changing expectations
- minimise the environmental impacts of its activities [1].

This involves familiar marketing activities such as research on customers' needs, preferences and expectations. However, to address sustainability, new types of information are required on socio-ecological problems and consumer attitudes to sustainability issues [2].

Apply a Strategic Approach

Individuals and businesses do not just buy brands; they also 'buy' corporate philosophies and policies [3, p. 12]. A strategic decision must be made on how, and to what extent, sustainable development goals, achievements and challenges are communicated publically or directly to consumers. This will depend, for example, on whether the business is appealing to committed 'green' consumers or to those who don't regard themselves as green but might be influenced by environmentally-improved products. It should also be asked whether individual products or the business as a whole are being promoted.

Figure 3.1 outlines an integrated approach to sustainability marketing, building on the work of Belz and Peattie [2]. At a corporate level, sustainable development and marketing strategies need to be responsive to stakeholder concerns and expectations. They also need to address consumer attitudes and behaviour, including their responsiveness to 'green' marketing and their interest in particular socio-ecological issues. Marketing strategies and the sustainability marketing mix (including packaging) should be selected to support corporate values and objectives. While stakeholder and consumer perspectives will inform corporate strategy, it is equally important to base decisions on a sound understanding of the life cycle environmental impacts of the product and its packaging.

This process is not always straightforward. For example, consumer perceptions may be counter-intuitive to decisions about packaging based on an understanding of environmental life cycles of products. Sustainability marketing therefore provides new challenges for marketing the business, brands and products—particularly as consumer perceptions of the environmental impacts of packaging are primarily linked to disposal issues (see Sect. 3.2.2).

Avoid Greenwash

> **Greenwashing** is actions … 'to make people believe that your company is doing more to protect the environment than it really is' [4].

Marketing sustainability must avoid 'greenwash'. This requires due diligence and additional investment in claims validation enabled by a life cycle management approach to product design. Guidance on how to avoid greenwashing can be found in Sect. 3.3.2.

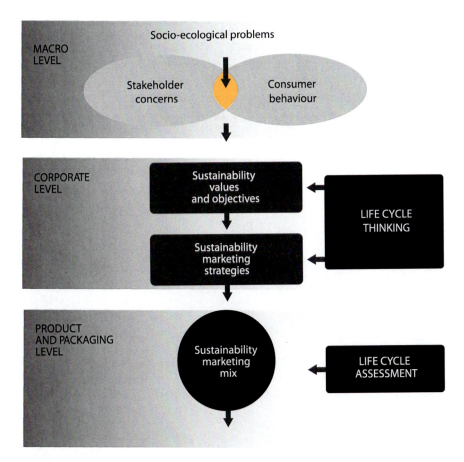

Fig. 3.1 A multi-level approach to sustainable development marketing. *Source*: Builds on the work of Belz and Peattie [2]

Confirm the Role of On-Pack Labelling

The role of on-pack labelling, including the selection of information used to support the business's sustainable development goals, brand and product positioning, must be confirmed. The relative importance of labelling to support sustainable development is assessed in circumstances where competition for 'label space' is high. There is an increasing number of potential logos and consumer information that must or can be applied to packaging. Some are required by regulation, such as food ingredient and nutritional labels, while others are optional for marketing purposes, such as organic and fair trade labels.

3.2 Understanding the Citizen Consumer

3.2.1 Responsiveness to 'Green' Marketing

Businesses interested in promoting sustainable development can learn from more general research on 'green' or 'ethical' consumers.

Add Brand Value

Deloitte notes that sustainability is emerging as an important brand differentiator as long as consumer expectations for price and performance are met [5]. Consumers do not expect green products to cost any more than conventional alternatives. The most effective green products are those that offer broad value including price, quality and sustainability.

Research commissioned by the Co-operative Bank in the United Kingdom in 2000 [6] found that purchasing decisions are based primarily on quality and value for money, but other factors such as how companies treat employees and their impact on the environment are also important. 'Global watchdogs' (5% of the population) place a high level of importance on the ethical performance of industry. They actively seek information on a business's performance, are less concerned than other groups about value for money and tend to be socio-political activists. 'Conscientious consumers' (18% of the population) are less likely to campaign on ethical issues but are fairly active in recycling. They have avoided unethical products and recommend ethical businesses to friends or family [6].

The Rise in the Green Consumer

The Co-operative Bank's follow-up survey in 2008 [8] noted an increase in the percentage of people reporting green or ethical purchasing: 57% of all respondents had chosen a product or service at least once in the previous year on the basis of the business's reputation, and 29% had bought primarily for ethical reasons, compared to 51 and 29% respectively in 2000 [8] (p. 3) (see Table 3.1).

Similar behaviour is evident in the United States, where *Time* magazine noted the rise of the 'citizen consumer' [9] in a poll it commissioned: about 60% of respondents said they had purchased an organic product in 2009, while 40% had purchased a product because they liked the social or political values of the business that manufactured it (p. 24).

In other research, LOHAS consumers ('lifestyles of health and sustainability') or 'cultural creatives', have also been identified through a segmentation model which cuts across demographic criteria using a more lifestyle-based approach [10].

Consumers' attitudes and behaviour are not necessarily fixed. Deloitte characterised consumers based on their purchasing and lifestyle behaviour into five 'levels of development' in green purchasing (see Table 3.2) [5]. Consumers move up the 'learning curve' as their awareness and commitment to sustainability

Table 3.1 Consumer behaviour, United Kingdom, 2000 and 2008

	% Done at least once (2000)	% Done at least once (2008)
Recycled	73	96
Supported local shops/suppliers	61	83
Talked to friends/family about a business's behaviour	58	59
Chose a product or service based on a business's reputation	51	57
Avoided a product or service because of a business's behaviour	44	57
Bought primarily for ethical reasons	29	51
Felt guilty about unethical purchase	17	38
Actively sought information on a business's behaviour/policies	24	36
Actively campaigned about environmental/social issue	15	26

Source: The Co-operative Bank [8, p. 3]

Table 3.2 Breakdown of shoppers by green purchasing development level, United States, 2009

Category	% of shoppers
Unaware: Sustainability is not a conscious purchasing value	13
Unsure: Sustainability is not considered as a major purchasing value	33
Influenced: Sustainability is a tie-breaker when other purchase requirements are met	34
Proactive: Sustainability is an equally important primary purchasing value	18
Committed: Sustainability is the dominant purchasing value	2

Source: Deloitte [5, p. 7]

increases, and their purchasing behaviour may vary between product categories. This development mimics the evolution in sustainable development of organisations discussed in Sect. 1.3.1. There is an opportunity for marketing therefore to accelerate sustainable development by encouraging purchases in each of the middle categories to move consumers from the 'Unsure' to the 'Influenced' and then to the 'Proactive' stage.

Environmental Concerns Do Not Necessarily Translate Into Purchasing Decisions

Marketers cannot assume that environmental concerns and a commitment to buy green products will influence purchasing decisions. Research consistently shows a gap between environmental and social awareness, commitment and behaviour. This is often referred to as the attitude-behaviour or value-action gap (e.g. [11, 12]). For example, 69% of respondents in an Irish survey said that a business's commitment to social responsibility is important when they make a decision about buying a product or service [13, p. 2]. However, when asked about activities they had undertaken in the previous year, only 11% said they had

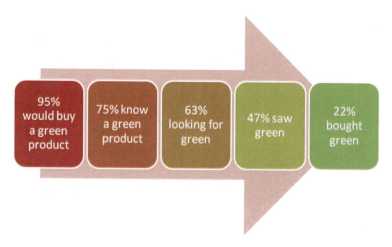

Fig. 3.2 The lost opportunity on the 'path to purchase'. *Source*: Based on research from Deloitte [5, p. 11]

purchased a product that was labelled as social, ethical or environmentally responsible (p. 3).

One of the most insightful reports into the apparent gap between environmental awareness, commitment to buy and actual purchasing behaviour was prepared by Deloitte for the Grocery Manufacturers Association (GMA) in the United States [5]. The survey found that while sustainability is considered by 54% of shoppers in their selection of products and stores, only 22% had actually bought a 'green product' on their visit to the store. Deloitte's list of characteristics for 'green products' included low water use, reduced packaging, organic/locally grown, fair trade, energy efficient, biodegradable, non-toxic and low volatile organic compounds, and recyclable (materials and content).

Improve Visibility of Green Products

The GMA research also identified that potential sales of green products are being lost on the 'path to purchase' (see Fig. 3.2). For example:
- While 63% had searched for a green product, only 47% had found one. Initiatives to improve the visibility of green products and draw shoppers' attention to them are likely to increase sales
- 47% had seen a green product, but only 22% had actually purchased one. It's important to influence consumers at point-of-sale by providing information on environmental credentials and providing them with more inspiration to buy.

Match Marketing Strategy to Target Market

An Australian study found that committed green consumers tend to be more receptive to environmental promotions [14].

3 Marketing and Communicating Sustainability

Table 3.3 How consumers regard the importance of different green claims, Australia, 2005

	High involved (%)	Low involved (%)
Ads should show product recycling symbols	86	14
Ads should emphasise how consumers will obtain environmental benefits	75	25
Ads should show environmental labels, based on third-party accreditation	75	25
Ads should promote corporate image	56	44
Ads should promote donating to environmental groups or causes	81	19
Ads should be clear about environmental claims	75	25

Source: D'Souza and Taghian [14, p. 59]

Consumers whose purchasing behaviour is strongly influenced by environmental concerns ('high involved') regard advertisements that promote a corporate image as the least important, compared to other forms of communication such as recycling labels or third party certification.

On the other hand, consumers whose purchasing behaviour is minimally influenced by environmental concerns ('low involved') are more responsive to environmental claims that promote the environmental achievements of a business. Hence this marketing approach is likely to be more successful in appealing to a broad market than specific claims about the product or packaging (see Table 3.3).

3.2.2 Perceptions of Packaging

Negative Perceptions Outweigh Positive Perceptions

Surveys conducted in the United Kingdom [15–17] and New Zealand [18] highlight the contradictory views that consumers hold about packaging.

While consumers understand and can mention (unprompted) many of the advantages of packaging, such as product protection, hygiene and ease/convenience, they also mention a number of perceived disadvantages. These include its environmental impacts, difficulties of disposal and resource use. In a UK survey [17] packaging was rated as the issue of greatest environmental and ethical concern (see Table 3.4). On balance, negative associations outweigh positive associations by 74–64% (p. 3) and tend to focus on particular packaging materials and products.

Negative Perceptions are Increasing

INCPEN has tracked UK consumer attitudes to packaging over 15 years and has observed an increase in negative perceptions [17, p. 3]:
- 79% of respondents agreed with the statement that products are over-packaged in 2008 compared with 68% in 1997
- 82% agreed that 'packaging is a major environmental problem' in 2008 compared with 71% in 1997.

Table 3.4 Concerns about environmental and ethical issues, United Kingdom, 2008

Environmental or ethical issue	% that mentioned issue
The amount of packaging	51
The chemicals/pesticides used in food	40
Fair trade—making sure farmers get a fair deal for their products	37
Animal welfare—making sure animals are well treated	33
The number of plastic bags used at the till	27
Food miles—the distance the product has travelled	24
The 'carbon footprint' of the product	16
Do not know/none of the above	8

Source: IPSOS Mori [17, p. 3]

Consumer Perceptions of Environmentally Friendly Packaging

Market research suggests that packaging improvements that consumers would regard as more environmentally friendly reflect packaging disposal issues:
- biodegradability
- recyclability
- reusability
- perceived over-packaging.

Dutch [19] and New Zealand [20] studies have demonstrated close correlations between perceptions of environmental friendliness of packaging types with perceptions of reusability and recyclability of the packaging or the materials used. In the New Zealand survey, consumers were asked to rate the environmental friendliness of packaging types and then rate their recyclability, and there was a strong relationship between the two. For example, paper bags and cardboard boxes rated amongst the most environmentally friendly and the most recyclable, while polystyrene rated lowest for both questions [20, p. 14]. This may be due to the high visibility of waste to the consumer and the enthusiasm with which most people participate in recycling programs. These results are reinforced by a survey from the United Kingdom [17, p. 4] which found:
- over-packaging, particularly of Easter eggs, electronics equipment, pre-packed fruit and vegetables and ready meals is a primary concern (see Table 3.5)
- consumers are more likely to think about packaging at home, when they have to dispose of it, than at the point of sale (41 and 19% respectively). This is reflected in consumers' priorities for packaging redesign (see Table 3.6).

However, the strong link in consumers' minds between the amount of packaging, recyclability and environmental friendliness poses a challenge for marketers. As shown in Sect. 2.4.2, in some cases more rather than less packaging may

3   Marketing and Communicating Sustainability

Table 3.5 Specific products that are perceived to be 'over-packaged', United Kingdom, 2008

Type of product	% that mentioned product
Easter eggs	59
Electronics equipment	57
Pre-packed fruit and vegetables	41
Ready meals	36
Pre-packed sandwiches and lunch meals	33
Cosmetics	32
Pre-packed meat and fish	20
Breakfast cereals	14
Chocolate	11
Drinks in bottles and cans	10
Household cleaning products	9
Food in tins/cans	6
Bread	4
Wine	2
Do not know/none of the above	11

Source: IPSOS Mori [17, p. 4]

Table 3.6 Consumer priorities for packaging redesign

Environmental improvement	% of respondents
Biodegradable material	44
Recycled content	42
Reducing the amount of material	42
Making materials easier to recycle	41
More compostable material	26
More refillable options in stores	20
More degradable materials	10

Source: IPSOS Mori [17, p. 5]

be required to minimise product waste. In some cases, lightweighting may also deliver more environmental benefit from a life cycle perspective, even if it means that the packaging cannot be recycled.

But Perceptions Have Little Impact on Purchase Decisions…

The environmental attributes of packaging are not 'top of mind' when a consumer selects a product in the supermarket.

A relatively small percentage of respondents to the UK survey [17, p. 4] took specific actions to reduce the environmental impacts of packaging when shopping:
- 39% said that they always take a reusable bag to the store
- 17% said that they always buy loose products where possible
- only 10% look for information on the label about recycling
- 9% avoid products with too much packaging.

In one Australian survey [21], shoppers were intercepted as they left a supermarket and asked about their purchasing behaviour. They were also asked to complete a follow-up questionnaire with more detailed behavioural and attitudinal information. When asked why they had purchased particular products, consumers mentioned attributes such as price, taste, convenience and habit. Only 4% mentioned anything to do with packaging, such as the fact that it was recyclable, biodegradable or contained recycled material. When they were specifically asked what they liked about the packaging, 16% mentioned its recyclability, but most responses related to functionality or appearance. A very small percentage of respondents (no more than 2% for any type of packaging) claimed that they avoided particular packaging materials for environmental reasons, despite the fact that most people claimed that the environment is important, and that they always try to behave in an environmentally friendly way. In their conclusion, the consultants noted that there is '[v]irtually no connection between attitudes to recycling, waste or the environment, and purchasing behaviour' (p. 5).

…Except Perhaps for Green Consumers

Green consumers may be prepared to trade-off selected functionality or convenience benefits of packaging to improve its environmental performance. A global survey by Nielson [22] found that almost half of all consumers who rate 'recyclable bags and packaging' as important when deciding where to do their shopping would consider giving up features that improve storage, convenience or ease of transport, but only a third would be prepared to compromise on shelf life, product information or hygiene.

3.3 Sustainability Marketing Principles

3.3.1 Reflect the Organisation's Sustainable Development Values and Goals

Many businesses that manufacture, package or sell consumer products have an environmental or sustainability policy that outlines their corporate values and objectives. This usually includes a list of commitments towards the natural environment, employees, consumers and other stakeholders (see the Hewlett Packard Case study 3.1). Marketing has an important role to play in ensuring that sustainable development and corporate social responsibility objectives are reflected in:
- marketing, communication and promotion activities
- the product/service mix
- the product development and design process.

3   Marketing and Communicating Sustainability   119

> **Case Study 3.1 Hewlett Packard's Global Citizenship Policies**
>
> Hewlett Packard (HP) has an overarching commitment to global citizenship:
>
> > 'Global citizenship is integral to the success of HP's business. From how we develop products, run our operations, manage our supply chain and engage with stakeholders, it drives us to accept challenges and pursue solutions that are the lifeblood of continuous innovation and growth. It also guides our decision making, ensuring that we uphold the values that have helped distinguish HP as a leader committed to using technology to benefit people, businesses, society and the environment'. Shane Robison, Executive Vice President and Chief Strategy and Technology Officer
>
> HP's global citizenship policy confirms its life cycle approach to environmental improvement:
>
> > 'HP is committed to providing products and services that are environmentally sound throughout their life cycles, and to conducting our operations in an environmentally responsible way'.
>
> HP also has specific policies or codes of conduct for accessibility, business ethics, corporate governance, diversity, environment, human rights, labour practices, privacy, products ('HP general specification for the environment') and supply chain. Packaging design and labelling requirements are included within the product policy. This provides comprehensive guidelines for suppliers on restricted materials, heavy metals, recyclability, polymer codes and other HP requirements.
> *Source*: Hewlett Packard [23, 24]

3.3.2 Tell the Truth: Honesty in Marketing

Be Transparent, Accurate, and Consistent

Sustainability marketing requires the communication of environmental claims and messages to be:
- transparent
- accurate
- consistent, irrespective of the mode in which they are communicated; e.g. sustainability report, website, on-pack labelling or other point of sale material.

Clear, truthful and relevant claims can:
- reduce the environmental impacts of consumption by encouraging consumers to purchase products with less or environmentally improved products

- strengthen a business's reputation or credibility
- demonstrate to regulators that a business is working to meet legal requirements
- ensure that it is able to meet labelling requirements of specific markets
- enhance the appeal of the business's products [25, p. 5].

Comply With Best Practice

Claims based on scientific research and independently verified are more likely to achieve commercial benefits and contribute to sustainable development (see Case Study 3.2). This is generally achieved through the use of life cycle thinking to identify and verify relevant claims. Table 3.7 presents best practice standards for environmental claims based on the international standard ISO 14021, Environmental claims and labels: self-declared environmental claims [26] and other government guidelines for regulatory compliance.

⇒ See Chap. 5 for more on life cycle thinking and LCA

Avoid Greenwashing

If environmental claims are inaccurate, misleading or not transparent, a business may be accused of 'greenwash'.

A 2008 survey of non-food consumer products in supermarkets by the Australian Consumers Association found:
- 637 claims on 185 items (see Fig. 3.3)
- general or vague claims, such as 'environmentally friendly', 'natural' or 'pure' were in common usage—contrary to the requirements of the Australian Government's guidelines on environmental claims [29] and ISO 14021 [26]
- a large number of 'CFC-free' claims, which are not relevant.

The survey concluded that greenwash is getting worse; in fact, it's 'out of control' [30].

A US and Canadian survey of 1,018 retail consumer products identified 1,753 environmental claims and concluded that all but one of these were 'false or at risk of misleading intended audiences' because they committed one of the 'six sins of greenwashing' [31, p. 1] (see Table 3.8). However, this survey used criteria over and above legal requirements. For example, claims were classified as:
- false or potentially misleading if they suggested a product was green based on a single attribute, such as recycled content, and failed to address other issues that may be more important, such as global warming
- potentially misleading if they were not substantiated by 'easily accessible' information at the point of purchase or on the product web site.

At the present time, single-attribute claims that are truthful and relevant are generally not regarded as greenwash. Most government guidelines simply require

Case Study 3.2 Orange Power: Learning From Experience

An article in The Sydney Morning Herald described the journey that one company undertook after it was accused of 'greenwash' [25]. Orange Power cleaning products are based on waste orange oil, a natural solvent, which is discarded from the juicing process. In 2008 one of their products was identified on national television as being guilty of 'greenwash' because the label contained about 17 claims that weren't substantiated. For example, they claimed that the product was biodegradable, when in fact all of the products in this category were required to be biodegradable.

The company took steps to address the problem. They sought an independent audit from Good Environmental Choice Australia (GECA), which licenses the use of an eco-label for products meeting its own standards, and changed their labelling. Orange Power became the first product on Australian supermarket shelves with the eco-label, but sales remained stagnant despite an A$750,000 advertising campaign promoting its environmental benefits. A paid endorsement from environmental group Planet Ark did result in increased sales. According to Managing Director, Iain Chaney, the experience has changed their corporate philosophy and they are now greening their entire manufacturing process [25].

Photo: Helen Lewis

Source: Sydney Morning Herald [25]

Table 3.7 Best practice standards for environmental claims

The claim needs to be accurate and not misleading
It is important to consider how an ordinary member of the public (not an expert) might interpret the claim. The UK Department for Environment, Food and Rural Affairs [27, p. 8] suggests that marketers: • avoid claims that indicate that an environmental benefit, while literally true, is unlikely to happen in practice • avoid restating a single environmental benefit in different terminology to infer multiple benefits • avoid claims that imply a recent improvement or enhancement when in actual fact it refers to a pre-existing or previously undisclosed benefit.
The claim needs to be substantiated and verifiable
The US Federal Trade Commission (FTC) [28] advises that an environmental claim should be substantiated with 'competent and reliable scientific evidence' (p. 2). The ISO standard [26] goes further by suggesting one of three test methods to evaluate a claim, in the following order of preference: • an international standard • a recognised national standard with international acceptability (for example CEN standards) • a method developed by industry as long as it has been subjected to peer review • if no methods currently exist, a method developed by the organisation as long as it is based on ISO 14021 and available for peer review. Relevant documentation should be retained, including identification for the relevant standard or test method and test results [27].
The claim should be specific and unambiguous
Claims that broadly imply that a product is environmentally beneficial or environmentally benign, such as 'sustainable', 'environmentally friendly', 'green' or 'ozone friendly' should not be used [26]. The FTC guidelines [28] state that broad environmental claims should be either avoided or qualified because of their potential to mislead consumers. For example, a claim such as 'environmentally preferable' is deceptive if it implies that the product is environmentally superior to other products and the claim cannot be substantiated by the manufacturer. However, '[t]he claim would not be deceptive if it were accompanied by clear and prominent qualifying language limiting the environmental superiority representation to the particular product attribute or attributes for which it could be substantiated…' (p. 5). A qualifying or explanatory statement should be of a reasonable size and close to the environmental claim it refers to [26]. A claim should specify exactly why a product is better for the environment and the level of environmental performance that has been achieved [27]. If a comparative claim is made (e.g., comparing performance with the previous product or a competitor's product), then a number of issues need to be considered [26]: • A published standard or recognised test method should be used • Comparisons should only be made with products serving a similar function in the same market place • Claims should be calculated using the same units of measurement • Claims may be based on percentages (e.g. 50% less packaging material) or absolute measures • Claims should be calculated over an appropriate time interval, typically 12 months.
The claim needs to be relevant
This means that customers understand the context within which the claim is made, for example [27, p. 9]: • It must be relevant to the particular product

(continued)

3 Marketing and Communicating Sustainability

Table 3.7 (continued)

The claim needs to be relevant
• It must be relevant to the place where the environmental impact is likely to occur (e.g. claiming that a package is technically recyclable when there is no recycling infrastructure in the place of sale is misleading—see Sect. 3.5.2)
• It should be clear whether the claim applies to the product, part of the product or the packaging
• The claim should not imply that the product is exceptional when all products in the same category share the same characteristic
• A claim about something being 'free' of a particular substance should not be made if the substance has never been used in this type of product or not for a long time (e.g., a claim of 'CFC-free' is no longer relevant to aerosols because their use is banned in most countries).

Sources: ISO [26], DEFRA [27], FTC [28]

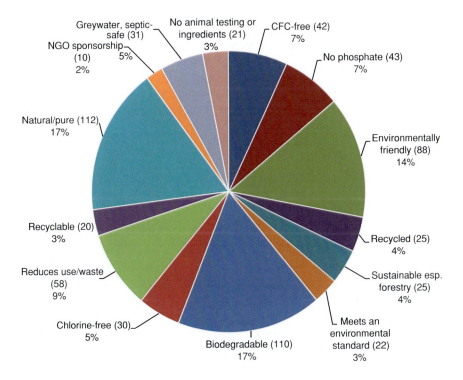

Fig. 3.3 Green claims on non-food supermarket products, Australia, 2008. *Source*: Data from Australian Consumers Association [30, p. 13]. (The number of claims is shown in brackets)

Table 3.8 The six sins of greenwash

1. *Sin of the hidden trade-off*: a claim that is based on a single attribute, such as recycled content for paper, or an unreasonably narrow set of attributes. Such claims are not usually false but may create a greener image than would be supported by a more complete environmental analysis. 2. *Sin of no proof*: a claim that cannot be substantiated by easily accessible supporting information, for example at the point of purchase or on the product web site. 3. *Sin of vagueness*: a claim that is so poorly defined or so broad that it is likely to be misunderstood by the intended consumer. Examples include 'all natural', 'green' and 'chemical-free'. 4. *Sin of irrelevance*: a claim that may be true but is unimportant or completely irrelevant. An example is 'CFC-free': CFCs have been banned since the 1970s. 5. *Sin of fibbing*: a claim that is false. 6. *Sin of lesser of two evils*: a claim that might be true within the product category but might distract consumers from the greater environmental impact of the category as a whole' for example so-called green insecticides and herbicides.
Source: TerraChoice [31]

companies to be able to substantiate their claims with rigorous and credible information.

Negative effects of greenwashing include [31]:

- encouraging well-intentioned consumers to make a purchase that does not deliver on its promise, with the result that the environmental benefit of the purchase is squandered
- taking market share away from products that offer more legitimate benefits and therefore slowing the penetration of environmental innovation in the market
- contributing to cynicism and doubt about all environmental claims, which may discourage consumers from purchasing genuinely green products
- a fall in sales when claims are found to be misleading or inaccurate (see Case Study 3.3).

⇒ Government guidelines provide more detail on the types of claims that are acceptable and legally compliant. They include:
- Federal Trade Commission (US) [28, 32]
- Department for Environment Food and Rural Affairs (UK) [33]
- Australian Competition and Consumer Commission [29]

3.3.3 Comply With Regulations

Environmental claims include 'any statement, symbol or graphic that indicates an environmental aspect of a product, a component or packaging' [26, p. 1]. They are found on product labels, point-of-sale promotions, advertising, corporate reports, technical bulletins, telemarketing and web sites.

> **Case Study 3.3 Mobil's Degradable Bags**
>
> In 1989, Mobil Chemical Corporation introduced a line of 'Hefty' garbage bags in the United States that they claimed were degradable. One year later they were sued by attorneys general in seven US states as well as the Federal Trade Commissioner, on claims that they were misleading the public with deceptive advertising. The attorneys general claimed that the bags would only degrade very slowly in landfill, if at all, and may cause more environmental damage than conventional plastic bags. [7]
>
> According to a report in Harvard Business Review, sales of the garbage bags soared when the company announced that they were degradable but fell when it emerged that they only partly decomposed [34].
>
> Mobil responded by deleting any references to degradability on the packaging.
> *Sources*: Spears and Larson [7], Kleiner [34]

Businesses need to ensure that environmental claims comply with local laws relating to false or misleading claims under trade practices law. A number of government agencies such as the US Federal Trade Commission [28] and the UK Department for the Environment, Food and Rural Affairs [33], have published guidelines to help businesses comply with regulations. In 2010 the Federal Trade Commission published proposed changes to their 'Green Guides' for environmental claims [32].

> ⇒ See Chap. 4 for more on regulations

3.4 The Sustainability Marketing Strategy

Building a sustainable brand is not achieved overnight. It requires a marketing strategy that reflects:
- the business's commitment to sustainable development
- the business's sustainable development goals
- a good understanding of current and future regulatory requirements.

How sustainable development objectives, information and outcomes are then applied to and communicated for particular products will depend on a range of factors, including:
- the nature of the product and the demographics of its market (mass market, niche etc.)
- attitudes and behaviour of consumers in response to sustainability/green marketing

- social and environmental impacts and benefits of the product/packaging life cycle and how these compare to industry benchmarks and competitors
- what improvements in sustainable development can be and have been achieved.

3.4.1 Developing the Strategy

Figure 3.1 provides an overview of the process involved in developing a sustainability marketing strategy and the marketing mix, including packaging. The aim here is not to provide a complete introduction to sustainability marketing but to show how packaging sustainability can be integrated within marketing and communication.

⇒ Frank-Martin Belz and Ken Peattie's book, *Sustainability Marketing* [2], shows how sustainability can be integrated within all aspects of marketing and communication.

Target the Marketing Campaign

Marketing and communication strategies must be carefully targeted and appropriate to the market segment. If the target consumer is not receptive to environmental claims and the method of communication, such as product labels, they are unlikely to succeed (see the Cascade Green Case Study 3.4).

Strategies to address a non-receptive market include:
- engaging and communicating with consumers (see Sect. 3.4.4)
- communicating sustainable development commitments and achievements indirectly; for example, via corporate reports, web sites, business rather than product promotion.

3.4.2 Understand and Engage With Stakeholders

A company's sustainability marketing strategy should be informed by a stakeholder analysis to identify:
- which organisations or groups can influence or are influenced by its operations [40]
- what concerns they have about the sustainability of its operations and products
- what impact (positive or negative) their views and/or actions could potentially have on the business
- opportunities for collaboration to improve sustainable development
- strategies for engagement to improve sustainable development
- key messages and appropriate channels for the dissemination of information for different stakeholders.

Case Study 3.4 Cascade Green

In 2008, Foster's launched a new low-carbohydrate beer called 'Cascade Green'.

The product achieved Greenhouse Friendly™ certification by the Australian Government because the annual greenhouse gas emissions of the product and packaging life cycle were 100% offset (initially by purchasing offsets for gas recovery at Hobart landfill). Other environmental benefits were achieved through the packaging design by using two-colour biodegradable vegetable inks and 100% recycled material for the corrugated shipper.

The name was chosen to communicate the environmental credentials of the brand and was considered to be 'straightforward' and 'authentic' [35].

Sales exceeded targets by 30% in its first month in the market and achieved a repeat purchase rate of 20% [35]. Since then, sales have levelled off and fallen below targets.

One critic attributed this 'failure' to the fact that:

> 'Greens don't like beer and beer drinkers don't like Green ideology...drinking beer and driving cars is about personal pleasure, not about the greater good [36]'.

According to a survey by Quantum in 2008, 67% of Australians agree that it is more important for brands to deliver good quality and value than to support charities and good causes, and this is a growing trend (58% agreed to the same question in 2006) [37]. Using green credentials to market this product to its target market may not have been the best approach.

Another potentially influencing factor on this green marketing failure is Australian's 'green fatigue' which has developed in response to the multitude of environmental claims on products and scepticism about environmental claims. In the same year that Foster's launched Cascade Green, one of its competitors, Coopers, received negative publicity about its claim to be 'Australia's greenest beer', a claim that was referred to the Australian Competition and Consumer Commission by the Australian Consumers Association [38].

Sources: Landor [35], Boase [37], Lee [38], Foster's [39]

The main stakeholder groups to consider are detailed in Table 3.9. Each group has a different potential impact on the business and requires different strategies for engagement. Government agencies, for example, have the ability to increase the business's costs through the imposition of regulations such as mandatory taxes or deposits. It may be in a business's interests to communicate its sustainable development initiatives to this group. Environmental groups have the ability to damage a business's reputation through boycotts and other

Table 3.9 Corporate stakeholders and communication strategies

Stakeholder group	Why they should be engaged	Communication channels
Customers	To gain their support for sustainable development initiatives, e.g. design changes to reduce environmental impact	Direct engagement
End-consumers	To encourage them to purchase the business's products and dispose of them correctly at end of life (e.g. by recycling)	Advertising, on-pack labels, point-of-sale displays, social media
Suppliers	To gain support for collaborative initiatives that reduce environmental and social impacts in the supply chain	Information for trade partners, product specifications
Employees	To encourage them to identify or implement opportunities to improve sustainable development To improve staff morale and productivity by appealing to personal values	Staff newsletters, intranet site, policies and procedures
Shareholders	To promote the benefits of sustainable development initiatives for long term shareholder value	Annual reports, sustainability report, business web site, sustainability indexes (e.g. DJSI)
Government agencies	To demonstrate environmental and social responsibility and convince governments that additional regulations are not required	Direct engagement and lobbying, public reports, participation in government inquiries
Environment and consumer organisations	To maintain corporate reputation by avoiding negative publicity associated with campaigns against the business or to gain their co-operation and support	Direct engagement, public reports
Recyclers	To better understand the recyclability of the business's packaging and/or collaborate to improve recycling systems	Direct engagement or through recyclers' industry associations

campaigns and may also have a strong influence on government regulations in some jurisdictions. As the McDonald's example in Case Study 3.3 demonstrates, however, they are often willing to work collaboratively with businesses to achieve positive outcomes.

Communicating with other businesses in the supply chain is essential as a focus on sustainable development will:
- change the nature of existing relationships
- require new areas of collaboration to achieve desired results
- create new partnerships and corporate relationships (see the Nude Food Movers Case study 3.6).

Case Study 3.5 McDonald's and the 'Great Clamshell Debate'

In the late 1980s McDonald's was under pressure from environmental groups to change their packaging to reduce packaging and solid waste.

In 1987, Friends of the Earth in the United Kingdom boycotted McDonald's to force it to abandon the use of chlorofluorocarbons (CFCs) in hamburger 'clamshells' made from polystyrene foam. The company switched to an alternative foaming agent and started to move back to cardboard cartons [41]. There was also a national letter-writing campaign by school children in the United States to protest against the use of polystyrene packaging [34].

In 1989, the Executive Director of the Environmental Defense Fund invited the president of McDonald's USA, Edward Rensi, to a meeting to discuss their packaging [42]. The two organisations established a task force to find ways to reduce McDonald's solid waste through source reduction, reuse, recycling and composting.

The task force firstly developed the criteria that would be used to evaluate waste reduction options [42]:

- consistency with the waste management hierarchy of reduce, reuse, recycle/compost and incinerate/landfill
- the magnitude of environmental impact resulting from a change
- public health and safety concerns
- practicality
- economic costs and benefits.

After extensive research and consultation with its staff, franchisees and suppliers, McDonald's developed a Waste Reduction Policy and Waste Reduction Action Plan.

Changes to packaging that resulted included:
- a switch from polystyrene to paper-based wraps for burgers
- a move to replace chlorine-bleached paper with unbleached paper or paper bleached with less damaging processes
- a commitment to use more recycled material and to design more recyclable packaging
- the establishment of recycling programs for in-house waste such as corrugated board [42].

The switch from polystyrene clamshells to paper-based packaging for burgers attracted some controversy. McDonald's had undertaken research 3 years earlier that showed that polystyrene was more recyclable than coated paper. According to Edward Rensi, the company changed its mind because its customers did not 'feel good' about polystyrene [34]. While there was a marketing benefit for McDonald's to respond to this view, the task force demonstrated that the change reduced packaging volume by 70–90% (and therefore waste to landfill) and led to substantial reductions in energy

(continued)

Case Study 3.5 (continued)

> consumption and pollution over the life cycle [42, p. 3]. The change in packaging annoyed some people in the restaurant industry, who felt that McDonald's had 'given in' to environmentalists [43].
>
> *Sources*: Kleiner [34], Elkington and Knight [41], Capatosta et al. [42], Oleck [43].

3.4.3 Understand and Engage with Consumers

Peattie [45] has attempted to explain the attitude-behaviour or value-action gap (see Sect. 3.2) and the success or failure of sustainability marketing strategies through a sustainability purchase matrix (Fig. 3.4). Two key factors impacting consumer behaviour are:

- the degree of compromise—such as having to pay more to buy a more sustainable product or to sacrifice some aspect of performance
- the degree of confidence—how sure the consumer is that the product addresses a genuine sustainability issue, that the product is superior in its sustainability performance, that the company behind it can be trusted and there will be a worthwhile benefit from the purchase [2, p. 79].

The sustainability purchase matrix is useful because it provides two simple strategies that marketers can use to successfully market products with sustainability benefits:

- reduce the compromises that consumers need to make when they purchase the product compared to alternatives
- build the confidence of consumers in the benefits of a more sustainable consumption choice [2].

The second strategy is very important for packaging. Most consumers, even those who are highly motivated 'green consumers', lack the knowledge to make informed judgements about the environmental impacts of different packaging types and tend to rely on perceptions or beliefs [19]. Where there is a discrepancy between environmental impacts and consumer perceptions, it may be necessary to invest in a communications program to convince consumers that the packaging has environmental benefits [19]. In other cases, businesses can benefit by adapting their packaging to meet consumer expectations, while at the same time communicating its environmental benefits.

Belz and Peattie identified a number of other factors that influence whether or not a consumer will reflect environmental or social concerns in their purchasing [2] (pp. 79–80):

- *Purchase value.* Consumers are likely to invest more effort in gathering and considering information relevant to a high value purchase, such as a house or car, than a low value product

Case Study 3.6 Nude Food Movers

In 2009 Smash Enterprises created the 'nude food movers' brand of portable food storage products in response to changing consumer expectations for healthier food and less packaging waste.

'We recognised consumers are changing their behaviour in response to global market trends as well as other key factors:
- heightened environmental awareness—climate change, pollution, deforestation, waste management
- increase in obesity—one in four kids are overweight or obese
- Gen Z—want to make a difference, environmentally aware
- global financial crisis—offer quality at an accessible price
- rubbish free lunch days—assist schools with their initiatives to reduce waste
- parental assistance—help to make preparing healthy lunches not only stress free, but healthy and nutritious'.

Sally Turno, Marketing Manager Smash Enterprises

With the support of Nutrition Australia, Nude Food Day was promoted to encourage Australians to bring a healthy, rubbish-free lunch to work or school on 13th October 2010. Nutrition Australia had previously implemented the flexible and tailored Healthy Eating Schools Program (HES). Co-branding with a government agency provided credibility and helped both organisations extend the reach of their messages.

'Nude Food Day' was considered a great success with 968 Australian organisations (schools and workplaces) and over 500,000 individuals participating. The company experienced engagement with customers and influencers through the revised (nude food mover) product offering.

'We have already seen a change in the behaviour of many of our customers. We are inundated with communication from not just parents, but teachers who are trying to educate the importance of healthy eating, teamed with a healthy environment. Schools have adopted the Nude Food Days into daily/weekly/monthly practice.

And we have definitely seen a change in the packaging of children's food in the past 12 months which stem from key market trends and more specially the 'movement' of schools requesting children a) reduce waste by not bringing wrapped (glad wrap, plastic bags, foil etc.) lunches with them to school and b) schools requesting children take home any waste from wrapped items, such as muesli bar wrappers.

We speak with many suppliers of these products and have been advised of the changes they are making to their ranges to adapt and appeal to the more environmental and health conscious consumer'.

Source: Nude Food Movers [44], Sally Turno, personal communication

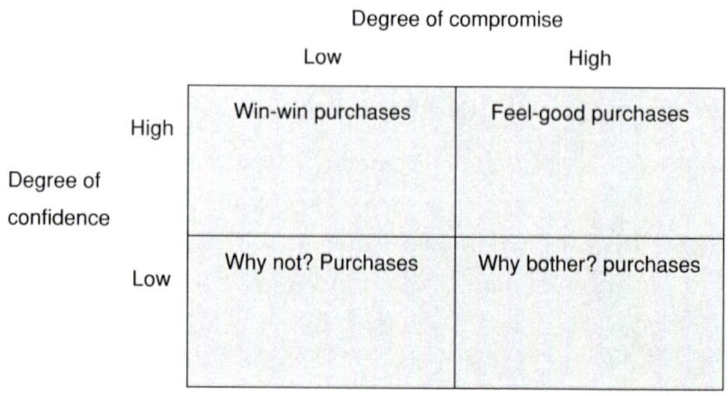

Fig. 3.4 Sustainability purchase matrix. *Source*: Belz and Peattie [2, p. 78]

	Degree of compromise	
	Low	High
High Degree of confidence	Win-win purchases	Feel-good purchases
Low	Why not? Purchases	Why bother? purchases

- *Frequency.* Frequent purchases are more likely to be habitual than carefully considered
- *Visibility.* Consumers may be more likely to consider the sustainability of products that are highly visible to others.

The importance of frequency and habit is supported by an Australian survey of supermarket shoppers. Packaging was a low priority when selecting products. While price, product and convenience are all important, most consumers exhibit 'habitual buying'—they keep buying the same products in the same packaging [21].

3.4.4 Identify Messages and Improvement Strategies

When combined with a life cycle map or life cycle assessment data, a sustainability impact matrix can be used to identify:
- which messages to communicate about the environmental or social attributes of a product
- potential strategies to improve sustainability, develop a more sustainable product and meet sustainable development goals.

⇒ See Chap. 5 for more information on creating a sustainability matrix

For example, a simplified matrix for roasted coffee is shown in Table 3.10 [2], followed by opportunities for improvement and labelling. One dimension of the matrix is the life cycle of the product and the other is the economic, environmental and social criteria of concern. The matrix highlights certain information:
- The highest environmental impacts occur at the cultivation stage of the life cycle. (Energy, water, soil and ecosystems bear a high impact, and air bears a medium impact.) These impacts arise from coffee growing that uses a system with high inputs of water for irrigation, fertilisers, pesticides and energy. The use of industrial fertilisers leads to eutrophication of groundwater and surface water

3 Marketing and Communicating Sustainability

- The primary processing stage of the life cycle (roasting and packaging) also produces significant environmental impacts but to a lesser extent overall than the cultivation stage—with the exception of waste where it is greater than and air (emissions) where it is similar to cultivation
- The greatest social benefits and potential negative impacts of the life cycle occur at the cultivation stage when coffee prices determine the viability of coffee cultivation and have an impact on the livelihood of coffee farmers, their families and communities—in particular in developing countries
- Consumption and disposal of coffee produce high impacts on energy consumption, emissions (including greenhouse gas emissions), water consumption and waste (both coffee grounds and packaging).

The matrix therefore helps to:
- identify areas where a business is having a positive triple bottom line impact
- identify 'hot spots' requiring strategies for improvement
- generate ideas for improvement and innovation throughout the complete life cycle
- assess whether communications and claims are greenwash. If a low impact area is promoted and diverts attention away from high impact areas this could be construed as one of the 'sins of greenwash', the 'hidden trade-off' [31]
- benchmark against best practice standards or competitors. The matrix can be developed and compared for competing products or services to identify areas of competitive advantage or disadvantage.

Table 3.10 Sustainability impact matrix for roasted coffee

	Cultivation/primary processing	Roasting, packaging	Distribution	Consumption	Disposal
Energy	High	Medium	Low	High	Low
Air	Medium	Medium	Low	Medium	Low
Water	High	Medium	Low	High	Low
Soil	High	Low	Low	Low	Low
Waste	Low	Medium	Low	Low	High
Ecosystems	High	Low	Low	Low	Low
Health	Medium	Low	Low	Low	Low
Equity	High	Low	Low	Low	Low
Opportunities for improvement and innovation	Organic growing Minimum prices paid to growers	Energy efficiency Cleaner production		Work with machine manufacturers to improve efficiency	Recyclable packaging Encourage composting of coffee grounds
Environmental claims or labels	Certified organic Fair trade				Recyclable packaging Mobius loop recycling symbol

Source: Builds on Belz and Peattie [2, p. 60]

For example, the coffee brand owner may decide to:
- work with its supply chain to source organic or more sustainably harvested coffee beans
- switch to more recyclable packaging
- develop strategies to support composting of used coffee grounds
- include 'certified organic' or 'recyclable packaging' logos and information on its packaging. Note: environmental labels should be considered where a product or package appears to be in a position of strength compared to its competitors. See Appendix B for information on labels and logos.

3.4.5 *Communicating the Messages*

Communicate Function and Provide Emotional Benefit

The most effective green positioning strategy is to provide clear information to consumers about the *functional* benefits of the product, environmental attributes such as waste reduction or recyclability, while also providing emotional benefits [46]. A strategy that focuses exclusively on the functional attributes will have limited success because it delivers no individual consumer benefit and provides insufficient motivation to change consumer purchasing behaviour.

Relevant emotional benefits include:
- the feeling of well-being that green consumers get from being altruistic; that is, from 'helping the environment'
- the personal satisfaction that consumers get from the socially visible consumption of green brands
- the benefits that consumers get from sensations and feelings normally experienced through contact with nature [46].

Visy, for example, uses nature-based images to promote its environmental commitment (see Case Study 3.7).

Use a Range of Communication Channels and Be Consistent

Communicating sustainable development credentials and positioning the business, brands and products involves many forms of communication, such as corporate reports, advertising and marketing materials, and websites. These must be consistent with each other and any on-pack labels. If, for example, a corporate mission statement explicitly states that a business will source paper-based materials from certified forests, these claims need to be factual and, where promoted, shown to be consistent and in-line with appropriate certification, regulatory or other standards (see Patagonia's Case Study 3.8).

Case Study 3.7 Visy: marketing the company

Packaging and recycling company Visy uses images of rainforests with the words 'Packaging and Recycling for a Better World' on its trucks to promote itself as an environmentally responsible business. This advertising provides customers and consumers with emotional benefits through the perception that they are helping to achieve a better world. Visy's website and other promotional materials support this message by providing detailed information on its environmental and social initiatives.

Photo Visy

Sources: Visy Industries [47, 48]

Case Study 3.8 Patagonia: 'The Good and The Bad' of Products and Packaging

Patagonia, an outdoor adventure brand operating primarily in the apparel category, is recognised as a pioneer in sustainable lifestyles marketing and has experienced growth of 19% in a period of market decline [11]. Patagonia has a consistent and transparent approach to communicating its environmental impact and journey to reduce impact and create positive change.

A sustainable packaging case study is featured on Patagonia's website, entitled 'the good and the bad of packaging the Capilene'. This case study shares with stakeholders and customers the understanding the company has gained on the impacts of the packaging, the modifications it has made over time

(continued)

Case Study 3.8 (continued)

and the commitment to ongoing improvement to the product and packaging system (including packaging that is invisible to the retail customer).

As Patagonia notes [49]:

'The first Capilene® Underwear (circa 1985) came packaged in red, white and blue polyethylene bags with attached handles and red, white and blue graphic inserts. The objective was to protect the product and keep it clean, and to be able to reopen and close the package with ease. But the packaging took extra store space to display properly—and was environmentally wasteful. As a stopgap we printed suggestions on the back for reuse of the bag. This did not go over well with customers, who found it patronising. Next we switched sometime in the early nineties to recycled paper bags, which tore quickly and without provocation. Next, we worked with an outside consulting team to develop the first generation of our current retail racking system and a minimal "sushi roll" package of basic information and two rubber bands. Other packaging ideas that we've explored over the years include self-bags, mesh bags, napkin rings, an enviro eggshell, paperhangers and webbing straps with buckles. But in the end, the sushi roll has proved the most effective. Some countries, culturally, won't accept sushi-rolled underwear and for those countries we ship base layers in recycled-cardboard boxes. But the majority of the units we ship go out with minimal packaging.

However, one large packaging problem goes unsolved as yet—and largely unnoticed by consumers. Every unit we produce goes into its own plastic shipping bag to keep it clean from factory to end consumer. We still do not have a good alternative. The bags do not even have, as yet, recycled content. We do try to recycle as many as we can, by shipping them to companies that make lumber from recycled plastic, but this is not enough. We continue to look for a solution.'

The deployment of the strategy to reduce the packaging impacts of the Capilene is linked to several corporate and communication strategies including Patagonia's code of conduct, paper use and procurement policy, procedures and life cycle assessment research through the Footprint Chronicles®. These documents can be viewed online and offer a good example of a business gaining a good understanding of impacts and communicating effectively.

Source: United Nations Environment Program [11] and Patagonia [49]

3.4.6 Confirm the Role of Packaging

Packaging is an important component of sustainability marketing, and a business must confirm within its strategy the role that packaging will play. As outlined in Chap. 1, packaging will contribute to achieving the business's sustainable development goals, and these will include reductions in the impacts of the packaging itself.

Packaging, however, is also a communicator of the brand, and on-pack labelling and packaging design can be used to facilitate changes in consumer purchases, use of products and disposal of packaging. New challenges in communication will also arise, as packaging and product-design decisions may be counter-intuitive to consumer perceptions.

⇒ Appendix B Labels and logos, provides a reference list of some potential packaging logos

Environmental claims and labels on packaging might be used to influence purchase decisions, reinforce the corporate, brand or product positioning and be one of the strategies used to meet sustainable development goals such as increased recycling of packaging. However, these claims and labels must compete for label space, which is already under pressure to communicate other messages or regulatory information to purchasers. The number of potential logos is also increasing. A website has been created in Germany to provide a reference for consumers to identify and understand over 300 different labelling schemes [50].

3.5 Environmental Claims and Labels

3.5.1 Types of Claims and Labels

ISO has divided environmental claims into three categories (Table 3.11).

Type I claims are often used to verify the source or production stages and are less likely to be used for the end-of life stage. These tend to have more credibility with consumers because they have been checked and verified by an independent organisation. Some of these certification schemes are for performance in a specific area, such as those for 'chlorine-free' paper and sustainable forestry. More general eco-labels, such as the German 'Blue Angel', indicate that the product has been evaluated across its entire life cycle against a range of environmental criteria.

Type II claims are currently the most common and are more likely to be used to communicate aspects of the use or end-of-life phase of a product's lifecycle.

Type III claims include quantitative environmental data and are therefore more difficult for the consumer to interpret. They are not widely used to communicate packaging benefits.

The selection of claims and their use on packaging needs to be aligned with the sustainable development goals of the business and the positioning of the business, brands and products in line with this. Examples of logos used to communicate specific aspects of sustainable development are provided in Table 3.12. These are linked to the packaging sustainability framework presented in Chap. 2 and briefly discussed in the following sections.

Table 3.11 Types of environmental labels, per ISO 14024, 14021 and 14025

Type of label	Description
Type I labels (ISO 14024: 1999)	Voluntary, third-party certified labels that can be displayed on products and services that meet or exceed a set of criteria established by the certifying body. They are commonly referred to as 'eco-labels' and indicate that a product is environmentally preferable within its category
Type II labels (ISO 14021: 1999)	Based on the 'self-declarations' of manufacturers, importers, distributors or retailers
Type III labels (ISO 14025: 2006)	Provide quantitative life cycle environmental data based on certain criteria, such as natural resource use, air emissions or solid waste

Source: Charter et al. [3]

3.5.2 Material Recycling

The cradle-to-cradle model introduced in Chap. 1 outlines two material recovery options: a 'technical' or industrial' metabolism (including material recycling) and a biological metabolism such as composting (organic recycling).

Environmental claims and labels inform and educate consumers about the correct way to recover packaging to support sustainability by:
- reducing waste
- ensuring materials are recovered and cycled through the correct metabolism
- eliminating cross-contamination between the two metabolisms.

Promote Recyclability

Recyclable packaging should be labelled with a claim and/or a symbol to indicate that the packaging can be recycled after use.

Many consumers are confused or unsure about which packaging is recyclable. A survey of householders in the United Kingdom [51, p. 58] found that only 48% of respondents claim to understand the use of their recycling system 'very well'. Current recyclers sometimes or often:
- put things in the bin because they are not sure if they can be recycled (48%)
- throw recyclable bathroom waste in the residual bin (41%)
- put things in the recycling bin even if they're not sure they can be recycled (36%) (p. 83).

Packaging labels, together with other forms of marketing and communication, can better inform consumers about recyclability and prompt them, at the point of disposal, to separate recyclable packaging from general waste. This will improve recovery rates for packaging in line with regulatory and corporate waste reduction goals.

Effective recyclability labels can also help to reduce the amount of non-recyclable material placed in recycling bins, which reduces recycling costs and may improve the quality of the recycled material.

3 Marketing and Communicating Sustainability

Table 3.12 Packaging sustainability principles (cyclic and safe) and relevant labels

Packaging sustainability principles	Design strategy	Recommended labels to communicate the strategy
Cyclic packaging	Use of recyclable materials/encourage consumers to recycle through a material recycling system	Recyclable Mobius loop symbol
	Use of biodegradable materials/encourage consumers to recycle through an organic recycling system	Compostable (if certified to a recognised national or international standard)
	Polymers used in the packaging are identified to support recycling	Plastics identification code
	The material contains recycled content	Mobius loop with percentage recycled content
Safe packaging	Encourage consumers not to litter	Please do not litter 'Tidyman' symbol
	Paper is manufactured using a less environmentally damaging bleaching process	Process chlorine free[a] Totally chlorine free[a]
	The fibre used to make paper or board is from a sustainably managed forest	Forest Stewardship Council certified[a]
	The company is taking steps to reduce greenhouse gas emissions	Carbon Trust label[a]
	Greenhouse gas emissions have been offset	Carbon neutral
	Renewable energy is used in production	Green-e marketplace[a]

[a] These labels are trademarks and can only be used by companies meeting rigorous requirements for environmental performance—contact details are provided in Appendix B

Confirm That it is Recycled

A claim that a package is 'recyclable' should only be made if there is a well-established system available for collection and recycling. According to ISO, this means that collection or drop-off facilities are 'conveniently available to a reasonable proportion of purchasers, potential purchasers and users of the product in the area where the product is sold' [26, p. 13]. If this is not the case, then a qualified claim of recyclability may be used.

A qualified claim must adequately convey the limited availability of collection facilities. A general claim such as 'recyclable where facilities exist' is inadequate because it doesn't provide sufficient information about the availability of collection facilities [26].

Proposed changes to the US Federal Trade Commission guidelines [32, p. 2] include more specific advice on disclosing the limited availability of recycling programs:

- If the 'substantial majority' of consumers or communities have access to recycling facilities, the marketer can make an unqualified recyclable claim
- If a 'significant percentage' of consumers/communities have access to recycling facilities, the marketer should make a qualified recyclable claim (for example, 'package may not be recyclable in your area')

- If less than a 'significant percentage' of consumers have access to recycling facilities, the marketer should make a qualified recyclable claim (for example, 'product is recyclable only in the few communities that have recycling programs').

The US Society of the Plastics Industry has asked the Federal Trade Commission to consider the use of alternative qualifiers that refer consumers to informative, accurate websites about recyclability [52]. In their view, this would provide businesses with a way to communicate more information about the limited recyclability of packaging than could be provided on a typical container, where label space is limited:

> 'We urge the Commission to assess whether advertising that encourages consumers to visit a website for accurate, up-to-date information on recycling options available to them might both empower consumers to educate themselves about recycling options (which they are more likely to do if affirmatively reminded on the product or its packaging) *and* provide them with the necessary roadmap by which to find recycling information quickly and readily, without a significant risk of prompting undesirable consumer behaviour (e.g., putting an item that cannot be recycled locally into the [kerbside] recycling bin without checking)' [52, p. 4].

Use Symbols Consumers Recognise and Understand

If packaging is 'recyclable' according to the ISO standard, then a recycling symbol should be considered. As outlined in Sect. 3.2.2, recyclability is one of the measures that consumers use to judge the environmental friendliness of packaging. The aim should be to use a symbol that is most likely to be recognised and interpreted correctly by consumers.

As a general rule, the Mobius loop should be used (following the requirements of ISO 14021) unless there is a mandatory labelling standard or a more recognisable symbol in the target market. ISO states that the use of a recycling symbol is optional, but if a symbol is to be used they recommend the Mobius loop. The meaning of this symbol was correctly identified by 56% of respondents to one UK survey—more than any of the other recycling symbols covered by the survey [17, p. 4]. There are other codes available, and there may be a more appropriate symbol for use in a particular country. For example the use of specific recycling labels is mandatory in Japan.

For packaging made up of more than one material, it may be necessary to advise consumers about the recyclability of each material and the need to separate them (if possible). A new voluntary label has been introduced in the United Kingdom (see Appendix B—Labels and logos and Case Study 3.9).

Understand the Resin Identification Codes

The resin identification code (Fig. 3.5) was developed by the US Society of the Plastics Industry in 1988 for marking rigid plastic containers to identify the plastic resin from which they are made. Their purpose was to support the introduction of

> **Case Study 3.9 On-Pack Recycling Label**
>
> The British Retail Consortium and WRAP (Waste & Resources Action Programme) have developed a label to provide more specific information to consumers about recyclability. Packaging and packaging components are labelled to distinguish between three levels of recyclability [53]:
> - 'Widely recycled': 65% or more of local authorities have collection facilities for that packaging type in their area
> - 'Check local recycling': 15–65% of local authorities have collection facilities for that packaging type in their area
> - 'Not currently recycled': less than 15% of local authorities have collection facilities for that packaging type in their area.
>
> The service is supported by the site http://www.recyclenow.com, which allows consumers to search for recyclable materials by entering their postcode.

collection and sorting systems for plastics packaging by helping consumers to identify and separate particular plastics from household waste.

The code is now mandated in 39 US states (as of 2009) and is also promoted as a voluntary identification standard by plastics industry associations in many other countries. For example, the European Plastics Association has stated that resin identification codes, based on the Society of the Plastics Industry codes, 'should be used when it is considered to aid the identification and sorting of used plastic packaging' [54, p. 2].

Use of the code now extends beyond rigid plastics packaging, although its use on non-recyclable flexible plastics has the potential to mislead consumers. The code is very similar to the 'Mobius loop' recycling symbol and therefore may incorrectly imply recyclability if it is clearly visible on a non-recyclable pack.

In 2008, the Society of the Plastics Industry began working with the American Standard for Testing and Materials to develop a new standard that would expand the system by providing for additional codes for resin types not covered by numbers 1–6 [55]. This process aims to accommodate new types of plastics, such as biodegradable starch based thermoplastics, and combinations of materials. According to the Society of the Plastics Industry, the guidelines presented in Table 3.13, should be followed for correct use of the resin identification codes.

The code may not be suitable for packaging of dangerous goods, including chemicals embossed with 'POISON' or labelled to meet the requirements of dangerous goods legislation [57]. There may be some risks associated with the inclusion of these products in collection, sorting or reprocessing systems. It is therefore important to consult with recyclers to check whether or not they should be included.

Fig. 3.5 Resin identification code for plastics

Table 3.13 Correct use of the SPI resin identification codes

How to use the Society of the Plastics Industry resin identification codes correctly
• For products sold in the United States, use the code on bottles and rigid containers in compliance with the 39 state laws currently in effect
• Make the code inconspicuous at the point of purchase so that it will not influence the consumer's buying decision
• The code should be moulded, formed or imprinted on all containers large enough to accept the ½ inch minimum-size symbol and all containers between 8-ounce size and 5 gallons
• The code should be on the bottom of the container, as close to the centre as possible
• Do not modify the code in any way
• Do not make recycling claims in close proximity to the code, even if such claims are appropriately qualified.

Source: Society of the plastics industry [56]
Notes: Technical bulletins with engraving masters and artwork are available from the Society of the Plastics Industry [http://www.plasticsindustry.org/AboutPlastics/content.cfm?ItemNumber=825&navItemNumber=1124 (accessed 27 January 2010)]

3.5.3 Organic Recycling

Promote Biomaterials Recovery Responsibly

Claims about 'degradability', 'biodegradability' and 'compostability' cause confusion for packaging developers and consumers. Many consumers believe that biodegradable materials are better for the environment than non-biodegradable materials and may be influenced by environmental marketing strategies that focus on this attribute (Photo 3.1).

⇒ Explanations of degradability, biodegradability and compostability are provided in Chap. 6

Unfortunately, many self-declared claims about degradability and its environmental benefits (particularly in landfill) are incorrect or misleading. There is also potential for biodegradable plastics to be disposed in recycling bins, as consumers frequently fail to understand the difference between recycling and composting recovery systems [58].

Comply with ISO

The ISO standard [26] defines the terms 'degradable' and 'compostable' and provides guidelines for these types of claims on products or packaging.

Photo 3.1 Example of self-declared claims about degradability. Photo: Helen Lewis

> **Degradability** is a: 'characteristic of a product or packaging that, with respect to specific conditions, allows it to break down to a specific extent within a given time' [26, p. 10].
>
> A '**compostable**' material or product is one that meets the more stringent requirements for degradation in a home composting or commercial (industrial or municipal) composting environment. The ISO standard defines compostability as: 'a characteristic of a product, packaging or associated component that allows it to biodegrade, generating a relatively homogenous and stable humus-like substance' [26, p. 9].

To ensure that any claim about degradability or compostability can be substantiated and verified, the packaging material should be certified as meeting the requirements of recognised international or national performance standards (see Table 3.14).

Validate or Test for Degradability

According to the standard, claims about degradability should only be made in relation to a specific test method that includes maximum levels of degradation and must be relevant to the circumstances in which the product is likely to be disposed. For example, a biodegradable polymer requires oxygen, moisture and microorganisms to enable it to completely break down. These conditions exist in a well-managed composting facility but are unlikely to be present in a conventional landfill. Modern landfills are designed for slow degradation by compressing waste and removing leachate.

Claims about degradability should not be made for packaging that releases substances in concentrations harmful to the environment. As with all claims, methods for evaluation should be based on international standards, regional or national standards with international recognition, or industry methods subjected to peer review (in that order).

A number of businesses in different jurisdictions have been charged with making false or misleading claims about degradability (see Case Study 3.10).

> **Case Study 3.10 Misleading Degradability Claims on Packaging (US)**
>
> In June 2009 the US Federal Trade Commission (FTC) charged three companies with making deceptive and unsubstantiated biodegradability claims:
> - Kmart for its American Fare brand of disposable plates
> - Tender for its Fresh Bath brand of moist wipes
> - Dyna-E for its Lightload brand of dry towels. [59]
>
> The FTC's marketing guidelines state:
>
> 'An unqualified claim that a product or package is degradable, biodegradable or photodegradable should be substantiated by competent and reliable scientific evidence that the entire product or package will completely break down and return to nature, i.e. decompose into elements found in nature within a reasonably short period of time after customary disposal' [28, p. 5].
>
> In their charges against the three companies, the FTC alleged that the defendants' products are typically disposed of in landfill, incinerators or recycling facilities, where it is impossible for them to degrade in a 'reasonably short time' [59].

Confirm Composting Facilities are Available

A claim of compostability should not be made for packaging that negatively affects the overall value of the compost as a soil amendment, releases substances harmful to the environment during decomposition or subsequent use, or reduces the rate of composting. The ISO standard also states that the claim needs to specify whether the packaging is suitable for home composting or commercial composting, unless it is suitable for both processes.

Claims about compostability should only be made if composting facilities are 'conveniently available to a reasonable proportion of purchasers, potential purchasers and users where the packaging or product is sold' [26, p. 10]. If this is not the case, the claim needs to be qualified with advice about the limited availability of composting facilities.

The compostability of packaging certified against European standard EN 13432, US standards ASTM D6400 and ASTM D6868 or Australian standard AS 4736 can be communicated through the use of an industry label (see Appendix B—Labels and logos—for information on the European 'seedling' logo). Certification schemes provide a simple and recognisable label and add credibility to any claim of compostability. They still need to be used carefully, however, to avoid misleading consumers. Certification indicates that a package is *technically* compostable, but if systems for the recovery and processing of compostable materials are not widely available then the claim may be considered misleading (if this is the case then additional qualifying information may need to be provided to consumers). Table 3.14 provides a list of the standards relating to degradable plastics.

3.5.4 Recycled Content

> **Recycled content** is the proportion, by mass, of recycled material in a product or packaging [26, p. 13].

Materials that can be claimed as recycled content are those defined as 'pre-consumer material' or 'post-consumer material' [26, p. 14]:
- *Pre-consumer material* is material diverted from the waste stream during a manufacturing process, but excludes rework, regrind or scrap that can be reclaimed in the same process that generated it. For example, plastic scrap generated during the manufacture of a plastic product, and then reused in the same process, is not considered to be 'recycled'
- *Post-consumer material* is material generated by households or by commercial, industrial and institutional facilities in their role as end-users of the product.

State the Percentage of Recycled Material

The ISO standard states that where a claim of recycled content is made, the percentage of recycled material should be stated. The use of a recycled content symbol is voluntary, but they recommend the use of the Mobius loop with the percentage value inside or immediately adjacent to the symbol. If the percentage varies over time, the claim should be something like 'at least X%' or 'greater than X%'. The use of an explanatory statement is also optional, but could include information on the type of material (pre-consumer or post-consumer).

3.5.5 Litter Reduction

There is some evidence that people need continuous anti-littering information to take up responsible behaviour [60]. Packaging labels provide one opportunity to

Table 3.14 Standards for degradable plastics

Standard	Region/country	Disposal environment
EN 13432: 2000, Requirements for packaging recoverable through composting and biodegradation	Europe	Municipal and industrial aerobic composting facilities
EN 14995: 2006, Plastics—evaluation of compostability—test scheme and specifications	Europe	Municipal and industrial aerobic composting facilities
ASTM D6400: 2004, Standard specification for compostable plastics	United States	Municipal and industrial aerobic composting facilities
ASTM D6868: 2003, Standard specification for biodegradable plastics used as coatings on paper and other compostable substrates	United States	Municipal and industrial aerobic composting facilities
ASTM D7081: 2005, Standard specification for non-floating biodegradable plastics in the marine environment	United States	Aerobic marine waters or anaerobic marine sediments
ASTM D5511:2002, Standard test method for determining anaerobic biodegradation of plastic materials under high-solids anaerobic digestion conditions	United States	High-solids anaerobic digester for the production of compost from municipal solid waste
ASTM D5526: 2002, Standard test method for determining anaerobic biodegradation of plastic materials under accelerated landfill conditions	United States	Biologically active landfill[a]
AS 4736: 2006, Biodegradable plastics—biodegradable plastics suitable for composting and other microbial treatments	Australia	Municipal and industrial aerobic composting facilities
AS 5810: 2010, Biodegradable plastics—biodegradable plastics suitable for home composting	Australia	Home composting

[a] Most landfills are not biologically active because they have been intentionally designed to minimise biodegradation. Some modern landfills are designed to accelerate degradation and extract the methane for energy generation. These are often called 'bioreactor landfills'

reinforce the anti-litter message, particularly for products consumed away from home. The most common anti-litter symbol is the 'Tidyman', and there are many variations of this (see Appendix B—Labels and logos). The symbol is often accompanied by a message such as 'Please dispose of thoughtfully'. This is fairly ambiguous, and a more direct message such as 'Please don't litter' might be more effective.

3.5.6 Chlorine Free

The Chlorine Free Products Association was established in 1994 to encourage paper manufacturers to move away from chlorine chemistry. (For more

information on pollution associated with chlorine bleaching, see Sect. 2.4.4.) Paper and paper products can be certified as either 'totally chlorine free' or 'process chlorine free'. Totally chlorine free certification is for virgin paper and must fulfil the following conditions [61]:
- No chlorine or chlorine compounds were used in the paper manufacturing process
- The mill has no current or pending environmental violations
- The paper does not contain any fibre from old growth forests.

Process chlorine free certification, which applies to recycled paper only, means the following [61]:
- The paper contains at least 30% post consumer waste
- No chlorine or chlorine compounds were used in the paper manufacturing process for virgin fibre or the re-bleaching of recycled fibre
- The mill has no current or pending environmental violations
- Any virgin fibre is not sourced from old growth forests.

3.5.7 Forest Stewardship

There are a number of certification schemes for paper and timber products that meet minimum standards for ecological stewardship and social sustainability in forest management. The Forest Stewardship Council (FSC) is a non-government organisation established by business, environmental and human rights groups in 1993 in response to concerns about global deforestation. Suppliers of products certified by the FSC have demonstrated that they comply with strict environmental and social principles such as [62]:
- compliance with all relevant laws and treaties
- recognition and respect for the rights of indigenous communities
- protection of workers' rights
- reduction of environmental impact of logging activities and maintenance of the ecological functions and integrity of the forest.

An alternative certification scheme for paper and timber products is available through the Program for the Endorsement of Forest Certification (PEFC). This is an umbrella organisation for approximately 30 national standard setting bodies, although other stakeholders can apply to join as full members. Products are certified against national standards that meet the program's 'Sustainability Benchmark'. According to the program, this provides international recognition in a global market: products are 'certified once, accepted everywhere' [63]. Certification is available for forest management or 'chain of custody', which outlines requirements for tracking certified material from the forest to the final product. A list of certified products, including packaging materials, is available on its website [64].

3.5.8 Carbon Reduction

Many businesses now measure and report on greenhouse gas emissions associated with individual products, but there is no international standard for claims or labels yet.

A national labelling scheme was introduced in the United Kingdom by The Carbon Trust, an organisation established by the government in 2001 as an independent company tasked with accelerating the move to a low carbon economy. The label shows greenhouse gas emissions generated during the product's life cycle, including production, transportation, processing, use and disposal. It is based on UK standard PAS 2050, developed by BSI Standards Solutions in partnership with the UK Department for Environment, Food and Rural Affairs and the Carbon Trust. When a business displays the carbon label, it is committing to reduce the carbon emissions associated with the product over the following 2 years, and if this commitment is not met the business will no longer be able to use the label.

One of the first businesses to use The Carbon Trust label was PepsiCo for the Walkers brand of potato chips. The company's market research found that consumers are positive about the label, with 79% saying that it makes them more aware of the environmental impact of the products and services they buy, and 44% saying that it made them more positive about Walkers as a business [65].

Many businesses aim to become 'carbon neutral' by generating sufficient renewable energy for their needs or by buying carbon offsets. The Federal Trade Commission advises that marketers should have 'competent and reliable scientific evidence to support their carbon offset claims' [32].

The transport of food products is becoming more prominent as an environmental issue through the debate on 'food miles', which calls on consumers to buy food locally [66]. Food miles are the distance travelled by food from the 'paddock to the plate'. In response to customer concerns about food transport, both Tesco and Marks & Spencer in the United Kingdom have announced that they plan to put an airplane symbol on products that have been imported by airfreight. The environmental impact of freight increases with distance travelled, and airfreight has higher environmental impacts than sea or road transport. However, a recent report for the UK Government [67] concluded that food miles are inadequate as a single measure of sustainable development. While the environmental and social impacts of food transport are significant, they need to be weighed up against other impacts such as the energy costs of growing food in colder climates like the United Kingdom compared to warmer countries such as Spain.

3.5.9 Renewable Energy Use

The Federal Trade Commission [32, p. 2] suggests that marketers should qualify renewable energy claims by specifying:
- the source of renewable energy such as wind or solar

3 Marketing and Communicating Sustainability

What business sustainable development goals, achievements and challenges will be communicated?

Will the sustainability marketing campaign be for individual products or the company as a whole?

What issues are consumers concerned about?

Where will the messages be communicated?
- On-pack
- Point of sale promotion
- Advertising
- Corporate reports
- Technical bulletins
- Telemarketing
- Websites

If on-pack labeling is used:
- Which business sustainability messages will be communicated?
- What will be the brand and product positioning?
- Will messages relate to the product and/or packaging?

Are the messages truthful, transparent and clear?

Fig. 3.6 Checklist for determining the sustainability marketing and communication approach

- whether less than all of the manufacturing processes involved in making the product/packaging were powered with renewable energy or conventional energy offset by renewable energy certificates.

The Centre for Resource Solutions in California has developed the 'Green-e Marketplace' logo to help businesses communicate their commitment to renewable energy. To be eligible to use the logo, businesses need to measure their annual electricity use and purchase (or generate) a qualifying amount of certified renewable energy. One of the first brands to use the logo on a packaged good was Frito-Lay, which started to use it on its SunChips multigrain snacks in 2007 [68].

3.6 Conclusion

Consumer surveys about environmental values and attitudes, including attitudes to packaging, highlight the opportunities and risks associated with marketing sustainability. Consumers tend to have negative associations with packaging and have limited awareness of packaging-specific environmental issues other than disposal issues (recyclability, reuse, biodegradability and over-packaging).

Businesses need to respond to consumer concerns, preferences and expectations if they want to remain competitive and grow their market. However, environmental marketing strategies also need to be based on a good understanding of legislative requirements (Chap. 4) and life cycle environmental impacts (Chap. 5). They should be carefully considered and integrated within the business's policies, corporate strategies and management systems (Chap. 8). The promotion of a product or its packaging as green or environmentally preferred, without a genuine and proactive commitment to sustainable development, is likely to be counterproductive. It opens a business to greenwash accusations and a consumer backlash.

A checklist for issues to consider when determining the sustainability marketing and communication approach of the organisation is presented in Fig. 3.6.

References

1. Polonsky MJ, Rosenberger PJ (2001) Re-evaluating green marketing—a sophisticated strategic marketing approach, Business Horizons 44(5): 21–30
2. Belz FM, Peattie K (2009) Sustainability marketing: a global perspective. Wiley, Chichester
3. Charter M, Peattie K, Ottman J, Polonsky M (2002) Marketing and sustainability. Centre for business relationships, accountability, sustainability and society (BRASS) in association with the Centre for Sustainable Design
4. Cambridge Dictionaries Online (2010) Cambridge advanced learner's dictionary (cited 9 September 2010)
5. Deloitte (2009) Finding the green in today's shoppers: sustainability trends and new shopper insights. Report to the Grocery Manufacturers Association (GMA)
6. Ipsos-Mori (2000) Ethical consumerism research. http://www.ipsos-mori.com/researchpublications/researcharchive/poll.aspx?oItemId=1496 (cited 14 February 2010)

7. Spears M, Larson A (1992) Mobil Chemical Corporation. World Resources Institute
8. The Co-operative Bank (2008) The ethical consumerism report 2008. Manchester, UK
9. Stengel R (2009) The responsibility revolution, Time, pp 24–27
10. Ray P, Anderson SR (2000) The cultural creatives: how 50 million people are changing the world. Harmony, New York
11. United Nations Environment Program (2005) UN Global Compact and Utopies. Talk the walk. United Nations Environment Program, Paris
12. Worldwide Fund for Nature (2008) Weathercocks and signposts: the environment movement at a crossroads. Godalming, Surrey, UK
13. Business in the Community—Ireland (2003) The first-ever survey of consumer attitudes in Ireland towards corporate responsibility. Dublin, Ireland
14. D'Souza C, Taghian M (2005) Green advertising effects on attitude and choice of advertising themes. Asia Pac J Mark Logist 17(3):51–66
15. Pegram Walters associates (1993) Consumer attitudes to packaging: summary report of findings. Report prepared for INCPEN, London
16. Pegram Walters associates (1997) Project packaging II: report of findings. Report prepared for INCPEN, London
17. IPSOS Mori (2008) Public attitudes to packaging 2008. Report to INCPEN and Valpak, London
18. New Zealand Paperboard Packaging Association (2005) Code of practice for the New Zealand paperboard packaging industry, Lower Hutt, NZ
19. van Dam Y (1996) Environmental assessment of packaging: the consumer point of view. Environ Manag 20(5):607–614
20. New Zealand Paperboard Packaging Association (2005) Attitudes to packaging, recycling and the environment, Lower Hutt, NZ
21. Taverner Research Company (2004) Consumer demand for environmental packaging. Report to the NSW Jurisdictional Recycling Group, Sydney
22. Nielsen (2008) Packaging and the environment: a global Nielsen report, New York
23. Hewlett Packard (2010) HP standard 011 general specification for the environment
24. Hewlett Packard (2010) Global citizenship strategy. http://www.hp.com/hpinfo/globalcitizenship/commitment/gcstrategy.html
25. Murphy M (2010) Proving cred is worth the effort. In: The Sydney Morning Herald, Fairfax, Sydney p 5
26. ISO (1999) ISO 14021:1999, Environmental labels and declarations—type II environmental labelling—principles and procedures. International Standards Organization (ISO), Geneva
27. DEFRA (2003) Green claims—practical guidance. http://www.defra.gov.uk/environment/business/marketing/glc/pdf/genericguide.pdf (cited 12 July 2009)
28. Federal Trade Commission (1998) Part 260—guides for the use of environmental marketing claims, Washington
29. ACCC (2008) Green marketing and the Trade Practices Act. http://www.accc.gov.au/content/index.phtml/itemId/81576 (cited 8 August 2008)
30. Australian Consumers Association (2008) Supermarket green watch, Choice, pp 12–16
31. TerraChoice Environmental Marketing (2007) The 'six sins of greenwashing': a study of environmental claims in North American consumer markets
32. Federal Trade Commission (2010) Green guides: summary of proposals. http://www.ftc.gov/os/2010/10/101006greenguidesproposal.pdf (cited 12 February 2011)
33. DEFRA (2011) Green claims guidelines: how to make a good environmental claim. Department for Environment Food and Rural Affairs, London
34. Kleiner A (1991) What does it mean to be green? Harvard Business Review, pp 38–47 (July–August)
35. Landor (2010) Cascade green: greening an authentic brew. http://www.landor.com/?do=ourwork.casehistory&cn=6101 (cited 27 December 2010)
36. Atkins G (2001) Why green marketing fails, Asian Correspondent, 31 October 2001. http://asiancorrespondent.com/42084/why-green-marketing-fails/. Accessed 26 Dec 2010

37. Boase F (2008) Walking the talk—it ain't easy being green. The Leader, November, Australian Business School
38. Lee J (2008) ACCC to probe green beer claims. The Age, 25 August 2008. http://www.theage.com.au/national/accc-to-probe-green-beer-claims-20080824-41f0.html. Accessed 12 Feb 2011
39. Foster's (2008) Cascade launches 100% carbon offset beer. Media release, 4 March 2008 (cited 26 December 2010)
40. Freeman RE (1984) Strategic management: a stakeholder approach. Pitman, Boston
41. Elkington J, Knight P (1992) The green business guide. Victor Gollancz, London
42. Capatosta T, Langert R, Magnuson K, Sprehe D, Denison R, Prince J, Ruston J (1991) McDonald's Corporation—Environmental Defense Waste Reduction Taskforce. McDonald's Corporation and Environmental Defense, New York
43. Oleck J (1992) The great clamshell debate. Restaur Bus 91(16):68–70
44. Nude Food Movers (2011). http://www.nudefoodmovers.com (18 March 2011)
45. Peattie K (1999) Trappings versus substance in the greening of marketing planning. J Strateg Mark 7:131–148
46. Hartmann P, Ibanez VA, Sainz FJF (2005) Green branding effects on attitude: functional versus emotional positioning strategies. Mark Intell Plan 23(1):9–29
47. Visy Industries (2010) Visy environment. http://www.visy.com.au/?id=99 (cited 2 January 2011)
48. Visy Industries (2009) National Packaging Covenant Report 2009, Melbourne
49. Patagonia (2007) Baselayer packaging: the good and the bad. http://www.patagonia.com/pdf/en_US/packaging_info1.pdf (cited 19 March 2011)
50. Verbraucher Initiative (2011) Label online. Undated. http://www.label-online.de (cited 18 March 2011)
51. Pocock R, Stone I, Clive H, Smith R, Jesson J, Wilszak S (2007) Barriers to recycling at home. WRAP, Bunbury
52. Society of the Plastics Industry (2008) Letter from the SPI to the Federal Trade Commission: green packaging workshop, Washington
53. OPRL Retailers and brand owners FAQ. Undated. http://www.onpackrecyclinglabel.org.uk/default.asp?section_id=2&content_id=13#5 (cited 29 January 2010)
54. APME, Report on polymer identification codes for packaging. Undated, Plastics Europe
55. ASTM (2008) ASTM international working to adapt SPI resin identification codes as new standard. Media release, July 2008. http://www.astmnewsroom.org/default.aspx?pageid=1449 (cited 27 January 2010)
56. Society of the Plastics Industry (2009) SPI resin identification code—guide to correct use. http://www.plasticsindustry.org/AboutPlastics/content.cfm?ItemNumber=823 (cited 27 January 2010)
57. PACIA (2003) Plastics identification code. Plastics and Chemicals Industries Association (PACIA), Melbourne
58. WRAP (2007) Consumer attitudes to biopolymers. http://www.wrap.org.uk/retail/materials/biopolymers.html (cited 17 February 2009)
59. Federal Trade Commission (2009) FTC announces actions against Kmart, Tender and Dyna-E alleging deceptive 'biodegradable' claims. Press release, 9 June 2009. http://www.ftc.gov/opa/2009/06/kmart.shtm (cited 1 February 2010)
60. DEFRA (2004) Reducing litter caused by 'food on the go': a voluntary code of practice for local partnerships. http://archive.defra.gov.uk/environment/quality/local/litter/documents/fastfoodcop.pdf. Accessed 9 Oct 2009
61. CFPA (2010) Chlorine Free Products Association (CFPA). Undated. http://www.chlorinefreeproducts.org (cited 7 January 2010)
62. FSC (2010) The FSC principles and criteria for responsible forest management. Undated. http://www.fsc.org/pc.html (cited 7 January 2010)
63. Program for the Endorsement of Forest Certification (2010) National standards (cited 12 January 2010)
64. Program for Endorsement of Forest Certification (2007) PEFC council information register. http://register.pefc.cz/search4.asp (cited 24 December 2010)

65. The Carbon Trust (2008) Working with PepsiCo and Walkers: product carbon footprinting in practice, London
66. Ellis H (2007) Food matters, 3 November. http://www.bbc.co.uk/food/food_matters/foodmiles.shtml (cited 4 November 2007)
67. AEA Technology (2005) The validity of food miles as an indicator of sustainability, July 2005. http://statistics.defra.gov.uk/esg/reports/foodmiles/default.asp (cited 5 November 2007)
68. Environmental Leader (2007) Sun Chips adds Green-e label to package. 14 September 2007. http://www.environmentalleader.com/2007/09/14/sunchipspackaging-includes-green-e-label/. Accessed 15 Feb 2010

Chapter 4
Complying with Regulations

Helen Lewis

Abstract Around the world, packaging is subject to a wide range of environmental regulations, voluntary standards and codes of practice. While these vary greatly, there are common themes and expectations to be considered in the packaging design process: environmental design, resource efficiency (materials and energy), reduction in toxic substances, end-of-life recovery, use of recycled materials, restrictions on plastic bags and some takeaway food containers, and responsible environmental labelling. These should be addressed in the packaging design process, as well as specific regulatory requirements in end-markets, to be well prepared to meet current and future regulatory obligations. Appendix C: matrix of international regulations, policies and standards provides a detailed guide to packaging regulations and standards, by region and country, as a resource to support the packaging design process.

Contents

4.1	Introduction	156
4.2	Common Themes and Expectations	158
	4.2.1 Environmental Design ★ $ ()	158
	4.2.2 Resource Efficiency ('Optimisation') $	160
	4.2.3 Reduce Toxic Substances ✚	162
	4.2.4 End-of-Life Recovery ()	164
	4.2.5 Use of Recycled Materials ()	164
	4.2.6 Restrictions on Plastic Shopping Bags and Takeaway Food Packaging ()	165
	4.2.7 Responsible Environmental Labelling ★	166
4.3	Conclusion	167
References		168

H. Lewis (✉)
Centre for Design, RMIT University, GPO Box 2476, Melbourne, VIC 3001, Australia
e-mail: lewis.helen@bigpond.com

Tables

Table 4.1 Policy instruments 157
Table 4.2 Heavy metals in packaging 163
Table 4.3 Regulatory themes 167

Photos

Photo 4.1 Painted message on the floor of a Woolworths store in South Australia, where non-biodegradable plastic bags were banned in 2009 166
Photo 4.2 Voluntary label promoting the recyclability of cardboard packaging on a Woolworths Private Label product, Australia 167

Case Studies

Case Study 4.1 Design requirements in the Australian Packaging Covenant 159
Case Study 4.2 The Chinese Excessive Packaging Law 160

4.1 Introduction

Around the world, packaging is subject to a wide range of regulations, environmental voluntary standards, and codes of practice. These have evolved from regulations in the 1970 and 1980s to manage litter and waste, through to voluntary standards and guidelines that promote 'sustainable packaging' in the 2000s (see Sect. 1.4).

⇒ See Appendix C: matrix of international regulations, policies and standards for packaging by region and country

Regulations Target Waste Reduction and Recycling

Most current regulations and standards are intended to promote environmental responsibility for packaging. Most, however, target waste reduction and recycling rather than broader sustainability goals. For example, 'bottle bills' (container deposit legislation), the European 'Green Dot' and Ontario (Canada) Blue Box programs involve the payment of a redeemable deposit or a recycling fee in order to promote packaging recovery.

These approaches contribute to:
- increasing packaging recycling rates
- reducing waste to landfill
- meeting country or regional recycling and waste reduction targets.

Some Regulations Do Not Promote Optimal Outcomes

An assessment of regulatory and policy approaches according to the principles for packaging sustainability introduced in Chap. 2 shows that many regulations do not promote triple bottom line sustainable development (see Table 4.1). Those that focus on a single issue achieve limited environmental outcomes. Waste reduction and recycling should therefore be components in a broader plan.

Table 4.1 Policy instruments

Type of policy or regulation	Description	Sustainability principle(s)
Bottle bill/Container deposit legislation (CDL)	A mandatory deposit on beverage containers paid by the consumer and redeemed when they return the container to the retailer or another authorised agent	○
Code of Practice (CoP)	A Code of Practice is a voluntary standard, generally developed by an industry or professional association to guide the behaviour of members	Depends on the code
Design requirements	Standards covering issues such as integration of environmental design in new product development, or specific requirements (e.g. packaging layers, void space etc.)	★ $ ○ ✚
Extended producer responsibility (EPR)	A regulation that makes producers (and possibly other industries/businesses in the supply chain) physically or financially responsible for the recovery of packaging at end–of-life	○
Labelling	Labelling to promote recycling is mandatory in some jurisdictions, e.g. the resin identification codes or recycling logos	○
Product stewardship (voluntary agreement)	A voluntary commitment by businesses in the packaging supply chain to reduce the environmental impacts of packaging. Voluntary agreements may be supported by back-up regulation to catch free-riders ('co-regulation')	★ $ ○ ✚
Packaging ban	A ban on the sale or issue of a particular type of packaging, e.g. plastic shopping bags or expanded polystyrene takeaway packaging containers	○
Packaging tax	A tax imposed by government on the sale of certain types of packaging. The tax may differentiate packaging material, weight and/or carbon dioxide emissions	$ ○ ✚
Packaging levy	While 'tax' and 'levy' are sometimes used interchangeably, a levy generally goes into a special fund (e.g. an environmental fund) rather than consolidated revenue	$ ○ ✚
Recycling requirements	Some governments specify minimum recycling rates or recycled content	○
Trade practices legislation	Trade practices regulations that restrict false and misleading claims applying to environmental claims and labels	★
Standards	Standards are developed by an independent national or international organisation, e.g. International Standards Organisation. Compliance with standards may be voluntary or regulated	Depends on the standard

★ (effective), $ (efficient), ○ (cyclic), ✚ (safe)

Life Cycle-Based Approaches are Emerging

A new generation of packaging guidelines, standards and policies recognise that waste reduction is only one of the strategies required to achieve better environmental outcomes. These combine waste reduction and recycling strategies with considerations about water and energy consumption and packaging efficiency at every stage of the product environmental life cycle.

⇒ Learn more about LCA in Chap. 5 and the life cycle of common packaging materials in Chap. 6

Flexibility is Required to Achieve Optimal Outcomes

A more flexible approach enables strategies:
- to be identified and optimised on a case-by-case basis
- that balance environmental issues with economic and social objectives (triple bottom line)
- that do not consider the environmental impacts of packaging in isolation from the product it contains [1].

This approach is evident in the packaging sustainability indicators and metrics framework developed by The Consumer Goods Forum [2] (see Table 8.7). Other examples include the Sustainable Packaging Coalition's guidelines and indicators [3, 4] and the Packaging and the Environment standards (TC 122/SC 4—Packaging and Environment) currently under development by the International Standards Organisation.

4.2 Common Themes and Expectations

Despite the different regulatory approaches applied around the world there are many common themes and expectations including:
- environmental design
- resource efficiency (optimisation)
- reduction of toxic substances
- end-of-life packaging recovery
- use of recycled materials
- restrictions on the use of plastic shopping bags and takeaway packaging
- responsible environmental labelling.

Each of these is discussed separately below and linked to the relevant principles in the packaging sustainability framework.

4.2.1 Environmental Design ★ $ ()

Many regulations, policies and standards require environmental impacts to be considered at the design stage, which is when decisions are made about materials and packaging formats. These decisions determine the environmental impacts of packaging at every stage of the life cycle.

4 Complying with Regulations

The promotion of environmental design is central to a number of national packaging strategies. The United Kingdom's packaging strategy [5] is to be achieved through partnerships with organisations such as WRAP and Envirowise, both of which provide guidelines, tools and advisory services for businesses in the packaging supply chain. The Australian Packaging Covenant (see Case Study 4.1) and the New Zealand Product Stewardship Scheme require signatories to adopt and implement design guidelines. Businesses

⇒ Learn more about life cycle based environmental tools and guidelines in Chap. 7

Case Study 4.1 Design Requirements in the Australian Packaging Covenant

The Australian Packaging Covenant (APC) is the voluntary component of a co-regulatory policy to reduce the environmental impacts of packaging. Brand owners that choose not to participate can be regulated through state-based EPR regulations.

The first goal of the APC relates to design:

'To optimise packaging to use resources efficiently and reduce environmental impact without compromising product quality or safety.'

APC signatories are required to implement the Sustainable Packaging Guidelines for design and procurement. This requires signatories to evaluate all new and existing packaging against opportunities to:
- maximise water and energy efficiency
- minimise materials (source reduction)
- use recycled materials
- use renewable materials
- minimise risks associated with toxic and hazardous materials
- use materials from responsible suppliers
- design for transport
- design for reuse
- design for litter reduction
- design for consumer accessibility
- provide consumer information.

The review process must be documented and the evidence relating to decisions retained on file for independent auditing.

Case studies on the implementation of the guidelines are available from the APC website: http://www.packagingcovenant.org.au/.

Source: Australian Packaging Covenant [8]

that market packaged products in Europe need to demonstrate compliance with the 'Essential Requirements' of the Packaging and Packaging Waste Directive:
- Packaging weight and volume should be minimised to the amount needed for safety and acceptance of the packed product
- Noxious and other hazardous constituents of packaging should have minimum impact on the environment at end-of-life
- Packaging should be suitable for material recycling, energy recovery or composting or for reuse if reuse is intended.

The European Committee for Standardization (CEN) has produced six voluntary standards for implementing the Essential Requirements (CEN Standards 13427 to 13432). The influence of these standards extends beyond Europe. For example, the Asian Packaging Federation modelled their Guidelines for Environmentally Conscious Packaging on the CEN Standards, and they provide the basis for the Packaging and the Environment Standards being developed by the International Standards Organisation [6].

4.2.2 Resource Efficiency ('Optimisation') $

Resource efficiency, including the efficient use of materials, energy and water throughout the packaging life cycle, is promoted through a range of policies. Waste avoidance and minimisation of packaging is required, for example, under the European Union (EU)'s Essential Requirements and the Chinese Government's Excessive Packaging Law (Case Study 4.2).

Case Study 4.2 The Chinese Excessive Packaging Law

The Excessive Packaging Law (2009) introduced mandatory legal standards and controls for packaging including:
- instructions on the number of primary and secondary packaging layers allowed
- limits on the cost of packaging, which is to be no more than 15% of the sale value of the product
- formulae for the calculation of allowable free space enclosing the product and the permitted product-packaging cost ratio.

The government argued that excessive packaging wastes resources, pollutes the environment and disadvantages consumers. Specific mention was made of luxury items and gift products that are popular during festive periods, such as Chinese New Year.

A National Standard has been developed entitled 'Excessive restrictions on merchandise packaging requirements for food and cosmetics'.

Source: I-Grafix.com [7]

Within the context of international efforts to address global warming, policy makers and practitioners in many countries are starting to consider the energy use and greenhouse gas emissions associated with every stage of the packaging life cycle. The packaging strategy released by the UK Government in 2009 has two goals which reflect the new carbon agenda [5, pp. 21–22]:
- 'Optimise' packaging by reducing waste at source. The strategy claims that this is the 'most effective way to pave the way for a low-carbon economy and to drive resource efficiency'
- Increase recycling, and where this is not possible, 'find other carbon- and economically efficient recovery routes (including energy from waste)'.

The strategy includes a proposal to shift from weight-based recovery targets, currently mandated through the EU's Packaging Directive, to targets based on carbon dioxide emissions. The rationale is that some of the actions taken by industry and local authorities to achieve weight-based recovery targets are not achieving an overall reduction in environmental impact. For example, much of the effort to increase recovery rates has focused on heavier materials, such as cardboard and glass, and recycling them does not always result in a reduction in life cycle greenhouse gas emissions. 'Closed loop' recycling of glass containers (back into new glass containers) reduces energy and greenhouse gas emissions as well as cost, but when glass is recycled into aggregate for road base it does not generate the same level of environmental benefit. There are also perverse impacts in other areas of the packaging market; for example, when the use of recycled material increases packaging weight and greenhouse gas impacts. In the government's view, carbon is a good proxy measure for other environmental impacts [5].

The United Kingdom's approach is consistent with similar developments in other countries. While packaging taxes are not uncommon in Europe, the Dutch

Government introduced an innovative tax in January 2008 based on the amount of carbon dioxide generated in the manufacture of each type of packaging material. The Belgian Government proposed a carbon-based tax on all packaging in 2007, but following widespread opposition the tax was confined to plastic carrier bags, plastic films, aluminium foil and disposable cutlery.

Economy-wide policies designed to reduce greenhouse gas emissions, such as a carbon tax or emissions trading scheme, are also likely to have a direct impact on the packaging supply chain. By increasing the cost of energy they provide a stronger financial incentive for manufacturers to improve the efficiency of production and distribution. The cost of raw materials with high-embodied energy, such as glass and aluminium, may increase relative to other materials, and raw material processors will have a stronger incentive to implement closed loop recycling.

4.2.3 Reduce Toxic Substances ✚

Avoiding or reducing the use of hazardous substances such as heavy metals is important for packaging sustainability. This is in line with the principle of safe packaging, one of the four principles introduced in Chap. 2. The EU Packaging and Packaging Waste Directive Essential Requirements and the US Toxics in Packaging Bill have become the default international standards for heavy metals in packaging.

⇒ Learn more about toxic substances in Sect. 2.4.4

The EU Essential Requirements state that the total weight of cadmium, mercury, lead and hexavalent chromium in packaging or packaging components must not exceed 100 parts per million (ppm). There are exemptions (derogations) for lead crystal glass, enamelled glass, glass that may have been contaminated with lead from old glass in the recycling process, and plastic pallets and crates manufactured with at least 80% recycled content where no heavy metals have been intentionally added during the production process.

The heavy metal limit is not enforced in all EU countries. In the United Kingdom, Trading Standards Officers may assess compliance by asking for technical documentation on any item of packaging [9]. A report on implementation of the European Packaging and Packaging Waste Directive found several packaging formats that commonly exceeded the 100 ppm limit (Table 4.2).

The US Toxics in Packaging Bill was introduced in 1990 by the Coalition of Northeastern Governors and has since been adopted by 19 US states. The Bill calls for:

- a ban on the intentional use of lead, cadmium, mercury and hexavalent chromium in packaging
- a limit on the sum of the concentration of incidentally introduced lead, cadmium, mercury and hexavalent chromium to 600 ppm 2 years after the law is introduced, 250 ppm after 3 years and 100 ppm after 4 years.

Table 4.2 Heavy metals in packaging

Packaging Component	Heavy Metal	Source
Coloured nets	Lead and chromium	Pigments linked to the colours of yellow and orange
Plastic caps	Cadmium	Pigments linked to the colours of yellow, orange, red and green
Plastic bags	Lead and chromium plus some hexavalent chromium	Pigments linked to the colours of yellow, gold, orange, red and green
Plastic non-food bottles	Lead, cadmium and chromium plus some hexavalent chromium	Pigments linked to the colours of yellow, orange and green

Source: Based on PIRA International Ltd and ECOLAS [10]

There are a number of exemptions including packaging made from recycled materials.

Chapter 2 highlighted other emerging issues for packaging safety, such as migration of Bisphenol A (BPA) and phthalates into foodstuffs (Sect. 2.4.4). Several jurisdictions in the United States have introduced or are considering regulations to restrict the use of BPA in packaging applications, particularly for products targeting children or infants. Connecticut, for example, bans

⇒ More information on BPA and phthalates can be found in Sect. 2.4.4

the use of BPA in containers for infant formula and baby food. BPA is a component of polycarbonate plastic and the epoxy lining of metal cans. Canada has banned the use of BPA in plastic baby bottles and in 2010 declared BPA to be a toxic chemical, paving the way for further restrictions [11].

Another group of chemicals that face increasing scrutiny is the phthalates used to soften PVC. In 1999 the European Union banned six phthalates in toys likely to be placed in the mouths of children under 3 years of age. A similar ban was introduced in California in 2007, and a number of other states followed. In 2008 the US Government prohibited the sale of children's toys or childcare items that contain more than 0.1% DEHP, DBP or BBP (for full chemical names see Table 2.16). An interim ban was imposed on DINP, DIDP and DNOP while more research was undertaken [12, pp 57–66]. The European Union has set very low limits for phthalates in food: 1.5 ppm for DEHP and 9 ppm for DIOP and DINP [13].

These examples highlight the need to understand chemicals used across the packaging life cycle in order to identify potential toxicity issues. One media commentator based in the US identified 'toxics' as one of the key strategic business issues in 2010 because of the accelerating level of regulatory activity in this area [14].

4.2.4 End-of-Life Recovery

Producer responsibility for packaging at end-of-life is regulated through EPR laws, voluntary agreements (for example, in Australia and New Zealand) and bottle bills.

Businesses that sell packaging in Europe can comply with EPR laws by joining one of the producer responsibility organisations that recover packaging for a fee. Many of these schemes license member organisations, which pay an amount per package based on weight and material type to display the 'green dot' logo on their packaging.[1] EPR laws for packaging have also been introduced in South Korea and Japan. In the United States, there are moves to introduce EPR regulations for packaging at a state level, driven by tight waste management budgets and the need to reduce greenhouse gas emissions [15].

Signatories to the Australian Packaging Covenant pay an annual fee to support recycling and litter reduction projects. There is no requirement for individual businesses to take back and recycle packaging, although brand owners that choose not to participate voluntarily in the Covenant or do not meet their Covenant obligations can be regulated under free-rider legislation. This requires them to ensure recovery of their own packaging.

Bottle bills are in place in many countries or states and apply to certain beverage containers. Brand owners are required to collect a specified deposit on each container and to refund consumers on the return of the empty bottle. This exchange normally occurs through a third-party recycling organisation, which charges businesses for the amount of money paid to consumers for the return of their branded packaging.

4.2.5 Use of Recycled Materials

Recycled content is not required by law in most jurisdictions, although it is important for packaging sustainability because it helps to 'close the loop' for recovered packaging. This is in line with 'cyclic' material flows, one of the four principles for packaging sustainability identified in Chap. 2.

It is regulated in two US states—Oregon and California—where a minimum level of post-consumer recycled content (25%) is one option for compliance with the Rigid Plastic Packaging Container statute. Mostly, however, increased use of recycled materials to help close the loop in recycling systems is encouraged through voluntary initiatives. For example, packaging manufacturers in the Netherlands have agreed to a voluntary target of 25% post-consumer recycled content in plastic bottles, while signatories to the Courtauld Commitment in the United Kingdom have agreed to increase recycled content in packaging.

---

[1] More information can be found at http://www.pro-e.org. In 2010 the green dot system was in use in 29 European countries.

The use of recycled materials in food packaging is controlled by health regulations or standards in each country. Compliance with these regulations is evaluated by testing for any migration that might occur when the material is in contact with food.

According to the US Food and Drug Administration (FDA), post-consumer glass and metals are not a concern because they are generally impervious to contaminants and are readily cleaned at the temperatures used in recycling processes [16]. Recycled pulp from recovered paper and paperboard and recycled plastics may be used for food contact packaging as long as they meet specific criteria in the Code of Federal Regulations. The test standard for recycled plastics is the FDA's Guidance for Industry: Use of Recycled Plastics in Food Packaging: Chemistry Considerations [16].

In the European Union, the use of recycled materials in contact with food is governed by the Framework Regulation (EC) 1935/2004. This is based on the principle that:

> 'any material or article intended to come into contact directly or indirectly with food must be sufficiently inert to preclude substances from being transferred to food in quantities large enough to endanger human health or to bring about an unacceptable change in the composition of the food or a deterioration in its organoleptic properties' [17, p. 1].

Recycled plastics for use in food contact packaging needs to be approved by the European Food Safety Authority.[2]

4.2.6 Restrictions on Plastic Shopping Bags and Takeaway Food Packaging ()

Plastic bags and takeaway food packaging are regulated in many jurisdictions, primarily because they are highly visible in litter. Some national and municipal governments have banned all lightweight plastic shopping bags (Photo 4.1), while others have only banned non-biodegradable bags. For example:

- China banned lightweight plastic shopping bags from 1st June 2008. Businesses are prohibited from manufacturing, selling or using bags less than 0.025 mm thick. More durable bags are permitted as long as they are sold to consumers
- Corsica was the first French region to ban non-biodegradable bags in 1999, and a similar ban was introduced in Paris in 2007. The French Senate approved a ban on non-biodegradable plastic supermarket bags from 1 January 2010
- In 2000 the Indian Government introduced a law banning the manufacture and use of plastic bags thinner than 20 microns in Bombay, Delhi and the entire states of Maharashtra and Kerala.

While plastic bags are highly visible in litter and waste, particularly in busy commercial precincts and popular recreational areas, they are not necessarily the

---

[2] http://www.efsa.europa.eu/en/ceftopics/topic/foodcontactmaterials.htm, accessed 20 November 2010.

Photo 4.1 Painted message on the floor of a Woolworths store in South Australia, where non-biodegradable plastic bags were banned in 2009

most commonly littered items. In Australia, for example, plastic bags make up less than 1% of all littered items [18]. There are concerns, however, about the risk of injury or death to wildlife from ingestion of plastic bags [19]. In Bangladesh and some Indian cities, plastic bags were banned because they contribute to blocked stormwater drains and flooding during the wet season.

Takeaway food packaging manufactured from expanded polystyrene packaging is banned in some jurisdictions in the United States. Foamed plastic packaging is also banned in a number of Chinese cities, including Beijing. South Korea and Taiwan have both restricted the use of disposable packaging in restaurants and stores.

4.2.7 Responsible Environmental Labelling ★

Environmental labels, which include written claims and logos that promote the environmental attributes of a product or packaging, are regulated through trade practices legislation in most countries (see Sect. 3.3.3). Unsubstantiated or exaggerated claims have prompted authorities in some countries to publish advice on the implications of trade practices law for environmental claims and guidelines on the use of specific terms such as

⇒ Chapter 3 provides information on the labels most commonly used on packaging, and explains how they can support a sustainability marketing strategy.

4 Complying with Regulations

Photo 4.2 Voluntary label promoting the recyclability of cardboard packaging on a Woolworths Private Label product, Australia

'recyclable' and 'biodegradable' [e.g. 20, 21]. The general advice provided to industry is that claims need to be:

- truthful and accurate
- relevant to the product
- specific and unambiguous.

Guidelines are also provided in the voluntary ISO standard 14021 Environmental labels and declarations—Type II environmental labelling—Principles and procedures [22]. Apart from trade practices law, the only other relevant legislation is the mandatory use of the resin identification codes in many US states (see Sect. 3.5.2). In most countries, the use of the codes to support recycling is voluntary. Photo 4.2 shows a recyclability label on packaging.

4.3 Conclusion

There are common themes and expectations in regulations that target environmental impacts of packaging. These should be embedded in the packaging design process and are reflected in the packaging sustainability framework described in Chap. 2 (see Table 4.3). By applying this or a similar framework for packaging

Table 4.3 Regulatory themes

Regulatory theme	Sustainability principle	
Environmental design	All	★ $ () ✦
Resource efficiency ('optimisation')	Efficient	$
Reduction in toxic substances	Safe	✦
Recovery at end-of-life	Cyclic	()
Use of recycled materials	Cyclic	()
Restrictions on plastic bags and takeaway packaging	Efficient	$
Responsible environmental labelling	Effective	★

design, a business will be well positioned to meet current and future packaging regulations.

Businesses also need to stay ahead of the game by monitoring regulatory trends and continually updating their design processes to reflect current best practice sustainable development strategies in product and packaging design. A new generation of packaging guidelines, standards and policies are emerging that combine waste reduction and recycling strategies with considerations about water and energy consumption and packaging efficiency at every stage of the product environmental life cycle. Chapter 5 provides an introduction to life cycle assessment and its application to packaging. This needs to be considered alongside regulatory requirements as part of a packaging for sustainability strategy.

References

1. EUROPEN and ECR Europe (2009) Packaging in the sustainability agenda: a guide for corporate decision makers. The European Organisation for Packaging and the Environment (EUROPEN) and ECR Europe, Brussels
2. The Consumer Goods Forum. Global packaging project (2010). http://globalpackaging.mycgforum.com/ (cited 21 January 2011)
3. SPC (2006) Design guidelines for sustainable packaging, version 1.0. Sustainable Packaging Coalition, Charlottesville, Virginia
4. SPC (2009) Sustainable packaging indicators and metrics. Sustainable Packaging Coalition, Charlottesville, Virginia
5. DEFRA (2009) Making the most of packaging: a strategy for a low-carbon economy. Department for Environment, Food and Rural Affairs, London
6. EUROPEN (2009) Work to begin on international packaging and environment standards. Media release (cited 5 June 2010; 8 December)
7. I-Grafix.com (2009) China restarts green packaging laws. 24 April 2009. http://www.igrafix.com/index.php/contributors/other/china-restarts-green-packaging-laws.html. Accessed 4 June 2010
8. Australian Packaging Covenant: a commitment by governments and industry to the sustainable design, use and recovery of packaging (2010), APC Secretariat, Sydney
9. BERR (2008) Packaging (essential requirements) regulations. Government guidance notes. http://www.berr.gov.uk/files/file49463.PDF (cited 14 July 2009)
10. PIRA International Ltd and ECOLAS N.V. (2005) Study on the implementation of Directive 94/62/EC on packaging and packaging waste and options to strengthen prevention and re-use of packaging, Final report. Surrey, UK
11. Egan L (2010) Canada declares BPA toxic, sets stage for more bans. Reuters. 14 October 2010. http://www.reuters.com/article/idUSTRE69D4MT20101014. Accessed 26 Jan 2010
12. Smith R, Lourie B (2009) Slow death by rubber duck. University of Queensland Press, St Lucia
13. Australian Consumers Association (2008) What's lurking under the lid? Choice. August 2008, pp 24–25
14. Makower J (2010) Toxics become a strategic issue: the state of green business 2010. Greenbiz.com. 11 February 2010. http://greenbiz.com/print/3249. Accessed 17 Feb 2010
15. Shoch L (2009) Legislation in packaging: a push for extended producer responsibility. Packaging Digest. 2 January 2009. http://www.packagingdigest.com/article/340571-Legislation_in_packaging_A_push_for_extended_producer_responsibility.php. Accessed 21 April 2009

4 Complying with Regulations

16. Food and Drug Administration (2006) Use of recycled plastics in food packaging: chemistry considerations. August 2006. Second edition. http://www.fda.gov/food/guidancecomplianceregulatoryinformation/GuidanceDocuments/FoodIngredientsandPackaging/ucm120762.htm (cited 9 January 2007)
17. European Commission (2004) Regulation (EC) No 1935/2004 of the European Parliament and of the Council of 27 October 2004 on materials and articles intended to come into contact with food, Brussels
18. KAB (2009) National litter index: annual results 2008–09 tabulations. Report by McGregor Tan Research for Keep Australia Beautiful (KAB): Frewville, SA
19. EPHC (2008) Decision regulatory impact statement (RIS): investigation of options to reduce the impacts of plastic bags. http://www.ephc.gov.au/pdf/Plastic_Bags/200805__Plastic_Bags__Decision_RIS__Options_to_Reduce_Impacts__incl_AppendicesCD.pdf (cited 26 September 2008)
20. DEFRA (2003) Green claims—Practical guidance. http://www.defra.gov.uk/environment/business/marketing/glc/pdf/genericguide.pdf (cited 12 July 2009)
21. ACCC (2008) Green marketing and the Trade Practices Act. http://www.accc.gov.au/content/index.phtml/itemId/815763 (cited 8 August 2008)
22. ISO (1999) ISO 14021:1999, Environmental labels and declarations—Type II environmental labelling—principles and procedures. International Standards Organization (ISO), Geneva

Chapter 5
Applying Life Cycle Assessment

Karli Verghese and Andrew Carre

Abstract Life cycle assessment (commonly known by its acronym LCA) is a useful sustainability tool and has many applications to support packaging decisions. In LCA, environmental impacts of product and packaging systems can be quantified in line with internationally accepted methodologies. With this knowledge, strategies can be identified and adopted that achieve the most effective and efficient environmental outcomes, and packaging design can be optimised. This chapter provides a general introduction to understanding and getting started with LCA, in particular for its application to product-packaging and packaging-system design. One important insight is that LCA studies have the potential to dispel common myths about packaging and sustainability issues. Packaging LCAs sometimes show that 'rules of thumb' or 'common sense solutions' simply do not apply or work in practice.

Contents

5.1	Introduction	173
5.2	Life Cycle Assessment Applications	174
5.3	Getting Started with LCA	180
	5.3.1 Create a Life Cycle Map	181
	5.3.2 Review LCA Studies	182
	5.3.3 Identify Level of LCA Required	183
5.4	Basic Principles of LCA	187
	5.4.1 The Goal (Purpose)	187

K. Verghese (✉) · A. Carre
Centre for Design, RMIT University, GPO Box 2476, Melbourne, VIC 3001, Australia
e-mail: Karli.Verghese@rmit.edu.au

A. Carre
e-mail: Andrew.Carre@rmit.edu.au

	5.4.2	The Functional Unit	188
	5.4.3	The System Boundary	188
	5.4.4	Inventory Analysis: Collecting and Calculating Data	189
	5.4.5	Impact Indicators	189
	5.4.6	Impact Assessment Methods	192
	5.4.7	Data Quality	193
	5.4.8	Uncertainty and Sensitivity Analysis	193
	5.4.9	Report Quality and Critical Review	194
	5.4.10	Limitations	195
5.5	Applying LCA to Packaging		196
	5.5.1	The Role of LCA in Packaging Design	196
	5.5.2	Some Basic Principles	198
	5.5.3	Accounting for Recycling	201
	5.5.4	Assessing Returnable Packaging	202
	5.5.5	Assessing Biodegradable Materials	205
5.6	The Future of Packaging LCA		208
5.7	Conclusion		209
References			209

Figures

Figure 5.1	Main steps in the four components of LCA	187
Figure 5.2	Simplified view of the life cycle stages, inputs and outputs	189
Figure 5.3	Example of an inventory table (extract)	190

Tables

Table 5.1	General terms used when describing components of life cycle assessment	176
Table 5.2	Applications of LCA in decisions about packaging sustainability	178
Table 5.3	System boundaries used in LCA: The cradle analogy	183
Table 5.4	Essential components of an LCA study	184
Table 5.5	Different levels of conducting an LCA study	185
Table 5.6	Example functional unit definitions	188
Table 5.7	Commonly used life cycle impact indicators	191
Table 5.8	Net savings for recycling for a typical Melbourne household per week	193

Case Studies

Case Study 5.1	How LCA works for Procter & Gamble	175
Case Study 5.2	Using streamlined LCA tools	181
Case Study 5.3	Reasons for engaging stakeholders	186

5 Applying Life Cycle Assessment   173

Case Study 5.4	Examples of carbon footprint initiatives	192
Case Study 5.5	Sensitivity analysis for two baby food packaging alternatives	194
Case Study 5.6	Example of a two-part critical review process...	195
Case Study 5.7	Disclosure of limitations: an example	196
Case Study 5.8	Implementing different packaging formats for different geographical regions...	197
Case Study 5.9	Beverage packaging LCA	199
Case Study 5.10	Example of true system analysis: the case of returnable packaging for fresh fruit and vegetables	203
Case Study 5.11	Example biopolymers and conventional polymers....................................	206

5.1 Introduction

> Life cycle assessment (LCA) can produce convincing evidence that intuition [is] no longer enough... 'Natural' products have been found to be not necessarily environmentally optimal. Many 'counter-intuitive' outcomes from LCA studies indicated the need for a closer systemic approach to identify and document impacts along the process chain and life time of products and services [1, p. 1].

To design packaging for sustainability it is necessary to understand the environmental impacts associated with all aspects of producing, using and disposing of a product and its packaging. A life cycle perspective, supported by LCA, provides the means to create a shared, comprehensive and informed understanding within a business of the environmental impacts of its operations and products. This can then be used to develop and communicate strategies for improvement. Using LCA tools reveals to all stakeholders the world behind the product: its potential environmental impacts and the decisions that may have led to them. This knowledge informs the business's sustainability strategy, packaging development and innovation plans.

Central to life cycle thinking is:
- the concept of a product life cycle
- methods to identify and quantify the environmental impacts associated with each life cycle stage and the life cycle system as a whole.

Life cycle maps provide a visual representation of the product life cycle, while LCA quantifies the environmental impacts in line with internationally accepted

methodologies enshrined in the ISO 14040 series of standards. When a life cycle approach is taken:
- a complete picture of the product-packaging system environmental impact becomes visible
- environmental impacts, such as material, energy and water consumption, emissions, waste production, and land use, and their sources are identified and understood
- the relative environmental impacts of the product and packaging are quantified
- the role of packaging in sustainable development is identified.

With this knowledge, strategies can be identified and adopted that achieve the most effective and efficient environmental outcomes.

Understand Basic Terms

The first packaging LCA study was undertaken in the late 1960s. The 'Resource Environmental Profile Analysis' was commissioned by Coca Cola Amatil to assess the resources and environmental profile of different beverage packaging materials (e.g., returnable glass vs. a range of single use packaging containers). These studies were the precursor of contemporary LCA which has now become a sophisticated assessment method for many materials and products, not just packaging.

There are some basic terms in life cycle thinking (see Table 5.1). Many of them appear similar but have significantly different meanings.

> Table 5.4 and Sect. 5.4 outline other specific terms used in LCA studies.

5.2 Life Cycle Assessment Applications

LCA is one part of a sustainability tool box [4] with many applications to facilitate packaging decisions (see Table 5.2 and the Procter and Gamble Case Study 5.1).

Embed LCA Throughout the Business

One of the main reasons for conducting an LCA is to increase knowledge and understanding about the interaction of materials and processes with the natural environment. Most importantly, LCA confirms whether changes will lead to environmental improvements or if environmental impacts will simply be shifted to other stages of the life cycle and/or other categories of impact. The LCA tools and results make this information visible and able to be shared throughout the business and, if desired, with customers, suppliers and other stakeholders.

Case Study 5.1 How LCA Works for Procter and Gamble

Procter and Gamble (P&G) has used LCA to guide decision-making since the late 1980s. Managers routinely use LCA to:
- analyse products from a system-wide perspective to
  - guide choices of raw materials
  - guide product innovation
  - design packaging with lower impact
- analyse the energy and resource use in its dishwashing detergent system
- analyse various emissions, wastes and resources using environmental themes
- benchmark products over time and report progress.

LCA is used to answer specific questions such as:
- How do two alternative manufacturing processes for the same product compare in terms of resource use and emissions?
- How do compact dishwashing detergents compare to regular dishwashing detergents in terms of resource use and emissions?
- What are the relative contributions of different stages of a product's life cycle to total emissions?

A study by Procter and Gamble in the 1990s identified possible savings in embodied energy in laundry detergent packaging from a range of improvement strategies [6]. By reconsidering the form and material make-up of the packaging system along with the concentration of product within, significant savings could be achieved.

Improvement strategies compared with existing bottle	Decrease in total energy consumption (%)
25% recycled content plastic bottle	6
25% consumer recycling	7
Triple concentrate product in existing plastic bottle	67
Single strength product in soft pouch (PET/LDPE)	32
Triple concentrate product in soft pouch	77
Triple concentrate product in paper gable	72
Source: Kuta et al. [6, p 188]	

Source: P&G [7].

Table 5.1 General terms used when describing components of life cycle assessment

Term	Definition	Example or notes
Life cycle	'[C]onsecutive and interlinked stages of a product system, from raw material acquisition or generation of natural resources to the final disposal' [2, p. 2]	For example, the life cycle of a plastic beverage bottle may involve the production of plastic from crude oil, the transformation of the plastic into a bottle, the transport of the bottle to the retailer, the purchase and storage of the bottle by the consumer, and the use and disposal of the bottle
Life cycle assessment (LCA)	'[C]ompilation and evaluation of the inputs, outputs and potential environmental impacts of a product system throughout its life cycle' [2, p. 2]. In a packaging context, an input refers to substances or energy going into a unit process and outputs refer to substances or energy leaving a unit process	An LCA of the bottle system described above would attempt to evaluate the complete environmental impact of the bottle, by considering all stages of its life cycle
Life cycle inventory (LCI)	[A] quantification of the 'relevant inputs and outputs of a product system. These inputs and outputs may include the use of resources and releases to air, water and land associated with the system. Interpretations may be drawn from these data, depending on the goal and scope of the LCA. These data also constitute the input to the life cycle impact assessment' [2, p. 2]	The main difference between an LCA and an LCI is that an LCI does not attempt to evaluate the environmental impacts of the inputs and outputs studied. For example, an LCI of the bottle system might conclude that over the bottle's life a certain quantity of methane is emitted, but the LCI would not evaluate this emission's contribution to global warming
Life cycle impact assessment (LCIA)	'[P]hase of life cycle assessment aimed at understanding and evaluating the magnitude and significance of the potential environmental impacts of a product system' [2, p. 2]	The LCIA takes the flows recorded in the LCI and assigns them to potential environmental impacts. For example, in the bottle system the LCIA takes the emissions of two greenhouse gases, carbon dioxide (which has a multiplying rate of 1) and methane (which has a multiplying rate of 21), and calculates their global warming potential
Life cycle management (LCM)	'Life cycle management is a business management approach that can be used by all types of businesses (and other organisations) to improve their products and thus the sustainability performance of the companies and associated value chains. A method that	For example, a beverage company would implement life cycle management through the business to identify areas for improvement within the business but also through the supply chain

(continued)

Table 5.1 (continued)

Term	Definition	Example or notes
	can be used equally by both large and small firms, its purpose is to ensure more sustainable value chain management. It can be used to target, organize, analyze and manage product-related information and activities towards continuous improvement along the life cycle' [3]	
Life cycle map	In this book, the life cycle map (see Sect. 5.3.1, Chap. 6 and Sect. 7.2.1) is defined as a diagram of the interrelated processes of the product or service system	In the bottle example, all processes would be drawn and linked to develop the life cycle map. The input of materials and energy and outputs of product, wastes and emissions would be identified
Product or service system	'[C]ollection of materially and energetically connected unit processes which performs one or more defined functions' [2, p. 3]	In the bottle example, the product is the bottle which performs functions including containment of the liquid beverage for retail sale. The product system refers to all processes that are studied to map the life cycle of the bottle's production and use (e.g., including raw materials, manufacturing, transport, use and disposal)
Life cycle thinking (LCT)	Life cycle thinking is not a term for which there is a commonly accepted definition. In this book it refers to a thought process that considers environmental impacts over the entire life cycle of a product and not just at one point (e.g., manufacturing or recovery)	

LCA Improves Decision-Making and Performance

LCA is used to assess alternative strategies or to optimise individual strategies. For example, studies can compare alternative product-packaging systems and manufacturing processes or technologies to determine which option will provide the greatest contribution to a business's sustainable development goals. This information can be used to inform strategic decisions regarding product mix or investment in new processing and packaging technologies. It also may highlight where a business contributes to the environmental impacts of other business's products, even though its own products have little direct impact. Strategies to reduce impacts can then be developed and factored into an organisation's sustainable development goals and relevant operational plans. The United Nations

Table 5.2 Applications of LCA in decisions about packaging sustainability

Business activity	Role of LCA	Outcomes
Developing the strategy	Identify sustainability metrics Set sustainable development goals and targets Confirm role of packaging Benchmark current products and packaging Set priorities for new products, product improvement, packaging development	LCA enables businesses to identify: • Sustainability metrics that are relevant to their product portfolio • 'Hotspots' in the life cycle for relevant metrics and strategies to address them • The environmental impacts of the product and its packaging • Opportunities to improve products and packaging
Designing packaging	Inform design strategy and priorities Design new packaging Improve current products and packaging	LCA enables packaging designers to identify: • How they will contribute to the business's sustainability goals • How to reduce the environmental impacts of current product-packaging systems • How to optimise the packaging system for specific products • New product and packaging innovations to create economic, social and environmental value
Marketing and communicating sustainability	Market and promote the business's sustainable development goals and achievements Position and market products based on environmental benefits Conduct competitive analysis and benchmarking of products and packaging from a sustainable development perspective Inform and substantiate environmental claims	LCA enables marketers to identify: • How they will contribute to the business's sustainability goals • Benefits of the business's sustainable development initiatives and achievements • The sustainability credentials of the products and packaging being studied • Competitive advantage or disadvantage of the products and packaging from a sustainable development perspective • Product development strategies to improve sustainability • Valid claims that are not 'greenwash'.
Regulatory compliance	Assist a business to meet or exceed its current regulatory requirements Position a business to meet and avoid future costs associated with future regulatory requirements	LCA enables a business to identify: • The impacts of product mix and design decisions on the environment • Where in the product's life cycle environmental impacts occur (hotspots)

Table 5.2 (continued)

Business activity	Role of LCA	Outcomes
	Assist a business to identify how it will facilitate compliance throughout the supply chain (product stewardship) Enable compliance with specific environmental standards or codes of practice e.g., PAS2050, National Carbon Offset Standard [5]	• The impact of current and future regulation on itself and supply chain partners • Its role in complying with or supporting compliance of environmental regulations • The benefits of complying with specific environmental regulations (e.g., waste and energy laws)

Environment Program and the Society for Environmental Toxicology and Chemistry [3] describe some excellent case studies showing how LCA is used to guide strategic decision-making.

The process of collecting data, as well as the results themselves, will lead to improvements in:

- setting relevant sustainable development goals
- focusing and aligning resources across the business and with supply chain partners upon the issues that matter
- decision making, in particular for product and packaging design, packaging material selection, process selection, procurement and investment strategies (including capital)
- measuring, monitoring, auditing and reporting sustainability performance.

Although regulations that mandate the use of LCA are not common, many regulations, standards and codes of practice require a life cycle-based approach. This is likely to increase in the future as LCA becomes more broadly adopted by policy makers.

⇒ See Chap. 4 for more on regulations

LCA Implementation is Challenging

LCA needs to be appropriately applied within an organisation to facilitate sustainable development. Just like the different phases of sustainability presented in Sect. 1.3.1, undertaking an LCA requires an organisation to be at a particular stage of awareness and interest in knowing more about the life cycle impacts of its products and processes. LCA can be time-consuming and expensive, with many studies taking more than 6 months to complete. This time lag can slow down the 'learning' process, so in many cases streamlined approaches are employed to keep feedback timely (see explanation in Table 5.5 in Sect. 5.3.3).

⇒ See Chap. 7 for more details on LCA based decisions support tools and how they can be deployed within the organisation

Some important issues should be addressed before embarking on an LCA:
- Clear objectives for the LCA must be identified (see explanation on goal and scope in Sect. 5.4.1). It should be more than just 'a good idea'. If the results of the study are to be used in marketing to communicate environmental benefits, a critical review will be required (see Sect. 5.4.9). Once the reasons for the study are clear, the functional unit of the study (see explanation in Sect. 5.4.2) should be defined correctly as this is the backbone of the LCA
- Appropriate people (internal and external to the organisation) should be engaged to become involved in the study to ensure that it can progress as smoothly as possible. Planning for sufficient time to collect the necessary data, including waiting for responses to data requests, is also important
- Appropriate personnel to do the LCA calculations should be engaged. In most cases, an external LCA professional/consultant will need to be engaged because of the specialist LCA knowledge required, which is unlikely to be available within the organisation
- It is also important to understand how the results of an LCA will be used to inform particular decisions. For instance, how and where can the results be integrated into the process for developing new products? At what stage should the LCA be undertaken? Who needs to be informed and kept up to date on the LCA process?

One approach to deploying LCA is presented in Case Study 5.2. This discusses how Nestlé uses the streamlined Packaging Impact Quick Evaluation Tool (PIQ-ET) (see Sect. 7.2.4) to screen different packaging designs. Nestlé has recognised the value of integrating life cycle thinking early within the new product development process, when the level of influence is relatively high and costs relatively low. PIQET is currently used globally in Nestlé by over 400 packaging technologists, designers and marketers:

> 'to understand the life cycle impacts associated with different packaging materials; to understand the actual impacts of recovery and recycling streams; to improve the environmental performance of their packaging designs over time with each new packaging format needing to have a lower environmental impact than previous designs; to identify any environmental showstoppers; and to be more consistent with communications to consumers' [8].

5.3 Getting Started with LCA

There are some preliminary steps to be taken before undertaking or commissioning an LCA. They include:
- creating a life cycle map
- reviewing previous LCA reports
- identifying the level of detail and rigour required.

5 Applying Life Cycle Assessment 181

> **Case Study 5.2 Using Streamlined LCA Tools**
>
> LCA has been adopted by many organisations worldwide as an integral part of their sustainability strategy. Day-to-day decision-making, however, is usually less influenced by LCA, due to the time required to undertake studies, the complexity associated with them and the need for specialists to conduct them.
>
> Nestlé have introduced a streamlined LCA tool for packaging design to incorporate LCA thinking into day-to-day activities.
>
> The Packaging Impact Quick Evaluation Tool (PIQET) is systematically used by Nestlé in its packaging decision-making process to ensure environmental impacts are considered at an early stage in new packaging design. The process anchors LCA to packaging development and allows development teams to assess the complete packaging life cycle with a representative set of environmental indicators when making packaging decisions.
>
> The governance model used for PIQET establishes roles and responsibilities in the packaging development process thus ensuring that environmental information is taken into account in a systematic manner in all decision-making concerning packaging development.
>
> The Nestlé approach gives management a rigorous assurance that packaging life cycle impacts are being considered and addressed but also helps to deploy and provide resources for LCA most effectively.
>
> Note: Permission received from Nestlé for publishing of case study.

5.3.1 Create a Life Cycle Map

A good place to start an LCA is with the construction of a life cycle map that provides a visual representation of the steps required to source and produce the product-packaging system as well as its use, disposal or recovery. The map will start to reveal 'the world behind' the product and to identify:

> ⇒ See Chap. 7 on how to create a life cycle map

- hot spots/impact categories or priority areas to focus on
- where in the life cycle improvements could and should be made
- issues within the control of the organisation
- issues outside the direct control of the organisation but which may be influenced indirectly through the organisation's choice of products, business model and supply chain structures
- issues of concern to other stakeholders (regulators, consumer groups and so on).

Ideally, a cross-functional and/or cross-supply chain team, or someone with access to cross-functional material, should be involved in constructing the LCA map to ensure a broad approach is taken and information is as complete and factual as possible. To construct the map, key processes such as sourcing of materials, manufacture, distribution, use and end-of-life waste management should be identified and documented in the form of a flowchart (an example is shown in Chap. 1, Fig. 1.3).

Developing and reviewing the life cycle map provides a mechanism to engage with suppliers, customers and waste management companies about their operations. It also leads to ideas and discussions to inform future strategies and actions. For example, litter impacts of beverage containers or fast food packaging might be identified and strategies adopted through design to reduce litter potential and through corporate sponsorship of litter prevention schemes.

5.3.2 Review LCA Studies

Reviewing previously published LCA studies is useful to:
- gain an understanding of LCA methods and how they are applied and interpreted
- source data that may inform the development of the sustainability strategy, assist creation of life cycle maps and accelerate an organisation's application of life cycle thinking
- develop the skills required to engage with life cycle practitioners including the commission of LCAs.

What to Look For

Different LCAs on the same product are often criticised for 'providing different answers'. This generally arises from differences in:
- the purpose of the study
- the defined functional unit
- the data used
- the defined system boundary—many studies discuss their scope using a 'cradle' analogy (see Table 5.3)
- other assumptions.

To assess different studies it is therefore important to firstly confirm critical information about the study as outlined in Table 5.4 and Sect. 5.4.

Table 5.3 System boundaries used in LCA: the cradle analogy

System boundary	Definition	Comments
Cradle-to-gate	Usually means an LCA has incorporated all the processes required to extract and transform materials from the environment and deliver a product to the factory or retail outlet gate	Exactly what is meant by the 'gate' is often unclear
Gate-to-gate	The term usually signifies that only an intermediate portion of the life cycle has been considered	Often just the processes that occur at a particular site are included, although this can be unclear
Cradle-to-grave	Usually infers that the entire product life cycle has been considered	Sometimes 'cradle-to-grave' LCAs will exclude important lifecycle phases such as use, so the statement usually needs to be investigated

5.3.3 Identify Level of LCA Required

There are several ways in which LCA can be used (Table 5.5), progressively increasing in complexity, time and cost:
- conceptual LCA
- streamlined LCA
- full LCA.

Data availability and the purpose of the study (for example, to support internal decisions or to be made public) are factors that need to be considered when determining which of these approaches should be taken.

Conceptual LCA

The conceptual LCA involves high level analysis of the processes that make up the life cycle of the product under study. It can be a useful way of introducing life cycle thinking into the organisation or design team and relies upon tools such as the life cycle map (Sect. 7.2.1) and sustainability impact matrix (Sect. 7.2.2) to collate understanding and knowledge of the interactions of the product across its life cycle.

⇒ See Chap. 7 on how to create a life cycle map and to complete a sustainability matrix (as well as Chap. 3)

Streamlined LCA

A streamlined LCA is a shortened form of LCA marked by a less rigorous approach (and perhaps less inclusive system boundary, more assumptions and less reliable data quality). It is mainly used for gaining insight into and understanding the major areas of focus in the life cycle of a product. It is not suitable for external

Table 5.4 Essential components of an LCA study

Component	Definition
Goal	'The goal of an LCA study shall unambiguously state the intended application, the reasons for carrying out the study and the intended audience, i.e., to whom the results of the study are intended to be communicated' [2, p. 5]
Scope	'The scope should be sufficiently well defined to ensure that the breadth, the depth and the detail of the study are compatible and sufficient to address the stated goal' [2, p. 5]
System boundary	'The system boundaries determine which unit processes shall be included within the LCA' [2, p. 6]
Functional unit	This is the 'quantified performance of a product system for use as a reference unit in a life cycle assessment study' [2, p. 2]
Data sources	'The results, data, methods, assumptions and limitations shall be transparent and presented in sufficient detail to allow the reader to comprehend the complexities and trade-offs inherent in the LCA study. The report shall also allow the results and interpretation to be used in a manner consistent with the goals of the study' [2, p. 8]
Assumptions	Many assumptions are usually required in order to undertake an LCA. As with data sources, it is imperative that assumptions are clearly documented in the report. Such communication contributes to the transparency of an LCA study
Transparency	This refers to 'open, comprehensive and understandable presentation of information' [2, p. 3]
Environmental impact categories	Impact categories are subsets of the characterisation of environmental impact undertaken by the LCA. Impact categories can include global warming, eutrophication, water use and so on
Critical review	'The critical review process shall ensure that • The study report is transparent and consistent • The interpretations reflect the limitations identified and the goal of the study • The data used are appropriate and reasonable in relation to the goal of the study • The methods used to carry out the LCA are scientifically and technically valid • The methods used to carry out the LCA are consistent with this International Standard Since this International Standard does not specify requirements on the goals or uses of LCA, a critical review can neither verify nor validate the goals that are chosen for an LCA, or the uses to which LCA results are put' [2, p. 9]

communication (marketing), although lessons learned can still be communicated and improvements prioritised and undertaken. As Parker states:

'If environmental performance measurement of packaging is to become more widespread, quicker, cheaper, streamlined systems will need to be developed and improved so that they are both feasible within a fast-paced business context as well as scientifically valid and accurate enough to form a sound basis for business decision making' [9, p. 234].

5 Applying Life Cycle Assessment

Table 5.5 Different levels of conducting an LCA study

Aspects	Level of LCA		
	Conceptual LCA	Streamlined LCA	Full LCA
Synonyms/terms	Matrix LCA Life cycle thinking	Screening LCA	Detailed LCA
Level of complexity	Low	Medium	High
Resources			
• Period of time	Short	Medium	Long
• Cost	Relatively low	Low to expensive	Costly
Expected outcome			
• Applications	Marketing New product development	Environmental labelling	Big decisions or policy
• Pinpoint components/materials	Yes	Yes	Yes
• Identify process(es) with emissions of specific interest in life cycle stages	No	Yes	Yes
• Reliability	Not as rigorous as streamlined/full LCA	Reliable	Reliable
Communication			
• Internal audience	Yes	Yes	Yes
• Public audience	No	Care needs to be taken. It depends on system boundaries and which impact categories are reported	Yes
• Formal reporting	No	No	Yes
Scope			
• Life cycle coverage	Limited	Limited indicators, data quality or life cycle stages	Entire life cycle
• Exclude phases in life cycle	Yes	Yes, but without compromise to overall result	No
Source data			
• Qualitative	Yes	Yes	Yes
• Semi-quantitative	Yes	Yes	Yes
• Quantitative	No	Yes	Yes
• New inventory data	No	No	Yes
Type of discussions/statements			
• Qualitative	Yes	Yes	Yes
• Quantitative	No	Yes	Yes
• Method	Manual	Computer-based	Computer-based

Source: Centre for Design, RMIT University, 2008

Full LCA

A full LCA is the most complete and detailed form of LCA that meets defined scoping requirements and can therefore be used for both internal and external communication purposes.

A full LCA, compliant with international standards, is essential if the purpose is to provide a public comparison of more than one product. This also requires a critical review of the methodology, assumptions, data collected and modelled, and conclusions. The ISO 14044 standard sets out the mandatory and non-mandatory elements to be included in an LCA and requirements for critical review of studies. However, not all full LCAs may undergo a critical review, if the study for instance is used internally within the company and results are not disclosed to the public.

An important consideration when conducting an LCA is the level of involvement required from stakeholders such as industry, government and non-government organisations. The degree to which stakeholders are involved can range from simple membership of an advisory committee, where the LCA practitioners present and discuss findings from the study, to the supply of inventory data or the review of draft and final findings (see Case Study 5.3 on stakeholders involved in a packaging waste LCA).

Case Study 5.3 Reasons for Engaging Stakeholders

In an LCA of packaging waste management in Australia, a Stakeholder Advisory Committee was used. It consisted of 'fifteen individual companies, eight industry associations, three Federal government agencies, five State government agencies and two others' [10, p. 155]. The committee was involved in meetings with the LCA project team to provide input to 'the study's approach and progress; barriers affecting progress and how to overcome them; methodological issues including system boundary decisions; and issues regarding confidentiality of data' [10, p. 152].

'The involvement of stakeholders was considered critical in regards to the envisaged credibility of the study and its outcomes. By involving as many stakeholders as possible in the study, the researchers aimed at meeting the following objectives:
- Increase industry and government awareness on the feasibility of the LCA approach for evaluating environmental impacts
- Achieving consensus among the various stakeholders regarding the study's approach including methodological decisions, system boundary decisions and cut-off rules
- Maximise cooperation in the provision of system information and data
- Obtain stakeholder critical review of the outcomes of the study
- Minimise the risk of stakeholder scrutiny towards the study's conclusions' [10, p. 151].

5 Applying Life Cycle Assessment

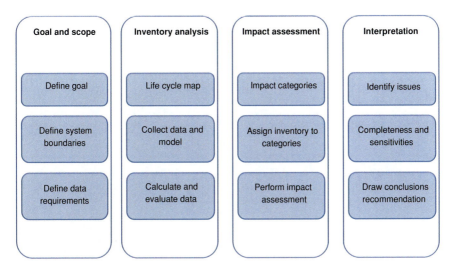

Fig. 5.1 Main steps in the four components of LCA. *Source*: Adapted from Verghese [11, p. 200]

5.4 Basic Principles of LCA

The technical framework for LCA consists of four phases (Fig. 5.1):
- *Goal and scope definition*: determines the purpose of the study, the system boundaries, functional unit and data quality issues to be confirmed
- *Inventory analysis*: involves the quantification of material and energy inputs and the outputs of emissions to air, land and water and generation of solid waste
- *Impact assessment*: assigns the inventory to environmental impact categories
- *Interpretation*: assesses the results of the inventory analysis and impact assessment to arrive at conclusions and recommendations.

Key aspects of the different stages are discussed below.

5.4.1 The Goal (Purpose)

A clear definition of the goal is required to clarify the reasons for the study and inform decisions about:
- the context of the study
- data quality
- assumptions
- system boundaries.

The international standard requires that statements of goal and scope be developed and refined as part of undertaking the LCA. Accordingly, the goal statement should include a description of the product-packaging system under study as well as the intended application of the results and the target audiences (for example, internal decision-makers or the public).

Table 5.6 Example functional unit definitions

LCA type	Examples of functional units used
Packaging system only	Provision of a proper vehicle for a child's baby food meal in France, Spain, and Germany in 2007 [13]
	Shipment of 1,000 tons ('907 tonnes or 2 million pounds') of each type of produce using reusable plastic containers and display-ready corrugated containers (DRCs) [14]
	Package and pallet for 1 m^2 of ceramic tiles [15]
	1,000 containers of capacity 0.4536 kg ('1 pound') each for the packaging (polylactic acid, polyethylene terephthalate and polystyrene) of strawberries [16]
Product-packaging systems	One cup of coffee ready to drink at home or in small offices. The packaging under study was flexible packaging consisting of a PET/aluminium/PE foil bag and cardboard box [17]
	The provision of 1 kg of butter ready to be eaten at home. The packaging under study was flexible packaging consisting of aluminium, wax and greaseproof paper [17]

5.4.2 The Functional Unit

To assess the potential environmental impacts of a product system, a unit of measure known as the 'functional unit' must be defined.

Clarify the Service Provided by the Product to be Studied

This is defined as the service provided by the product or system being studied. For example, the functional unit for an LCA on a beverage container would be based on the service provided by the beverage container, not the beverage itself (e.g., to facilitate containment, distribution and storage of beer and/or carbonated soft drinks from the breweries via retailers to consumers (packaging and distribution of 1,000 litres of beverages) [12]. Other examples of functional units are provided in Table 5.6.

This approach to defining the functional unit encourages alternatives to be considered that may take radically different forms. In a packaging context, this is particularly salient as it could allow a disposable packaging system to be compared with a reusable system, a large pack with a small pack, a box with a can, and so on. The concept of functional comparison often affects product design-thinking more powerfully than actual LCA study results.

5.4.3 The System Boundary

An important part of the scoping phase is to detail which life cycle stages, processes and data are included in or excluded from the study (which can be represented as a life cycle map (see Sect. 7.2.1) or process flow chart. The main processes, materials, energy, emissions, waste and products that flow across the life cycle are identified and documented (Fig. 5.2).

5 Applying Life Cycle Assessment

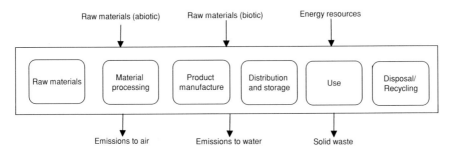

Fig. 5.2 Simplified view of the life cycle stages, inputs and outputs

5.4.4 Inventory Analysis: Collecting and Calculating Data

Inventory analysis is concerned with the collection, analysis and validation of data that quantifies the inputs (for example, materials, energy and water) and outputs (for example, emissions and waste) of the product's life cycle; that is, those that cross the system boundary (see Fig. 5.2).

For each life cycle stage, the material and energy flows (inputs and outputs) are collected and modelled. These flows are summed across the life cycle, as defined in the system boundary, and linked back to the functional unit. The outcome of the inventory analysis includes a process flow chart and a list of resources (materials, fuels and energy) and wastes and emissions (air, land, water) presented as an inventory table (Fig. 5.3). This table is also known as a life cycle inventory (LCI). This information can be useful on its own, for example, to show kilograms of carbon dioxide, although it generally needs further analysis and grouping into impact indicators to be useful in decision-making [18, p. 43]. This further analysis takes place in the impact assessment phase (see Sect. 5.4.6) where the environmental exchanges are grouped and compared.

5.4.5 Impact Indicators

Appropriate indicators are identified in the goal and scope section of the LCA, to provide the context for reporting.

One of the main reasons for conducting an LCA is to seek solutions that genuinely reduce environmental impact rather than shift the impact elsewhere. Therefore, indicators should be selected based on their relevance and to confirm that no adverse consequences arise. One of the strengths of LCA is that results are presented for a range of environmental impact indicators (see Table 5.7 for commonly used indicators). This makes all impacts visible and almost always leads to discussions about trade-offs and opportunities for design strategies such as those outlined in Chap. 2.

No	Substance	Category	Unit	
1	1,2-dichloroethane	Air	17.1	µg
2	Acetaldehyde	Air	383	µg
3	Acetic acid	Air	1.43	mg
4	Acetone	Air	256	µg
5	Acid as H⁺	Water	45.3	µg
6	Acrolein	Air	78.6	µg
7	Ag	Raw	4.5	µg
8	Ag	Water	1.30	µg
9	Al	Water	265	mg
10	Al	Air	89.56	mg
11	Alcohols	Air	15	mg
12	Aldehydes	Air	9.7	mg
13	Alkanes	Water	327	µg
14	Alkanes	Air	2.68	mg
15	Alkenes	Water	25	µg

Fig. 5.3 Example of an inventory table (extract)

Not Just Carbon

Carbon flows have traditionally been calculated and reported in LCAs as global warming potential, although recently the term 'carbon footprinting' has become more fashionable [20]. As a result there has been an increase in the number of LCAs that only report on 'carbon'. Examples of carbon footprinting initiatives are outlined in Case Study 5.4.

Although assessments are simplified when a single environmental indicator such as global warming potential is used, such simplification limits the conclusions that can be drawn. For instance, a simple packaging format with both paper and plastic components is likely to cause environmental impacts in areas of land use, water use, global warming and resource depletion, any one or all of which may be significant and can be improved through packaging design. Assessing only one indicator could result in decisions that adversely affect indicators that have not been assessed.

Not just LCA metrics

Even when a relatively broad suite of indicators is used, there is still a risk that one or more indicators of significant environmental impact will not be considered. Environmental effects of packaging that are not well served by existing indicators include the visual impact of packaging litter and the impact of waste packaging on plants and animals. For this reason LCAs need to be considered in conjunction with other sources of environmental information and packaging specific sustainability metrics to support packaging design decision-making.

⇒ See Table 7.3 for metrics used in various LCA and packaging decision support tools and Tables 7.6 and 7.7 for general lists of LCA and packaging specific metrics.

Table 5.7 Commonly used life cycle impact indicators

Impact indicator	Scale	Examples of LCI data (i.e., classification)	Common possible characterisation factor
Global warming	Global	Carbon dioxide (CO_2) Nitrogen dioxide (NO_2) Methane (CH_4) Chloroflurocarbons (CFCs) Hydrochloroflurocarbons (HCFCs) Methyl bromide (CH_3Br)	Global warming potential
Stratospheric ozone depletion	Global	Chloroflurocarbons (CFCs) Hydrochloroflurocarbons (HCFCs) Halons Methyl bromide (CH_3Br)	Ozone-depleting potential
Acidification	Regional Local	Sulfur oxides (SOx) Nitrogen oxides (NOx) HydrocHydroflouric acid (HF) hydrochloric acid (HCl) Ammonia (NH_4)	Acidification potential
Eutrophication	Local	Phosphate (PO_4) Nitrogen oxide (NO) Nitrogen dioxide (NO_2) Nitrates Ammonia (NH_4)	Eutrophication potential
Photochemical smog	Local	Non-methane hydrocarbon (NMHC)	Photochemical oxidant-creation potential
Terrestrial toxicity	Local	Toxic chemicals with a reported lethal concentration for rodents	LC_{50}
Aquatic toxicity	Local	Toxic chemicals with a reported lethal concentration for fish	LC_{50}
Human health	Global Regional Local	Total releases to air, water and soil	LC_{50}
Resource depletion	Global Regional Local	Quantity of minerals used Quantity of fossil fuels used	Resource-depletion potential
Land use	Global Regional Local	Quantity disposed of in a landfill or other land modifications	Land availability
Water use	Regional Local	Water used or consumed	Water-shortage potential

Source: Curran [19, p. 49]

Reporting Often Needs to be Simplified

While the outputs of an LCA are based upon commonly accepted scientific units (for example, 'CO_2 equivalent' or CO_2 eq. for global warming potential), it may sometimes be necessary to translate them into more recognisable units to aid

> **Case Study 5.4 Examples of carbon footprint initiatives**
>
> - 'An international standard ISO 14067 on carbon footprint of products (Part 1: quantification, Part 2: communication)
> - The World Business Council for Sustainable Development and the World Resources Institute developed two standards under their Greenhouse Gas Protocol/Supply Chain Initiative: A Product Life Cycle Accounting and Reporting Standard and a Corporate Accounting and Report Standard: Guidelines for Value Chain (Scope 3) Accounting and Reporting
> - The UNEP/SETAC Life Cycle Initiative launched a project group on carbon footprinting
> - The Japanese Ministry of Economy, Trade and Industry launched a carbon footprint trial project and a Technical Specification—General principles for the assessment and labelling of Carbon Footprint of Products' [20, p. 91]
> - A standardised approach known as the publicly available specification 2050:2008, Specification for the assessment of the life cycle greenhouse gas emissions of goods and services (PAS 2050) has been developed by the carbon trust and the UK department for environment, food and rural affairs that details how to assess the carbon footprint of goods and services [21, p. 203].

communication to non-experts in LCA. In Australia, it has been found useful in several LCAs to engage and communicate the technical results in 'layperson' terms. In 2001, for example, when the first comprehensive LCA of Australian packaging waste management was completed [22], the results were presented in both typical LCA units and equivalency units (see Table 5.8). The inclusion of the equivalency units assisted in explaining the LCA results to government and industry stakeholders.

5.4.6 Impact Assessment Methods

The impact assessment phase is used to identify and establish a link between the product's life cycle (inputs and outputs) and the environmental impacts associated with it. Impact assessment methods (for example, Eco-indicator 99, CML 2001 and IMPACT 2002+) [24] have been developed to assign inventory input and output flows to impact categories and calculate their contribution to each impact.

5 Applying Life Cycle Assessment

Table 5.8 Net savings for recycling for a typical Melbourne household per week (2001)

Impact	Totals	Unit	Equivalence
Greenhouse gases	3.2	kg CO_2 eq.	This equates to 0.25% of a household's total allocation of greenhouse gases from all sources
Embodied energy	32.2	MJ	Enough energy (9 kWh) to run a 40 W light bulb for 72 h (accounting for electricity losses)
Smog precursors	1.3	g C_2H_4 eq.	Equivalent to the emissions from 4.5 km of travel in an average post-1985 passenger car
Water use	92.5	litres	The equivalent of five sink-loads of dishes
Solid waste	3.6	kilogram	Depending on the material, 60–90% of the product put out for recycling will remain out of the solid waste stream

Source: Verghese [23, p. 60]
kg CO_2 eq carbon dioxide equivalents, *kWh* kilowatt hours, *MJ* megajoule, *g C_2H_4 eq.* grams of ethylene equivalent

To better understand the relative magnitude of impacts, the different types of data for each one can be normalised to one reference value (for example, CO_2 equivalent). Indicators with quite different units of measure can still be compared this way. The normalised results describe environmental impacts relative to a known baseline impact, although they do not describe which impacts are most important.

The impact categories can then be weighted based upon their environmental significance to arrive at a single point or 'eco-indicator'. This procedure involves subjective judgements and should be used with caution.

5.4.7 Data Quality

Data quality parameters are defined to give structure to the data collection and analysis and generally include:
- the age of the data
- the age of the technology to be modelled
- the geographical coverage of the data
- data variability, representativeness and reproducibility.

5.4.8 Uncertainty and Sensitivity Analysis

Uncertainty and sensitivity analyses are performed in the interpretation stage of an LCA to test key data sources and assumptions. This allows the validity of aspects of the study to be checked before conclusions are drawn.

An example of a sensitivity analysis is provided in the Nestlé Case Study 5.5 below.

> **Case Study 5.5 Sensitivity analysis for two baby food packaging alternatives**
>
> An LCA conducted by Nestlé compared 200-gram plastic pots and glass jars for baby food manufactured in three production sites, France, Spain and Germany, in addition to alternative logistical scenarios [13]. In this study, nine different sensitivity analyses were performed to test the assumptions made and assess their level of influence on the overall outcomes of the study:
> - impact assessment methodology chosen
> - steam consumption at the production site
> - collection rate of used packaging
> - efficiency of incinerators
> - polypropylene (PP) data consistency
> - production process for the ethylene–vinyl alcohol copolymer (EVOH) layer
> - recyclability of the PP–EVOH–PP multilayer plastic cup
> - type of fuel substituted by the polypropylene in steel and cement industries
> - the findings for the 200-g package size against two other package sizes.
>
> *Source*: Humbert et al. [13, p. 98]

5.4.9 Report Quality and Critical Review

Comply with ISO

A useful approach for consistency in application of LCA is to use reports compliant with the ISO 14044 standard. If there are deviations from the ISO standard for a particular study, these should be explained and justification given to why they are exceptions in the report. LCAs that do not comply with ISO14044 may also be valid, although they are more difficult to specify, as the ISO standard no longer applies.

Look for and Use Critical Review

The quality of a report is significantly improved if it is subjected to critical (or peer) review during the drafting process in line with ISO 14044. This provides:
- a mechanism for someone external of the study to check and ensure that it has been completed in accordance with the methodology
- credibility for the study and its findings.

> **Case Study 5.6 Example of a Two-part critical Review Process**
>
> 'The "stakeholder review" was undertaken with draft chapters for each material section being sent to relevant industries for comment prior to broader distribution of the draft report for review by all stakeholders represented on the Stakeholder Advisory Committee (SAC). This gave data suppliers the chance to correct any misreporting of data before the information was made available to other stakeholders. A draft report on the entire project was then distributed for comment to all stakeholders involved in the SAC. Parallel to this process a "technical critical review" was undertaken by the Centre for Environmental Studies (CML) at Leiden University, The Netherlands'.
> *Source*: James et al. [10, p. 153]

The review may be conducted by internal or external experts or interested parties. For instance, a panel of industry and government stakeholders may be established to review a study's methodology and findings (as in Case Study 5.6).

The international standard requires all reviewer comments to be noted and the author's responses documented. This transparency adds to the credibility of the study and can identify areas of weakness.

If a reviewer or review panel is selected appropriately, the critical review process is a check for whether the study adheres to LCA principles and is accurate in its interpretation of the process being studied. Sometimes the review requires a panel of experts in LCA as well as the subject matter. Independence and expertise are the hallmarks of good reviewers, both of which can be determined by looking at their list of publications and other relevant experience. Although critical review often comes at a financial cost, it is arguably one of the best things that can be done to improve the quality of the LCA.

5.4.10 Limitations

All LCAs have limitations, and it is important that these limitations are disclosed. If a study's credibility is related to its transparency then a thorough disclosure of limitations is essential to a high quality LCA report. A small-scale example from a beverage LCA is provided in Case Study 5.7.

> **Case Study 5.7 Disclosure of limitations: an example**
>
> An LCA in the United Kingdom was intended to assess different milk packaging systems, but the data quality was insufficient to provide a realistic direct comparison so the LCA practitioners reported the limitations. They admitted that the study could not be used 'in drawing specific conclusions relating to the relative performance of all milk packaging', but did show how the results could be used:
>
> '[T]he results give an insight into:
> - the type of impacts that the different milk packaging systems studied have on the environment
> - the magnitude of the selected environmental impacts for the different milk packaging systems studied
> - areas where knowledge of the different milk packaging systems is lacking
> - an indication of any environmental benefits of:
>   - incorporating recycled content in the containers
>   - lightweighting containers
>   - increased recycling of used milk containers.'
>
> *Source*: Fry et al. [25]

5.5 Applying LCA to Packaging

LCA studies have the potential to dispel common myths about packaging and sustainability issues. An important conclusion from LCAs on packaging is that many 'rules of thumb' or 'common sense solutions' simply do not apply or work in practice.

5.5.1 The Role of LCA in Packaging Design

It's crucial to understand the life cycle impacts of a product system in order to achieve the best environmental outcome from products and their packaging.

Use LCA to Inform Design Strategy

LCA tools should therefore be used as earlier as possible in a business's strategy development, design and innovation processes to:
- identify priorities and goals for sustainable development from product and packaging design, renovation or improvement

> ⇒ See Chap. 7 for more on LCA tools

- benchmark current products and packaging with the 'best in class' and competitors' products.

Embed LCA in the Design Process

LCA provides environmental data that identifies the life cycle stages affected by specific product and packaging design decisions: material choice, packaging shape and size, recyclability and so on. It therefore has an important role that can be used to inform:

> ⇒ See Sect. 1.2 for more on eco-efficiency and eco-effectiveness and Chap. 2 for more details on design strategies

- assessment of alternative product and product-packaging concepts (see Case Study 5.8)
- assessment of alternative packaging systems
- optimisation of specific product-packaging systems (both new ideas and current products).

Apart from its results, LCA can be used to facilitate business innovation by identifying short, medium and long term opportunities to contribute to sustainable development, addressing both eco-efficiency and eco-effectiveness strategies.

LCA can help reinforce the 'trial and error' cycle of the design process, by making sure designers receive feedback about the environmental performance of their designs. This provides a mechanism by which designers can begin to 'learn' strategies that deliver economic, social and environmental value.

Case Study 5.8 Implementing different packaging formats for different geographical regions

In the late 1990s Procter and Gamble (see Case Study 1.4) investigated six different options to reconfigure their laundry detergent packaging system. Two options were implemented:

- In the United States, the gable-top carton and a 25%-recycled content plastic bottle were introduced because the paperboard provided greater savings through source reduction than HDPE which wasn't recycled at the time of the study (early 1990s) and because consumers preferred the carton over the pouch format.
- In other geographical regions (e.g., Europe), the pouch format was introduced because of consumers' familiarity with that packaging for other product categories and to take account of existing waste management situations at the time (e.g., incineration).

Source: Kuta et al. [6]

Drive Continuous Improvement and Innovation

The environmental impacts of products and their packaging are not static. Innovation throughout the supply chain is constantly occurring, resulting in:
- new and more sustainable products entering the market
- improved efficiencies in processes (for example, in energy and water consumption and waste generation) and transport throughout the supply chain
- new and improved packaging materials (for example, recycled HDPE and recycled PET for food contact applications, as discussed in Chap. 2, and biopolymers from waste products) and packaging options (See case studies 2.9 and 2.10)
- new and more efficient end-of-life recovery, reuse and reprocessing options.

It is therefore important to update and review LCA data on an ongoing basis, and this should be integral to strategic and operational planning across the business, in particular for product and packaging design. Even if the scope to change existing packaging designs is limited, the following may be possible:
- short term environmental gains through changes to secondary and tertiary packaging, supply chain management, procurement strategies, and marketing and communication initiatives
- long term environmental gains, planned as part of future innovations coinciding with planned capital upgrades or new investments.

It is equally important to understand emerging innovations when using LCA that may alter the relevance or context of the conclusions. Studies and estimates of new technologies should be undertaken against existing processes and technologies to determine the appropriate way forward and how best to capitalise on new innovations that have a lower environmental impact than current processes.

5.5.2 Some Basic Principles

Case Study 5.9 presents a packaging-specific LCA conducted in 1998 [12]. Alternative beverage packaging formats were compared, and the alternatives were ranked based on their environmental impacts. The study is included because it represents an extremely thorough attempt to answer the question: 'which packaging system is the best?' Whereas many other studies tend to truncate documentation, this study is extremely transparent, and results could be readily reproduced.

Container Size is Important

The study compared packaging systems of a similar size. This may seem an obvious point, but when one considers that an LCA typically applies a functional unit (in this case 1,000 litres of beverage distribution), comparison across a range of sizes is technically straightforward only if one assumes that all packaging sizes perform the same function. The study pointed out that it was not valid

Case Study 5.9 Beverage packaging LCA

This LCA was conducted in Denmark in 1998.

Goal: to compare the potential environmental impacts of existing and alternative packaging systems for beer and carbonated soft drinks filled and sold in Denmark. The comparisons were of refillable and disposable glass and PET bottles, as well as aluminium and steel cans. Only packaging of the same size was compared, because consumption of the beverage was likely to be affected by the container size.

Functional unit: the functional unit was 'to facilitate containment, distribution and storage of beer and/or carbonated soft drinks from the breweries via retailers to consumers (packaging and distribution of 1,000 litres of beverages)'.

System boundary:

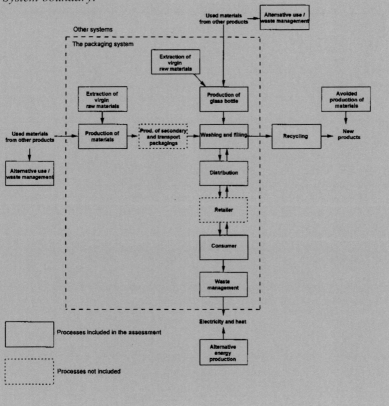

(continued)

Case Study 5.9 (continued)

Example of results

Only a small extract of results is shown here for one of the packaging systems considered. The table below shows the comparison of net potential environmental impacts of 50-cl (500-ml) soft-drink packaging systems in the base case.

Environmental impacts	Unit	Refillable PET bottle	Disposable PET bottle	Aluminium can	Steel can
Global warming	kg CO_2-eq	6.1E+00	2.0E+01	2.1E+01	2.6E+01
Photochemical ozone formation	kg C_2H_4-eq	1.1E−02	6.4E−02	8.4E−03	1.1E−02
Acidification	kg SO_2-eq	4.4E−02	2.2E−01	9.9E−02	1.2E−01
Nutrient enrichment	kg NO_3-eq	4.3E−02	1.4E−01	1.0E−01	1.2E−01

Notes: The table presents characterisation results. The functional unit is packaging and distribution of 72.3 l of soft drink

The above results were combined with overall study findings and translated into simple rankings. A ranking of 1 meant that the packaging system had the lowest impact in that impact indicator, and a ranking of 4 meant that the system had the highest impact.

Environmental impacts	Refillable PET bottle	Disposable PET bottle	Aluminium can	Steel can
Global warming	1	2–4	2–3	3–4
Photochemical ozone formation	1–3	4	1–2	2–3
Acidification	1–2	4	1–2	3
Nutrient enrichment	1–2	2–4	1–3	3–4

Source: Ekvall et al. [12]

The transparent nature of this study allows a number of specific issues to be highlighted when applying LCA to packaging.

in this case to compare different packaging sizes because each size performed a different function and was therefore not comparable.

For beverages this means that an LCA should not compare a large bottle (1 litres) for home consumption, with a small one (300 ml) for out-of-home consumption because their functions are different. Experience suggests that as beverage container sizes increase, packaging system impacts reduce per litre of beverage transported. Drawing comparisons between different packaging sizes can confuse this characteristic of scale with inherent system properties and so will potentially mislead the reader.

5 Applying Life Cycle Assessment

Material Production Impacts do not Determine Life Cycle Impacts of Packaging

Aluminium production had the highest global warming impact per kilogram when compared with the production of other materials assessed in the study. However, the global warming impact of the aluminium can system was similar to the single use PET bottle system and lower than the steel can system. There are important reasons for this difference:

⇒ See Chap. 6 for more details on the life cycle impacts of materials

- Aluminium uses less material to contain 500 ml of product (18.5 g compared with 28 g for the PET bottle and 40.2 g for the steel can)
- Reprocessing limits prevent closed-loop recycling. Aluminium tends to retain its original properties after reprocessing as this largely offsets its original production impacts
- Collection and recovery of recycled materials do not necessarily lead to closed-loop recycling. A large proportion of PET, for instance, is used to create fibres for fabric manufacture
- Reprocessing impacts differ for different materials.

5.5.3 Accounting for Recycling

Recycling is an important component of a packaging sustainability strategy (see Chaps. 1 and 2), and most packaging LCAs address recycling either directly or indirectly. However, the assessment of recycling benefits can be difficult because many systems are not closed-loop, and in that case it must be decided whether the recycling benefits should be attributed to the original product or the new one with the recycled content. (See below for a further explanation of the difference between closed-loop and open-loop recycling.)

⇒ See Sect. 2.3.3 for description of closed-loop and open-loop recycling

Closed-loop Recycling

The net environmental impacts of a closed-loop recycling system for an indicator can be calculated from the following equation:

Recycling = impacts of reprocessing−impacts of virgin material avoided−impacts of waste treatment avoided

Hence, if the impacts of producing the virgin material and avoiding waste treatment are greater than the reprocessing impact, recycling generates a net benefit.

Open-Loop Recycling

In an open-loop system, the waste product being recycled is used in a different product system, which makes the calculation of benefit more complex. For example, recycled PET resin is sometimes diverted into fabric production and made into clothing.

In this case, although the benefit of recycling can be calculated, a decision must be made as to which product system the benefit will be allocated to: the product system from where the waste material originated or the product system that used the recycled material. According to Boustead, 'there is no scientific way of partitioning the change in any parameter in an open-loop system between the two product flows' [26, p. 5].

As open-loop recycling is common for most packaging materials, it is therefore important to recognise that a subjective decision must be made to account for the benefits. Tools such as sensitivity analysis must be used to test the impact of recycling assumptions on study conclusions.

5.5.4 Assessing Returnable Packaging

Returnable or reusable packaging can be used to address the cyclic design principle discussed in Chap. 2 and reduce packaging waste. However, to understand the overall environmental impact or how to optimise the use of this type of packaging a life cycle approach is required that takes into account the transport and reprocessing stages associated with reuse.

Case Study 5.10 summarises an LCA undertaken to compare reusable plastic containers (RPCs) and display-ready corrugated containers (DRCs) to package fresh fruits and vegetables [14].

As the table shows, results were determined for a range of different products and generally reveal that RPCs have lower impacts than DRCs;

> 'For the produce shipping scenarios analysed within the defined scope of this study, findings indicate that, on average across all 10 produce applications, RPCs require 39% less total energy, produce 95% less total solid waste and generate 29% less total greenhouse gas emissions than do DRCs' [14, p. 292].

The impacts of DRCs are primarily associated with their manufacture, transport and disposal and could be improved by:
- reducing material use (packaging weight)
- improving transport efficiency
- improving end-of life-strategies such as recycling.

RPCs are more complex:
- As for DRCs, their manufacture causes impacts, although for most RPCs the impacts are small contributors because the packaging is reused. The number of times the container is used (cycles) is therefore a key assumption which directly affects the impacts

5 Applying Life Cycle Assessment

Case Study 5.10 Example of true system analysis: the case of returnable packaging for fresh fruit and vegetables

Goal: to identify and quantify the energy, solid wastes and atmospheric and water-borne emissions associated with reusable plastic containers (RPCs) and display-ready common footprint corrugated containers (DRCs) used for shipping fresh produce. Ten different high-volume fruit and vegetable (produce) applications were analysed.

Functional unit: in order to ensure a valid basis for comparison for the container systems studied, a common functional unit is essential. For this study, the functional unit for each system was shipment of 1,000 tons ('907 tonnes or 2 million pounds') of each type of produce using RPCs and DRCs.

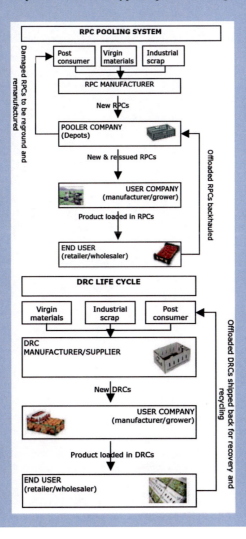

(continued)

Case Study 5.10 (continued)

System boundary:
Summary of LCI results for all produce container scenarios (all results reported on basis of 1,000 tons of produce shipped

Fresh produce	RPCs			DRCs	
	Average	Average with 80% BH	Conservative	Average	Conservative
Total energy (million BTU)					
Apples	853	789	900	1073	966
Bell peppers	1121	1040	1188	1818	1637
Carrots	531	504	567	981	883
Grapes	1080	1010	1141	1920	1729
Lettuce (head)	905	839	958	1485	1338
Oranges	650	601	692	1241	1117
Peaches/nectarines	671	621	707	1284	1156
Onions	533	501	566	1075	968
Tomatoes	797	736	864	1241	1117
Strawberries	1975	1858	2071	2455	2212
Total solid waste (tons)					
Apples	1.35	1.32	1.60	25.3	22.8
Bell peppers	1.99	1.96	2.37	43.2	38.9
Carrots	1.04	1.03	1.25	23.4	21.1
Grapes	2.15	2.12	2.50	45.5	41.0
Lettuce (head)	1.53	1.50	1.82	35.1	31.6
Oranges	1.23	1.21	1.47	30.2	27.2
Peaches/nectarines	1.25	1.23	1.45	30.5	27.5
Onions	1.09	1.07	1.28	25.7	23.1
Tomatoes	1.57	1.54	1.84	30.1	27.1
Strawberries	4.03	3.98	4.57	55.6	50.1
Total greenhouse gas (tons CO2 eq)					
Apples	62.7	57.5	64.3	67.1	60.5
Bell peppers	81.3	74.7	83.6	113.0	102.0
Carrots	37.8	35.6	39.0	61.1	55.1
Grapes	78.3	72.6	80.4	120.0	108.0
Lettuce (head)	65.9	60.5	67.7	92.8	83.6
Oranges	46.6	42.7	48.1	76.9	69.2
Peaches/nectarines	49.0	44.9	50.2	80.1	72.2
Onions	38.2	35.7	39.4	67.0	60.3
Tomatoes	57.5	52.5	59.3	77.0	69.3
Strawberries	145.0	135.0	148.0	155.0	140.0

Average scenario defined as RPC with average use/loss rates (separate results for maximum and 80% backhaul) and reported weight DRC.
Conservative scenario for RPC is use rate and loss rate 2 x the average loss rate. Conservative scenario for DRC is 10% light-weighting.

Source: Singh et al. [14]

- Transport impacts tend to be similar for both container types when moving produce from the farmer to the retailer, although the RPCs' impacts may increase on the journey that returns them to the farmer. The impacts of the return journey are heavily dependent on assumptions about the type and use of vehicle involved, which are in turn influenced by whether the RPCs can be folded or nested (Both strategies optimise the use of the return vehicle)

- RPC reprocessing, which includes washing, increases impacts and has no equivalent stage in the DRC system. (Reprocessing was not, however, addressed in the LCA.) Washing can be very energy-intensive, and there are examples where benefits associated with reuse are more than offset by washing
- RPCs are disposed at end-of-life, which causes impacts, but as in the manufacturing phase, these impacts are reduced through reuse of the packaging.

Improving the RPC system therefore requires attention to aspects such as RPC return rates (reducing loss rates, improving durability), improving transport efficiency through measures such as foldability, and strategies to improve reprocessing efficiency. Overall transportation distances can also become significant, especially where dedicated trucking loops are required to return reusable containers to producers.

5.5.5 Assessing Biodegradable Materials

Biodegradable polymers are now widely available as an alternative packaging material and need to be assessed for potential packaging applications (see Chap. 6). Rather than being produced from non-renewable, fossil-derived feedstock, like most typical plastics, most biodegradable plastics use renewable sources such as cornstarch, and many are biodegradable.

Case Study 5.11 summarises an LCA that compared the use of a biodegradable polymer (polylactic acid (PLA)) with more traditional plastic alternatives: PET and PS for strawberry clamshell containers [16].

The results indicate that the biodegradable material has similar impacts to the alternative fossil-based systems in most indicators. To understand the reasons for this, it is useful to look at the global warming indicator as an example.

Material production and packaging material conversion impacts (e.g., bottle blowing) are similar for all materials. This suggests that the packaging format is highly sensitive to transportation, and environmental impact may have more to do with trucking efficiency and supplier location than materials. It also suggests that it is difficult to draw a general conclusion for the whole system, because most strawberry producers actually use different supply chains to the one modelled.

The study discloses two aspects of production that have proved to be problematic in LCA studies:
- The production inventory includes a credit associated with the absorption of biogenic carbon dioxide from the atmosphere during plant (corn) growth. This reduces the global warming impact of the biopolymer. The problem with this approach, however, is that if carbon dioxide is absorbed it must also be emitted or sequestered at the packaging's end-of-life. If the LCA is rigorously conducted, this emission will also be counted and the credit will be nullified. It is unclear how this was addressed in the study

Case Study 5.11 Example biopolymers and conventional polymers

Goal: to compare the environmental impact of PLA (polylactic acid), PET (polyethylene terephthalate) and PS (polystryrene) thermo-formed clamshell containers used for the packaging of strawberries.

Functional unit: 1,000 containers of capacity 0.4536 kg ('one pound') each for the packaging of strawberries

System boundary:

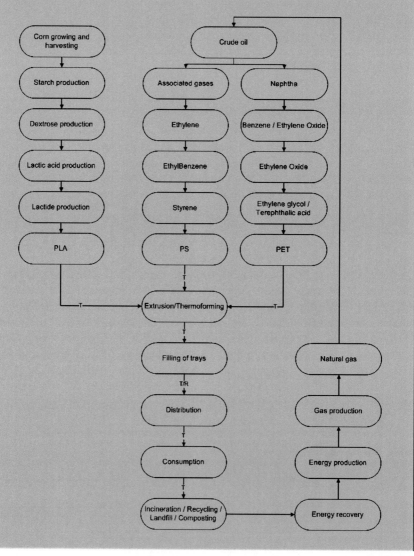

(continued)

5 Applying Life Cycle Assessment

Case Study 5.11 (continued)

Results: characterisations of environmental impact categories for the LCA of PLA, PET and PS clamshell containers for strawberries

Impact category	Stage	PLA		PET	PS
Global warming	Resin production	60		65	70
	Extrusion	15		16	12
	Thermoforming	22		24	18
	Electricity production	3		4	3
	Transportation (R)[a]	28.7		50.2	31.7
	Transportation (C)[b]	41.8		39.2	30.1
	Sub-total	**171**		**198**	**165**
	Transportation (S)[c]	564		565	565
	Total	**735**		**763**	**730**
Aquatic acidification, kg SO2	Resin production	1.17		0.36	0.47
	Extrusion	0.06		0.07	0.05
	Thermoforming	0.11		0.12	0.09
	Electricity production	0.01		0.02	0.02
	Transportation (R)	0.19		0.34	0.22
	Transportation (C)	0.28		0.27	0.20
	Sub-total	**1.82**		**1.14**	**1.04**
	Transportation (S)	3.84		3.83	3.83
	Total	**5.66**		**4.97**	**4.87**
		R	NR	NR	NR
Energy, MJ, surplus	Resin production	991/32.4[d]	1019/33.4[d]	2412/74.0[d]	2400/96.1[d]
	Extrusion	283		303	231
	Thermoforming	476		508	389
	Electricity production	41		54	42
	Transportation (R)	477		837	528
	Transportation (C)	697		655	501
	Sub-total	**991**	**2993**	**4560**	**4090**
	Transportation (S)	9416		9440	9410
	Total	**13,400**		**14,000**	**13,500**
Land occupation, m2org.arable	Resin production	0.04		0.37	0.001
	Extrusion	0.62		0.66	0.50
	Thermoforming	1.33		1.42	1.08
	Electricity production	0.0009		0.0015	0.0011
	Transportation (R)	0.38		0.66	0.42
	Transportation (C)	0.55		0.51	0.39
	Sub-total	**2.92**		**3.62**	**2.4**
	Transportation (S)	7.4		7.38	7.4
	Total	**10.3**		**11**	**9.8**

Notes:
[a] Transportation (R) – transportation of resin from resin supplier to container manufacturer by a 16 ton truck.
[b] Transportation (C) – transportation of containers from strawberry filler to distribution/market by a 16 ton truck.
[c] Transportation (S) – transportation of 1,000 lbs of strawberries (only food and no containers) by a 16 ton truck.
Variations between the PLA, PET and PS values are due to rounding error in the software.
[d] Energy consumption for 1 kg of resin. R = renewable; NR = non renewable

For presentation in this case study a selection of impact indicators were presented. The remaining indicators that were omitted from the presentation are: ozone layer depletion, aquatic eutrophication, respiratory organics, respiratory inorganics and aquatic eco-toxicity.

Source: Madival et al. [16]

- The PLA inventory assumes that electricity used in production comes from renewable energy sources. This assumption was made because the PLA producer purchased green energy from its local electricity producer. Arguably, all the material producers in the study could have purchased green energy, so this point of difference was simply determined by the PLA producer's choice rather than an inherent quality of the material.

These two issues are not uncommon when considering biodegradable plastics in an LCA and are typically addressed by sensitivity analysis. In this case, neither was addressed in detail, making it hard to work out how the PLA clamshell would compare if the assumptions were different.

5.6 The Future of Packaging LCA

Increased Use of LCA and LCA Tools

In the world of packaging, the growth in understanding and use of life cycle thinking and LCA tools will escalate in coming years as more businesses:

- come to terms with sustainable development and their role in it
- need to comply with product stewardship expectations articulated by regulations and codes of practice such as the European Packaging Waste Directive and the Australian Packaging Covenant.

This will increase demand for streamlined tools and standardisation of methodologies.

> Learn more about LCA tools in Chap. 7

> Regulations are discussed in Chap. 4

LCA will Continue to Evolve

LCA is expected to remain the methodology of choice to quantify the environmental impacts of products and packaging. However, methodologies will continue to evolve and potentially be standardised for specific applications.

Improvements to impact assessment methods that will provide better characterisation of environmental impacts, particularly for water and land use and potentially for litter, are likely. There is also a push to improve the regional specificity of datasets, allowing LCA practitioners to draw conclusions at a more local scale, also enabling global systems to be assessed and optimised.

> Learn more about international packaging standards in Chap. 4

Increased Adoption of the Product-Packaging System Approach

Better understanding of the relative impacts of the product versus the packaging, and clarity about the role of packaging in sustainability, are emerging as important challenges. Most packaging LCAs have only focused on the packaging material life cycle impacts. With more LCAs being undertaken on food systems, it is now possible to map the contribution of both to the product-packaging system. In the case of food products, studies have demonstrated that in many cases the packaging contributes a lower impact to the product-packaging system than the food item contained [27, 28], although there are exceptions.

It is therefore expected that models of packaging's role in the protection, containment and marketability of products within LCA will become more sophisticated and the following aspects will be incorporated in future [8]:

- functional environmental benefits of packaging; for example, extended shelf life
- advantages and disadvantages of different packaging formats and trade-offs within the supply chain; for example, laminate pouches that are currently not technically recyclable compared with heavier steel or glass containers that are highly recyclable
- the impact of different serving sizes to respond to changing demographics and work-life circumstances.

5.7 Conclusion

Life cycle thinking and the use of life cycle tools and information are necessary to inform any sustainability strategy. They are increasingly used to inform packaging design at every stage from the component to the packaging system level, and in the future they will be an essential part of the process to optimise product-packaging systems. Life cycle thinking should be applied in all stages of the product and packaging design process. Specifically, a range of variables such as size, shape, weight, thickness, cube utilisation and end-of-life recovery and reprocessing should be considered. Material choice influences many of these variables and therefore the environmental impact of the packaging system as a whole. The life cycle impacts of commonly used packaging materials are accordingly outlined in detail in Chap. 6. Commonly used LCA tools are described, and how they can be used is discussed, in Chap. 7.

References

1. Horne R (2009) Life cycle assessment: origins, principles and context. In: Horne R, Grant T, Verghese K (eds) Life cycle assessment: principles, practice and prospects. CSIRO Publishing, Collingwood
2. International Standards Organisation (1997) ISO 14040 Environmental management standard—life cycle assessment, principles and framework. Australian Standards—Published as AS 14040:1998, Sydney
3. UNEP, SETAC (2009) Life cycle management—how business uses it to decrease footprint, create opportunities and make value chains more sustainable
4. Guinée JB (2002) Handbook on life cycle assessment operational guide to the ISO standards. Int J life Cycle Assess 7(5):311–313
5. Department of Climate Change and Energy Efficiency (2010) National carbon offset standard. Commonwealth of Australia, Canberra
6. Kuta CC, Koch DG, Hildebrandt CC, Janzen DC (1995) Improvement of products and packaging through the use of life cycle analysis. Resour Conserv Recycl 14:185–198
7. P&G. *LCA at P&G*, unknown. http://www.scienceinthebox.com/en_UK/sustainability/lifecycleassessment_en.html#five (cited 29 April 2011)

8. Verghese K, Lockrey S, Clune S, Sivaraman D (2012) Life cycle assessment of food and beverage packaging. In: Yam K (ed) Innovative and sustainable food and beverage packaging, Woodhead Publishing Limited, Cambridge, England (in press)
9. Parker G (2008) Measuring the environmental performance of food packaging: life cycle assessment. In: Chiellini E (ed) Environmentally compatible food packaging. Woodhead Publishing Limited, Cambridge
10. James K, Grant T, Sonneveld K (2002) Stakeholder involvement in Australian paper and packaging waste management LCA study. Int J Life Cycle Assess 7(3):151–157
11. Verghese K (2008) Environmental assessment of food packaging and advanced methods for choosing the correct material. In: Chiellini E (ed) Environmentally compatible food packaging. Woodhead Publishing Limited, Cambridge
12. Ekvall T, Person L, Ryberg A, Widheden J, Frees N, Nielsen PH, Weidema BP, Wesnaes M (1998) Life cycle assessment of packaging systems for beer and soft drinks, in main report. Ministry of Environment and Energy, Danish Environmental Protection Agency, Denmark
13. Humbert S, Rossi V, Margni M, Jolliet O, Loerincik Y (2009) Life cycle assessment of two baby food packaging alternatives: glass jars vs. plastic pots. Int J LCA 14:95–106
14. Singh SP, Chonhenchob V, Singh J (2006) Life cycle inventory and analysis of re-useable plastic containers and display-ready corrugated containers used for packaging fresh fruits and vegetables. Packag Technol Sci 19:279–293
15. Bovea MD, Serrano J, Bruscas GM, Gallardo A (2006) Application of life cycle assessment to improve the environmental performance of a ceramic tile packaging system. Packag Technol Sci 19:83–95
16. Madival S, Auras R, Singh SP, Narayan R (2009) Assessment of the environmental profile of PLA, PET and PS clamshell containers using LCA methodology. J Clean Prod 17:1183–1194
17. Busser S, Jungbluth N (2009) The role of flexible packaging in the life cycle of coffee and butter. Int J LCA 14(1):S80–S91
18. Lewis H, Gertsakis J, Grant T, Morelli N, Sweatman A (2001) Design + environment: a global guide to designing greener goods. Greenleaf Publishing, Sheffield
19. Curran MA (2006) Life cycle assessment: principles and practice National Risk Management Research Laboratory, Office of Research and Development. U.S. Environmental Protection Agency, Ohio
20. Finkbeiner M (2009) Carbon footprinting—opportunities and threats. Int J LCA 14:91–94
21. Sinden G (2009) The contribution of PAS 2050 to the evolution of international greenhouse gas emission standards. Int J LCA 14:195–203
22. Grant T, James K, Lundie S, Sonneveld K (2001) Stage 2 report for life cycle assessment for paper and packaging waste management scenarios in Victoria, for EcoRecycle Victoria. Centre for Design at RMIT University, Centre for Packaging Transportation and Storage at Victoria University and the Centre for Water and Waste Technology at University of New South Wales
23. Verghese K (2009) Life cycle assessment and waste management. In: Horne RE, Grant T, Verghese K (eds) Life cycle assessment: principles, practice and prospects. CSIRO Publishing, Collingwood
24. Goedkoop M, Oele M, de Schryver A, Vieira M (2008) Simapro database manual. Methods Library, Pre Consultants, Amersfoort, The Netherlands.
25. Fry JM, Hartlin B, Wallén E, Aumônier S (2010) Life cycle assessment of example packaging systems for milk, in doorstep distribution system. Waste and Resources Action Programme (WRAP)
26. Boustead I (2001) Who gets the credit? http://www.plasticseurope.org/plastics-sustainability/life-cycle-thinking.aspx (cited 24 November 2010)
27. Erlov L, Lofgren C, Soras A (2000) Packaging—a tool for the prevention of environmental impact. Packforsk, Kista
28. Roy P, Nei D, Orikasa T, Xu Q, Okadome H, Nakamura N, Shiina T (2009) A review of life cycle assessment (LCA) on some food products. J Food Eng 90:1–10

Chapter 6
Packaging Materials

Karli Verghese, Enda Crossin and Margaret Jollands

Abstract Material selection is inherently linked to the economic, social and environmental value of a product-packaging system. While the properties of a packaging material and the process by which it is converted into a packaging component contribute to its effectiveness, strategies to optimise environmental performance are informed by an understanding of material life cycles. Paper and board products, polymers, glass, aluminium and steel are the most widely used packaging materials, although there is increasing use of renewable materials such as starch and cellulose. This chapter provides an overview of the life cycle and applications of commonly used and emerging packaging materials.

Contents

6.1	Introduction	213
6.2	Aluminium	214
6.3	Steel	217
6.4	Glass	220
6.5	Paper and Board	224
6.6	Non-renewable Thermoplastics	230
6.7	Renewable Thermoplastics	242
6.8	Conclusions	249
References		249

K. Verghese (✉) · E. Crossin · M. Jollands
Centre for Design, RMIT University, GPO Box 2476, Melbourne, VIC 3001, Australia
e-mail: Karli.Verghese@rmit.edu.au

E. Crossin
e-mail: Enda.Crossin@rmit.edu.au

M. Jollands
e-mail: Margaret.Jollands@rmit.edu.au

Figures

Figure 6.1	Life cycle of aluminium	216
Figure 6.2	Generic Materials Recovery Facility (MRF) model (dashed areas are less commonly used)	217
Figure 6.3	Life cycle of steel	219
Figure 6.4	Life cycle of glass	222
Figure 6.5	Life cycle of paper and board (generic)	228
Figure 6.6	Life cycle of non-renewable fossil-derived thermoplastic	236
Figure 6.7	The life cycle of post-consumer recycled non-renewable thermoplastics	239
Figure 6.8	Life cycle of a generic renewable thermoplastic	245

Tables

Table 6.1	Consumption of packaging materials across different regions	213
Table 6.2	Classification of paper and board	224
Table 6.3	Densities of common non-renewable thermoplastic polymers	231
Table 6.4	Properties of commonly used non-renewable - fossil derived thermoplastic polymers	232
Table 6.5	Applications of commonly used non-renewable - fossil derived thermoplastic polymers	234
Table 6.6	Manufacture of commonly used non-renewable fossil-derived thermoplastic polymers	238
Table 6.7	Compatibility of common non-renewable thermoplastics in mechanical recycling	241
Table 6.8	Environmental impact drivers of materials	248

Photos

Photo 6.1	Aluminium can	215
Photo 6.2	Glass packaging	221
Photo 6.3	Folding boxboard	225
Photo 6.4	Corrugated Carton	226
Photo 6.5	New applications for moulded paper include food service packaging	227
Photo 6.6	PET Packaging	235
Photo 6.7	Polypropylene closures	237
Photo 6.8	Plastic packaging can be reprocessed into different products such as plastic 'lumber': example of 'open loop' recycling	240
Photo 6.9	Example of a packahe made from a biodegradable starch-based polymer	243
Photo 6.10	Biodegradable Polylactic acid bottles	244
Photo 6.11	Limes nets	246

Case Studies

Case Study 6.1	Owens-Illinois Lean and Green	223
Case Study 6.2	Automatically sorting glass	223
Case Study 6.3	Automatic sorting of plastics: Visy Industries	240

6 Packaging Materials

6.1 Introduction

Material selection is inherently linked to the economic, social and environmental value of a product-packaging system. While the properties of a packaging material and the process by which it is converted into a packaging component contribute to its effectiveness, strategies to optimise environmental performance are informed by an understanding of material life cycles. For example, it influences:
- delivery of the packaging functional requirements
- production processes associated with that material and material form
- product cost and economic viability
- packaging recoverability
- safety of the product and its packaging.

Commonly used and Emerging Packaging Materials

Paper and board products, polymers, glass, aluminium and steel are the most widely used packaging materials (see Table 6.1).

Table 6.1 Consumption of packaging materials across different regions

Packaging material	Region		
	USA (%)	Australia (%)	Europe (27 EU members) (%)
Paper and board	50	62	38
Polymers	17	21	18
Glass	13	14	20
Metals	6	3	6
Other	14	Not reported	18

Sources: Lewis [5], EPA [27], EC [28]

These materials can be used alone or in combination, forming composite sandwich or laminated materials used in packaging applications such as multi-layered plastic films and aseptic packaging. Composites are used to combine the properties of two or more materials, often to counteract unfavourable properties of one or more of them.

Concerns about the environmental impacts of non-renewable materials have led to the development of materials derived from renewable sources and new hybrid composites, which include resin/cellulose for strength, mineral/fibre for barrier papers, and starch as a gas barrier. Other less commonly used renewable materials include calico, jute, hemp, kenaf, palm and sugar-cane bagasse. In addition, there is growing interest in edible films for food packaging derived from a range of plant and animal sources such as whey, collagen and gelatine [1, 2].

Plastic polymers used for packaging are generally thermoplastic (soften when heated) and made from non-renewable resources (such as oil and gas), renewable resources (such as starch from corn, wheat, tapioca) or a combination of both. Non-renewable thermoplastics include polyethylene (PE) and polypropylene (PP). Examples of renewable thermoplastics are starch- and cellulose-based polymers and polylactic acid (PLA or polylactide).

Effective Packaging is Determined by its Function

The properties of a packaging material, together with the processes used to convert the material into packaging, provide the packaging function and therefore contribute to fulfilling the 'Effective' packaging principle (see Sect. 2.3.1).

Efficient, Cyclic and Safe Packaging Requires Life Cycle Thinking

In order to assess if packaging is efficient, cyclic and safe, the life cycle environmental impacts of a product and its packaging need to be considered. Each stage of the packaging life cycle has environmental impacts, and each material has its own unique environmental impacts.

⇒ Learn more about life cycle thinking in Chap. 5

The remainder of this chapter provides an overview of the life cycle and applications of commonly used and emerging packaging materials.

6.2 Aluminium

Material Properties

Aluminium has excellent barrier properties and is impervious to liquids, gases, aromas, light and micro-organisms. These properties mean that it is commonly used as a barrier layer in laminated (composite) materials. It is corrosion-resistant in most packaging applications and has mechanical properties that are unaffected by the typical temperatures achieved in hot filling of beverages, for instance.

Aluminium used for packaging is generally pure and mechanically weak, although it can be alloyed if specific property requirements are required; for example, strength in cans. The density of pure aluminium is 2.70 g/cm^3.

Packaging Applications

Aluminium can be processed into different packaging formats including cans (Photo 6.1), trays, tubes and foil. Can applications include food products (such as soft-drink) and personal care products (such as deodorant). Sheet applications include food products (such as frozen food trays for reheating) and composite materials (such as jar seals, tamper-proof seals, aseptic packages). Technical constraints for aluminium in packaging include limitations to the thickness of foil and limitations in shape.

6 Packaging Materials

Photo 6.1 Aluminium can (Photo: Amcor Packaging (Australia) Pty Ltd)

Life Cycle

Figure 6.1 presents a simplified life cycle flow chart for aluminium. Aluminium is produced from bauxite ore (mined in open-cut mines) and two major additives: lime (limestone quarried in open-cast mines and calcined) and caustic soda (NaOH, produced from electrolysis of salt).

Aluminium oxide (alumina) is extracted from mined bauxite using caustic soda under heat and pressure. Hydrated aluminium oxide is precipitated from the caustic solution and calcined to produce aluminium oxide. The oxide is mixed with fluoride salts and smelted to produce liquid aluminium, which is drawn off and cast into ingots that are then cast and rolled.

In the cradle-to-cradle model (see Sect. 1.2), aluminium belongs to the 'technical metabolism' (according to McDonough and Braungart [3], a 'technical nutrient' is designed to be returned to the industrial metabolism or system from which it came).

Environmental Impacts

Significant environmental impacts associated with the production of aluminium are:
- land use impacts of mining, such as habitat destruction, threats to biodiversity, and soil erosion
- impacts of bauxite refining, such as generation of caustic effluents and bauxite-residue slurry (red-mud)
- impacts of smelting, such as production of chlorofluorocarbons and indirect emissions associated with electricity generation.

The high electricity requirement for smelting means that there are significant variations in environmental impacts, depending on location, energy mix and emissions associated with the electricity grid. To produce aluminium using recyclate requires about 95% less energy than using virgin material [4].

Recovery and Disposal

Aluminium is fully recyclable and in packaging is mostly used for beverage containers [4]. Non-rigid aluminium packaging, such as foil, is theoretically recyclable; however, losses through material recovery facilities (often known by their acronym

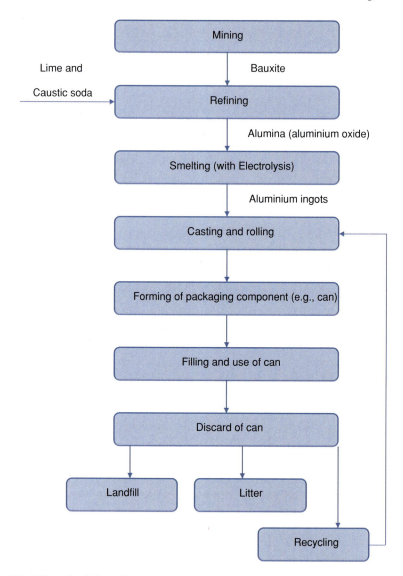

Fig. 6.1 Life cycle of aluminium

MRFs) (see Fig. 6.2) are high [5], and it is difficult to know how much non-rigid aluminium packaging is recycled. A MRF takes commingled packaging and paper from kerbside collections, sorts it into individual materials, and bales the different materials for transport. Global recycling rates are reported in Table 2.14 in Chap. 2. Aluminium is generally inert in landfill and does not break down.

Aluminium cans are collected through kerbside systems, drop-off centres and commercial scrap dealers, and sorted for reprocessing at an MRF. Cans are compressed into 'bricks' and transported to processing plants where they are

6 Packaging Materials

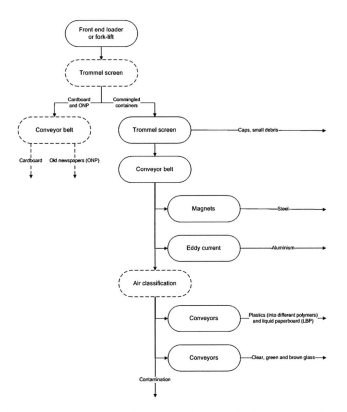

Fig. 6.2 Generic materials recovery facility (MRF) model (dashed areas are less commonly used). *Source*: Grant et al. [4]

processed in rotary furnaces and heated to 700°C. During heating, any lacquers applied to the cans during their manufacture are burnt off. Stronger alloys, which can be used for the can lids and tabs, melt at a lower temperature than the alloy used for the can body. The different alloys are separated before further processing.

Recovered molten aluminium alloys are cast into ingots, which are reused in a manufacturing process. The manufacturing process can be for the same application as the virgin material (closed loop recycling) and/or a different application (open loop recycling). The alloys used in aluminium can bodies is close to pure, and the recycled material can be used in a broad range of applications, including packaging, automotive components and building products.

6.3 Steel

Material Properties

Steel is an alloy of iron, carbon and other elements, such as manganese. It has a high mechanical strength and is a barrier for gas, liquid and light. The properties of steel are not affected by sterilisation and pasteurisation temperatures, which enables 'in package' processing to be used to extend the shelf life of foods. For food and beverage cans, a lacquer and/or tin coating can be applied to prevent corrosion of the packaging and food spoilage. The density of steel is approximately 7.8 g/cm^3.

Packaging Applications

There are many different packaging forms that can be produced from steel including cans for food and aerosols, pails and drums. Can applications include food products, such as vegetables and pet food, and personal care products, such as aerosols. Sheet applications include bulk packaging products, such as drums and pails.

Life Cycle

Figure 6.3 presents a simplified life cycle flow chart for steel. The primary materials used in its manufacture are pig iron and recycled steel.

To produce pig iron, iron ore (iron oxide) is extracted by strip mining, concentrated by crushing and grinding, and magnetically separated to remove granite. This ore is then sintered to make pellets and fines. The oxide from the sintered ore is removed in a blast furnace using coke and limestone to form liquid pig iron. The liquid pig iron is then removed, solidified, and used as a feedstock for steel production in a steel furnace.

There are two principal types of steel furnaces:
- basic oxygen furnace (BOF) that can use up to 28% recycled steel
- electric arc furnace (EAF) that can use up to 100% recycled steel.

The recycled content of steel is 20–30% in most packaging applications. Molten steel from the steel furnace is cast into ingots or transferred to other machines where it undergoes shaping processes such as hot rolling. Cold rolling further reduces the thickness of the steel. Additional processing steps are possible after forming, such as the electrolytic deposition of tin to form a protective barrier layer.

In the cradle-to-cradle model (see Sect. 1.2), steel belongs to the technical metabolism.

Environmental Impacts

The most significant environmental impacts associated with the production of steel are:
- mining impacts, including land-use impacts and the depletion of iron ore, coal and limestone resources
- emissions associated with the coking processes used in pig iron production.

Recovery and Disposal

Like aluminium, packaging applications for steel are dominated by cans. Although other product forms, such as pales, are technically reusable or recyclable, data on the fate of these forms is limited.

Steel cans are fully recyclable and can be reprocessed indefinitely with no loss of quality, although any impurities need to be removed. The magnetic properties of steel

6 Packaging Materials

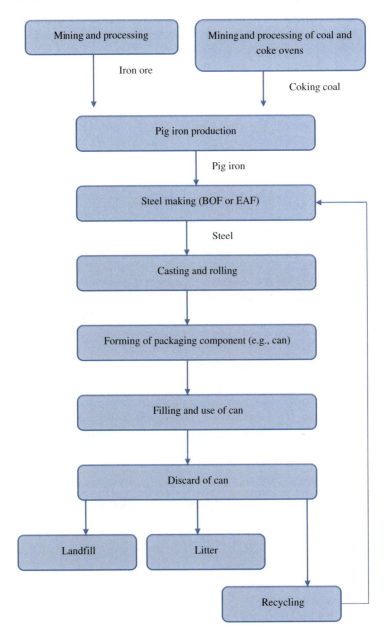

Fig. 6.3 Life cycle of steel

mean that it can be easily sorted from other materials at a MRF. The cans are crushed and baled for transport to refineries, where they are shredded. Tin, a contaminating element for steel, can be removed from the shredded material by an electrolytic process. The material is then melted in a furnace to make new steel, including tinplate for packaging. Relative to virgin steel produced via the basic-oxygen furnace process, recycling steel can reduce energy consumption by approximately 60% [6]. Global recycling rates for steel cans are reported in Table 2.14 of Chap. 2.

Steel slowly corrodes in landfill, the resultant oxides of which are inert and cause minimal environmental impacts.

6.4 Glass

Material Properties

Glass is chemically inert and impermeable to gas and liquid. Soda-lime glass is commonly used in packaging applications. Glass is strong but brittle, and its properties are not affected by the temperatures reached during hot-filling. Soda-lime glass is typically clear (white, flint or colourless glass) or coloured green, blue or brown (amber).

The spectrum of visible light transmitted through glass varies with the colour. Clear glass transmits a high proportion of the natural light spectrum, including ultraviolet light. Less light is transmitted for green and blue glass. Brown glass prevents the transmission of ultraviolet light and much of the visible light spectrum. The light transmission properties of coloured glass can be used to prevent product spoilage and control product shelf life; for example, in medicine bottles and bottled wine. The density of glass is approximately 2.4–2.6 g/cm^3.

Packaging Applications

Soda-lime glass can be processed into many different sizes and shapes. Bottles, jars and vials are used in a variety of packaging applications including food, such as wine, beer and soft-drinks, personal care products such as cosmetics and perfumes, and medical products such as medicines (Photo 6.2).

Life Cycle

Figure 6.4 presents a simplified life cycle flow chart for glass. Virgin glass is made from silica sand (SiO_2, 74%), soda ash (Na_2O, 13%), lime (CaO, 11%), feldspar (Al_2O_3, 1%) and other minerals and salts including iron oxide, chromium oxide and sulfur oxide, which are added to control processing properties and colour.

'Cullet' (recycled glass) can be substituted for virgin materials to reduce energy and virgin material consumption. For example, 50% cullet reduces energy consumption by 10–15% [7]. The cullet content of packaging glass is typically 30–70% [8, 9].

The raw materials are pre-mixed and transferred to a furnace where they are melted and fused in a continuous process at temperatures of 1500–1600°C. Glass fining and conditioning occurs in the later stage and ensures glass which is gas-free and has uniform composition. While in its melted form, the glass is formed into a 'gob' and then transformed into the required shape with the use of a mould.

6 Packaging Materials

Photo 6.2 Glass packaging
(Photo: Amcor Packaging
(Australia) Pty Ltd)

Two processes are common in packaging applications:
- blow-and-blow moulding (for narrow neck bottles). This involves blowing the glass gob with air into a pre-form (or blank) in a pre-form mould, transferring the pre-form into a different mould then blowing it with air to the final shape
- press-and-blow moulding (for jars and increasingly for narrow neck bottles). The pre-form (or blank) is pressed from the gob with a plunger and blown into the final shape.

Press-and-blow products have greater dimensional accuracy than blow-and-blow products and can result in a reduced wall thickness (refer to Case Study 6.1).

The formed glass packaging is finally annealed (heated and cooled slowly) to relieve surface stresses and improve the mechanical properties of the glass.

In the cradle-to-cradle model (see Sect. 1.2), glass belongs to the technical metabolism (Fig. 6.4).

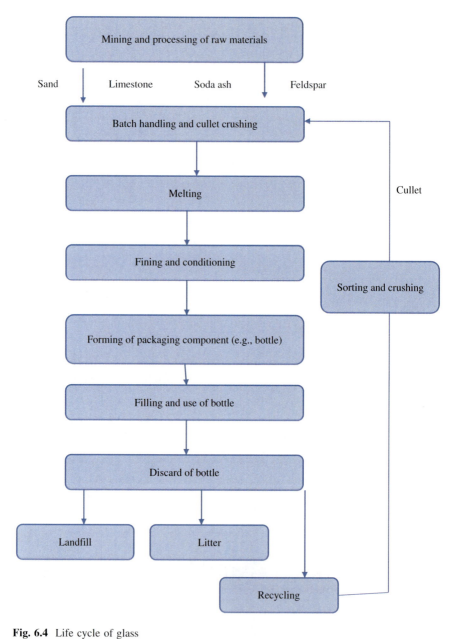

Fig. 6.4 Life cycle of glass

> **Case Study 6.1 Owens-Illinois Lean and Green**
>
> In response to increased demand for sustainable innovation, Owens–Illinois (O–I) have developed a range of narrow-neck wine bottles produced via a press-and-blow process. These thinner walls reduces the mass of a typical 750 ml bottle by up to 27% without compromising aesthetics [10]. This mass reduction reduces material required for production, as well as greenhouse gas emissions associated with manufacture and transport.

Environmental Impacts

The most significant environmental impacts associated with glass production include:
- extraction of raw materials and degradation of land associated with mining
- emissions from the large amounts of energy used in manufacturing
- emissions from processing
- impact of the weight of the glass packaging system on transport and distribution energy [11].

Recovery and Disposal

Glass is recyclable but is sensitive to contamination by other ceramic materials such as bone china and safety glass, metals such as lids, or differently coloured glass. It can be manually or automatically sorted for recycling. (See an example of automatic sorting in the Visy Recycling Case study 6.2.) A high level of glass breakage in collection vehicles leads to the mixing of glass colours. Such contamination results in the loss of critical properties, including processing, transmission and mechanical properties. Global recycling rates for packaging glass are reported in Table 2.14 in Chap. 2. Glass is inert and does not degrade in landfill.

Glass for packaging can be recycled in one of several ways. The most common involves the collection and crushing of glass packaging (bottles and jars) for use in the manufacture of new packaging. The containers are separated into different

> **Case Study 6.2 Automatically Sorting Glass**
>
> There are a variety of new technologies available to sort glass and remove contaminants automatically. For example, the Binder process used by Visy Recycling in Australia uses lasers to identify glass fragments as small as 6 mm at high speed. Air jets sort the glass fragments into the different colours and remove any non-recyclable materials such as steel caps and corks. This technology enables a higher percentage of broken glass to be colour-sorted to meet specifications for container manufacture.

colours (green, brown and clear), and contaminants such as metals, ceramics and plastics are removed. The glass is crushed into cullet and used as a feedstock to replace virgin materials in the manufacturing process. Due to its different composition, non-packaging glass, such as windows, generally cannot be used in the manufacture of packaging.

Recycled packaging glass can also be used to make new glass products such as tiles and ornaments using techniques such as fusing or slumping. Fusing involves melting crushed glass at high temperatures in a mould to form a new product. Slumping involves placing a single piece of glass on a mould to be heated in a kiln. As the glass melts, it 'slumps' around the shape of the mould.

Packaging glass can also be recycled by crushing it into granules or powder for use in a range of applications, including construction aggregate and filtration medium. These markets provide a secondary use for crushed or contaminated glass that is unable to be economically sorted and cleaned for use in new packaging. These applications reduce waste-to-landfill but do not generate the environmental benefits associated with closed loop recycling.

6.5 Paper and Board

Material Properties and Packaging Applications

Paper and board materials are the most widely used packaging materials, although their thickness, construction and applications vary (Table 6.2). The most common application of paper and board in packaging is as secondary packaging.

Table 6.2 Classification of paper and board

Type	Grammage (g/m^2 — GSM)	Construction	Example applications
Kraft paper	10–120	Single sheet	Bags, sacks, sheets, cartons, boxes, trays, labels, inserts
Boxboard (folding boxboard, cartonboard, paperboard)	120–800	Single or multi-layered	Folding cartons, milk and juice cartons
Corrugated Board (fibreboard)	250–1,500	Multi-layered, with fluting	Shipping boxes and cartons, pallets, edge protectors, trays, separators, corner blocks for bracing
Liquid paperboard	300–400	Multi-layered with polymer and optional aluminium foil	Fresh milk, soap and aseptic packaging, including long-life milk and juice
Moulded paper packaging (moulded pulp, moulded fibre)		Single layer	Egg cartons, takeaway drink trays, cushioning for electronic products, food service packaging

Kraft Paper

Kraft paper is strong, transulcent or opaque, and can be rigid or flexible, depending on the grade and thickness. In the uncoated form, it has limited gas-barrier properties and is sensitive to moisture. Grades include natural brown, unbleached, heavy duty and bleached white. The density of Kraft paper is typically 0.6–0.8 g/cm^3.

Kraft paper can be used as a barrier layer to prevent direct contact of a product with another packaging surface; for example, food with cartonboard. For applications in wet or humid conditions, agents can be added to improve tear resistance.

The liquid and gas barrier properties of Kraft paper can be improved by coating or laminating it with resins, wax, polymers such as polyethylene or aluminium foil. Grades of coated Kraft paper include greaseproof paper, glassine and parchment paper. Greaseproof papers are applied in the packaging of food, including biscuits, confectionary bars and other foods with a high oil content. Glassine is a strong greaseproof paper with a smooth, glossy finish. Food packaging applications of glassine include biscuit liners, fast foods and baked goods. Parchment paper is impervious to water and oil and is used in food packaging applications such as butter and margarine.

Boxboard

Grades of boxboard include folding boxboard, solid board (bleached or unbleached), folding boxboard and white-lined chipboard. Solid board is made from bleached or unbleached pulp. Folding boxboard typically consists of middle layers from mechanical pulp and outer chemical pulp layers. The outer layer of solid board and folding boxboard can be coated with multiple layers of a white pigment. White-lined chipboard consists of an inner layer made from recycled pulp and an outer layer made from either recycled or chemical pulp [12] (Photo 6.3).

Photo 6.3 Folding boxboard (Photo: Amcor Packaging (Australia) Pty Ltd)

Photo 6.4 Corrugated carton (Photo: Amcor Packaging (Australia) Pty Ltd)

Corrugated Board

Corrugated board is a layered assembly, consisting of an inner and outer layer of linerboard, made from Kraft or recycled paper, and a middle layer of fluted paper (fluting), which is wavy in appearance. Multiple corrugated layers, known as walls, are available. The structure of corrugated board provides varying degrees of resistance to crushing, shock and bulging, depending on the type of fluting and number of walls. The fluting also provides thermal insulation due to the air gap between the linerboards. Wax can be applied to the linerboard to improve barrier resistance against moisture and oils. Corrugated board is mostly used for secondary packaging, including boxes, trays and dividers (Photo 6.4).

Liquid Paperboard

Liquid paperboard packaging is a layered composite material consisting of solid bleached board and a polymer, typically low density polyethylene (LDPE). The LDPE layer is added to create a liquid barrier. An aluminium foil layer is sometimes applied to provide additional barrier properties. Liquid paperboard has excellent gas and liquid barrier properties. It commonly has a glossy outer and matt inner finish. Its uses include aseptic packaging for food applications, such as long life milk and wine cartons.

Moulded paper

Moulded paper packaging (Photo 6.5) is lightweight and often used for packaging products that need protection, such as eggs and electronic components or products. The thickness and density can be varied with the amount of pressure applied during the moulding process. It is more fire resistant that expanded polystyrene (EPS), which is its main competitor [13].

Photo 6.5 New applications for moulded paper include food service packaging (Photo: Helen Lewis)

Life Cycle

Figure 6.5 presents a simplified life cycle flow chart for paper and board materials. Paper and board are produced by extracting and pressing cellulose fibres. The cellulose fibres are usually derived from wood pulp (hardwood and softwood timbers), although other fibre sources such as bamboo and hemp are possible. In recent decades waste paper, not necessarily from packaging products, has become an important cellulose fibre source for the production of packaging paper and board.

In the cradle-to-cradle model, paper and board can be recovered through the biological metabolism (for example, in a commercial composting facility) or in a technical metabolism (paper recycling process).

The first process involved in wood-derived papermaking is the growing and felling of trees. The wood is then processed into chemical and/or mechanical pulp. The pulping process separates the cellulose fibres from other materials within the wood, including lignin, resins and oils. Chemical pulp (sulphate pulp) is produced by heating woodchips in an alkaline solution of sodium hydroxide and sodium sulfide. Semi-chemical pulps are used for the production of base papers of corrugated board. Mechanical pulp is produced by de-barking logs, chipping, refining and grinding, then screening, cleaning and thickening of the pulp.

Different bleaching processes are used to make the paper or board white (see Sect. 2.4.4), each with different environmental implications. These include chlorine bleaching, elemental chlorine-free, process totally chlorine-free and processed chlorine-free.

Solid Board

Solid board is produced from bleached or unbleached chemical pulp. Solid board, folding boxboard and white-lined chipboard can be made using a 'fourdrinier' or cylinder paper board process. The fourdrinier process deposits pulp onto a moving mesh. Production from a cylinder machine relies on the formation of a sheet on one

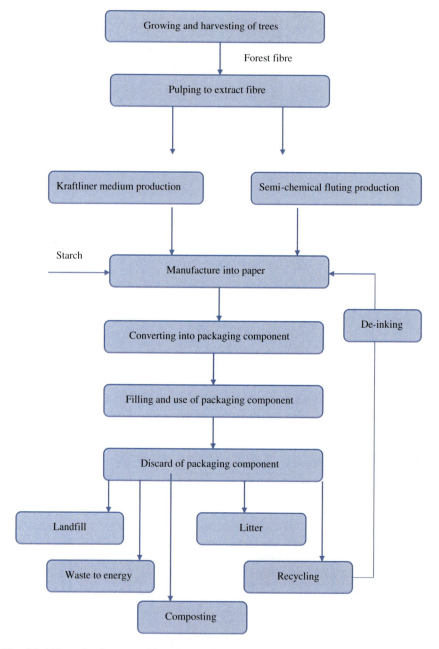

Fig. 6.5 Life cycle of paper and board (generic)

or more cylinders that rotates in a vat of pulp. Multi-layered sheets are formed by additional layers of pulp that are deposited throughout the process. Water is removed by gravity and vacuum processes before being pressed and dried.

Corrugated Board

Corrugated board is produced by firstly conditioning the feedstock papers with heat and steam. Paper is fed into corrugated rollers to produce the fluting material, after which glue is applied to adhere to one linerboard side. The other linerboard is then glued before more heat is applied to bond the glues. Boxes can be formed from the corrugated board by a series of cutting, folding and printing steps.

Liquid Paperboard

Liquid paperboard is constructed by coating paperboard with plastic (gable top) or plastic and aluminium (aseptic). The composite board is then cut, folded and printed before being distributed for filling.

Moulded Pulp

A moulded paper package is manufactured on a screen that has the desired shape. The pulp is either forced under pressure on to the screen mould or sucked on to it [13]. It can be made with up to 100% recycled feedstock.

Environmental Impacts

The most significant environmental impacts associated with the use of paper and board are:
- the loss of biodiversity through tree plantations, soil erosion and watershed destabilisation at the forestry level
- fertilisers used during the growing of the trees
- chemicals used in paper production
- emissions generated and water used during production.

Sustainable Forestry Initiatives

Sustainable forestry initiatives, such as the Forest Stewardship Council, aim to minimise the environmental, social and economic impacts of forestry activities by setting forestry management guidelines and providing third-party accreditation and auditing to those guidelines [14]. However, these schemes are not without published criticism. Pattberg [15] suggests that the sustainable forestry certification favours trade countries with well-developed forestry industries and that rival eco-labelling driven by commercial interests can clutter the marketplace and undermine the intent of sustainable forestry initiatives.

> Refer to Sects. 2.4.4 and 3.5.7 for more information on sustainable forestry initiatives and labelling schemes.

Recovery and Disposal

Paper and board are recycled along with other types of paper waste in a conventional paper-making process. Pre-consumer wastes include envelope trimmings, printer off-cuts, paper mill scraps and unsold magazines. Paper mill scrap (or mill broke) has always been used in paper manufacturing. Post-consumer waste paper includes non-packaging materials such as old newspapers and magazines.

In the recycling process, the recovered paper and board are mixed with water and chemicals to separate the fibres and form a slurry. This is then passed through a series of screens and centrifugal cleaners to remove contaminants such as wax, ink, glass, metal and plastic. After pulping, the slurry is dried to reduce water content in preparation for pressing. Pressing reduces the water content from 80% to approximately 5%. After pressing, the paper is wound onto rolls for shipment.

Recycling reduces the length of the cellulose fibres in the paper pulp, which reduces the final strength. This strength reduction is an important consideration for box applications, which requires long fibres for structural integrity. For this reason, the preferred source of fibre for box applications is old corrugated boxes. Strength can be improved by blending recycled with virgin pulp. New production technologies and adjustments to the paper-making process have allowed some manufacturers to produce corrugated boxes with up to 100% recycled fibre at a comparable or lower cost than those with 100% virgin fibre [16]. The inks in recycled pulp can lead to a loss of brightness and variation in colour.

The cellulose within paper and board is biodegradable, but the amount of degradation, its rate and emissions associated with it depend on the nature of the material and the degradation environment. For example, anaerobic (in the absence of oxygen) degradation of uncoated paper in landfill can generate methane, a greenhouse gas, whereas aerobic (with oxygen) degradation in landfill generates carbon dioxide. Coated layers, such as wax, polyethylene and aluminium, can reduce the degradation of paper and board by limiting the exposure of the underlying material to the environment.

Liquid paperboard is theoretically recyclable, although due to bonding with aluminium and/or polymer layers, the recovery of fibre content from the paper is low. Fibres can be recovered using a modified pulping process, but the aluminium and polymer are generally not recovered. Producers of liquid paperboard are developing processes to recover all materials used in liquid paperboard. One such process, developed by TetraPak, involves shredding and pulping of the paperboard to separate paper fibres, followed by plasma separation of the polymer and aluminium materials [17]. The commercial utilisation of this separation technology is currently limited to certain countries (i.e. Brazil), however there are developments to expand use.

6.6 Non-renewable Thermoplastics

Material Properties and Packaging Applications

Thermoplastics vary significantly in their molecular structure, which gives rise to differences in properties, including density, stiffness, strength, toughness, elongation

6 Packaging Materials

Table 6.3 Densities of common non-renewable thermoplastic polymers

Polymer			Density (g/cm^3)	Sinks/floats (in water)
PET	Polyethylene terephthalate	♳	1.38–1.40	Sinks
HDPE	High density polyethylene	♴	0.96	Floats
PVC	Polyvinyl chloride	♵	1.35–1.40	Sinks
LDPE	Low density polyethylene	♶	0.92	Floats
PP	Polypropylene	♷	0.90	Floats
PS	Polystyrene	♸	1.06	Sinks
PC	Polycarbonate	♹	1.20–1.22	Sinks

to failure, gas and water permeability, light transmission and degradation behaviour. This property range allows thermoplastics to be used in a wide range of applications and forms such as rigid containers (bottles, jars), flexible packaging (films, wrapping), coatings and lacquers, and adhesives.

Tables 6.3, 6.4, 6.5 summarise commonly used non-renewable thermoplastics used as packaging materials, together with their densities, resin identification code, common properties and applications (Photos 6.6 and 6.7).

Life Cycle

The raw materials used to produce non-renewable thermoplastics are mostly derived from petrochemicals (hydrocarbons), either natural gas or crude oil extracted from the earth. Other raw materials are derived from a variety of sources, such as the chloride in polyvinyl chloride (PVC), which is derived from salt.

Table 6.4 Properties of commonly used non-renewable—fossil derived thermoplastic polymers

Polymer name		Properties
PET	Polyethylene terephthalate	Clear, tough, chemical resistant, good gas barrier, fairly high temperature toleration, safe for food contact, opaque (CPET), easy to mould or extrude. Coloured PET has the same physical attributes as clear PET but provides an opaque appearance that can give light protection to packaged products
HDPE	High density polyethylene	Rigid to flexible, waxy surface, opaque, fits all forming processes, vapour barrier, poor gas barrier, chemical resistant, easy to mould or extrude. A number of polymers and coatings can be used to improve the appearance, physical and barrier properties of HDPE (i.e., multi layer HDPE). Most are used to improve the barrier properties of the packaging against gas, liquid and odour transfer as well as ultraviolet light and chemical resistance as a means to extend a product's shelf life. Common multi-layer and barrier materials used in packaging include silicon dioxide coatings, metallised film, EVOH (ethylene vinyl alcohol) and nylon. Barrier structure and properties can be tailored to application to provide extended shelf life
PVC	Polyvinyl chloride	Hard, rigid, can be clear, tough, can be solvent-welded, thermoformable, good impact strength, good chemical resistance, oil- and alcohol-resistant, easy to mould or extrude on modified equipment. A number of polymers and coatings can be used to improve the appearance, physical and barrier properties of PVC (i.e., multi barrier PVC). Most are used to improve the barrier properties of the packaging against gas, liquids and odour transfer as well as ultraviolet light and chemical resistance as a means to extend a products shelf life. Common multi-layer and barrier materials used in packaging include: PVDC (polyvinylidene chloride) and EVOH (ethylene vinyl alcohol)
LDPE	Low density polyethylene	Tough, semi-flexible, waxy surface, shock-resistant, moderate clarity, good seal medium, good water vapour barrier, poor gas barrier, easy to mould or extrude
LLDPE	Linear low-density polyethylene	
PP	Polypropylene	Tough, heat-resistant up to 165°C, excellent chemical resistance, moderate barrier, oriented films, easy to mould and extrude. A number of polymers and coatings are used to improve the appearance, physical and barrier properties of PP. Most are used to improve the barrier properties of the packaging against gas, liquids and odour transfer as well as ultraviolet light and chemical resistance as a means to extend a product's shelf life. Common multi-layer and barrier materials used in packaging include PVDC (polyvinylidene chloride), silicon dioxide coatings, metallised film, PEN (polyethylene naphthalate), epoxy-amine organic coatings, EVA (ethylene vinyl acetate), EVOH (ethylene vinyl alcohol), EAA (ethylene acrylic acid), ionomers

(continued)

Table 6.4 (continued)

Polymer name		Properties
PS	Polystyrene	Amorphous polymer, very stiff and brittle easy to extrude and thermoform, low cost, relatively poor barrier to water vapour and gases, excellent x-ray resistance, free from odour and taste, low shrinkage, easy to mould and extrude
GPPS	General purpose polystyrene	Clear, glossy, rigid, brittle, semi-tough, softens at 95°C, easy to mould or recycle
HIPS	High impact polystyrene	Tough, opaque, poor barrier, easy to mould or extrude. Contains butadiene rubber additive, which increases impact resistance and flexibility and reduces brittleness compared with unmodified PS
EPS	Expanded polystyrene	Moulded expanded beads, light-weight, impact-energy absorbing, thermal insulating, protective packaging, easy to mould. EPS is a form of amorphous PS treated with a blowing agent e.g.: a hydrocarbon or carbon dioxide. It has cellular structure which reduces brittleness and provides excellent cushioning and insulating properties. Produced within a mould. EPS beads range from large to very small depending on the amount of blowing agent used and processing conditions
PC	Polycarbonate	Transparent, good resistance to weather, flames, impact, easy to process. It's an amorphous polymer, rigid, transparent, with excellent impact resistance, good weather resistance, good resistance to sunlight, relatively poor water/gas barrier properties, good resistance to water, oil, alcohols, fruit juices, aliphatic hydrocarbons and aqueous ethanol solutions

Table 6.5 Applications of commonly used non-renewable — fossil derived thermoplastic polymers

	Polymer name	Example applications
PET	Polyethylene terephthalate	Carbonated soft drink bottles, trays, blisters, food jars, clamshells for takeaway food and prepared meals, cups, high-performance films, and coatings. PET has gained market share in carbonated beverages at the expense of glass and aluminium
HDPE	High density polyethylene	Rigid and flexible applications. Beverages include milk, juice, water. Personal and health care products include shampoo, deodorants, cosmetics, medicines. Household products include detergent, bleach. Food products include cereals, crackers, snack foods, delicatessen products, produce bags
PVC	Polyvinyl chloride	Rigid and flexible applications, including bottles/containers for juice, cordial, dairy products, edible oils, liquor, toiletries, cosmetics and labels
LDPE	Low density polyethylene	Bags for food and clothing, produce bags and liners, wrapping film, squeeze bottles, shrink/stretch film, tubs, caps, foam, coatings, laminates
LLDPE	Linear low-density polyethylene	Stretch/cling film, grocery sacks, heavy duty shipping sacks. LLDPE is often blended with LDPE. The use of barrier layers for some applications is essential to ensure product preservation
PP	Polypropylene	Ice cream tubs, takeaway food containers, dairy dip containers, paint tubs, bottle closures, juice bottles, margarine and yoghurt tubs, cosmetic jars
PS	Polystyrene	Bottles for tablets and capsules and thermoformed trays
GPPS	General purpose polystyrene	Blister, foamed meat trays, moulded trays, plastic cutlery, CD covers, fruit and dairy tubs
HIPS	High impact polystyrene	Thermoformed containers for dairy products, single service food packaging, bottles, blister-packs, overcaps, films, barrier trays for modified-atmosphere products
EPS	Expanded polystyrene	Hot drink cups, produce and fish boxes, protective packaging for fragile products
PC	Polycarbonate	Microwave-proof containers, ovenware, food storage containers and water containers

Photo 6.6 PET Packaging
(Photo: Amcor Packaging
(Australia) Pty Ltd)

Figure 6.6 presents a simplified life cycle flow chart for non-renewable, fossil-derived thermoplastics, and Table 6.6 summarises the materials and manufacturing processes used for different materials.

In the processing (refining) of fossil fuels (crude oil, natural gas), hydrocarbons are split (cracked) into smaller molecules such as ethane which may then be converted and purified into a large number of different monomers such as ethylene (C_2H_4), propylene (C_3H_6), various butylenes (C_4H_8) and butadiene (C_4H_6). These monomers are then polymerised using pressure and energy to form polymers such as polyethylene that are then converted to plastic packaging products through various processes, including extrusion and moulding.

In the cradle-to-cradle model, non-renewable fossil derived thermoplastics belong in the technical metabolism.

Environmental Impacts

Significant environmental impacts from non-renewable fossil derived thermoplastics include:
- use of non-renewable resources as raw materials
- emissions during the refining and cracking processes
- some by-products or emissions are carcinogenic such as vinyl chloride monomer and styrene monomer.

Recovery and Disposal

At end-of-life, non-renewable fossil derived thermoplastics can be recycled, disposed to landfill, or incinerated for energy recovery. The most suitable method of recovery depends on the type of thermoplastic.

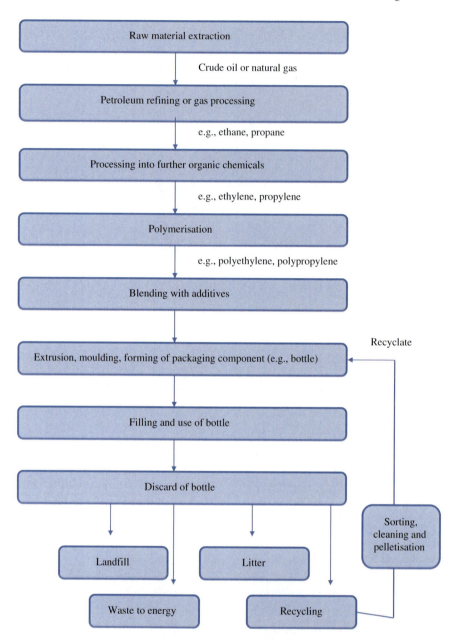

Fig. 6.6 Life cycle of non-renewable fossil-derived thermoplastic polymer

Photo 6.7 Polypropylene closures (Photo: Amcor Packaging (Australia) Pty Ltd)

Technologies and Infrastructure for Recovery and Recycling Exist

Most non-renewable fossil-derived thermoplastics are recyclable; that is, recovery and recycling technologies exist. However, recycling facilities may not be available for all packaging formats or in all locations. These include:

- mechanical recycling: plastics are shredded, washed and extruded to form pellets, or moulded directly into a new product
- feedstock recycling: plastics are converted back into a monomer or new raw materials by changing their chemical structure [18, p. 3]
- energy recycling: plastics are processed through controlled combustion to generate electricity or steam.

The infrastructure for mechanical recycling (collection and reprocessing) is well-established in most developed economies. It is also generally preferable to feedstock recycling because it maintains the economic value of the polymer at a relatively high level and significantly reduces the amount of energy required to manufacture new plastic products.

Mechanical recycling is most viable for plastics that are available in large quantities, in a clean and homogenous form, and in locations with reasonable access to recycling facilities. Some plastics can be problematic in the recycling process as they are not compatible with other polymers and therefore need to be separated.

Post-consumer packaging from households is generally sorted at a materials recovery facility (MRF) into individual polymers and/or mixed streams, depending on market requirements (Fig. 6.7).

Non-recyclable materials and other contaminants are sent to landfill or an incineration facility. Further sorting is undertaken during the mechanical recycling process. For example, non-compatible polymers and paper labels are separated

Table 6.6 Manufacture of commonly used non-renewable fossil-derived thermoplastic polymers

Polymer name(s)		Manufacturing process
PET	Polyethylene terephthalate	Produced by polycondensation from ethylene, terephthalic acid (TPA) and glycol
HDPE	High density polyethylene	Made from ethylene, a petroleum and natural gas derivative, and catalysts such as aluminium trialkyltitanium tetrachloride and chromium oxide. Through an adjusted polymerisation process and a higher crystallinity, high-density polyethylene (HDPE) is produced
PVC	Polyvinyl chloride	Vinyl monomer is manufactured from ethylene (from petroleum or natural gas) and chlorine (from rock salt). Produced by polymerisation of vinyl monomer with other co-monomers to obtain a wide variety of properties
LDPE	Low density polyethylene	Produced by polymerisation of ethylene, a petroleum and natural gas derivative. Peroxide can be used to initiate the polymerisation process. a flexible low-density polyethylene (LDPE) can be produced without plasticisers
LLDPE	Linear low density polyethylene	Butene and hexene are polymerised in addition to ethylene to produce a linear low-density polyethylene
PP	Polypropylene	Polypropylene is prepared by polymerisation of propene, a petroleum and natural gas derivative, along with a catalyst (Ziegler–Natta catalyst). Polypropylene films are often stretched (Oriented Polypropylene or OPP) to improve mechanical properties, including heat-seal strength and heat-shrink properties, printability, clarity and barrier properties for gasses and odours
PS	Polystyrene	Styrene monomer is manufactured from ethylene (from petroleum or natural gas) and benzene (from coal)
GPPS	General purpose polystyrene	Polymerisation of styrene, often by co-polymerisation with other monomers to obtain a wide variety of properties
HIPS	High impact polystyrene	HIPS is normally produced by co-polymerisation of styrene and a synthetic rubber
EPS	Expanded Polystyrene	EPS is produced by expanding PS beads, using low boiling hydrocarbon as a blowing agent (usually pentane)
PC	Polycarbonate	Polycarbonate is a polyester produced from carbonic acid

6 Packaging Materials

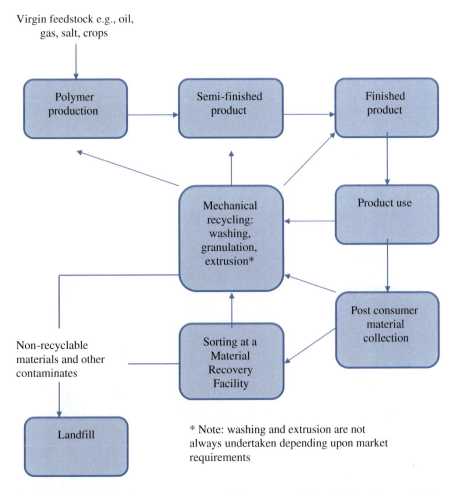

Fig. 6.7 The life cycle of post-consumer recycled non-renewable fossil derived thermoplastics

during the washing of the flake. The quality of the end product depends on both the nature of the original waste materials and the effectiveness of sorting processes at the MRF and during reprocessing. MRF operators are starting to sort plastics automatically; for example, with near-infrared optical sort machines (see Case Study 6.3).

Recovered plastics are baled and sent to a recycler, where they are granulated, washed (if required), extruded and pelletised. An example of an application of recycled plastic is in Photo 6.8.

Case Study 6.3 Automatic Sorting of Plastics: Visy Industries

Infrared optical sorting technology sorts plastics by their polymer type when illuminated. Each material reflects light in the infrared wavelength. The reflections are analysed by a fast scanning sensor installed over a conveyor belt. The sensor identifies the polymer and blows it into the correct polymer stream for further sorting.
Source: Visy Recycling [19]

Photo 6.8 Plastic packaging can be reprocessed into different products such as plastic 'lumber': example of 'open loop' recycling (Photo: Replas)

Recyclable Does not Necessarily Mean Recycled

Reprocessing is undertaken only when recycling is commercially viable or subsidised until it is viable. The cost, volume and quality of recyclate must be matched by a demand for that price, amount, and specification of material. This depends on many factors including economies of scale, the physical structure of the polymer and the degree of contamination from food waste, incompatible polymers and so on.

6 Packaging Materials

Table 6.7 Compatibility of common non-renewable thermoplastics in mechanical recycling

	LDPE	LLDPE	HDPE	PP	PS (GEN PURPOSE, HIGH IMPACT)	PVC	PC	PET
LDPE								
LLDPE	1							
HDPE	1	1						
PP	4	3	3					
PS (GEN PURPOSE, HIGH IMPACT)	4	4	4	4				
PVC	4	4	4	4	4			
PC	4	4	4	4	4	4		
PET	4	4	4	4	4	4	1	

Key
1 Excellent
2 Good
3 Fair
4 Incompatible

Notes: This table is adapted from [29, p. 24] and should only be used as a guide. Consultation with recyclers is recommended

Not all plastics are compatible when mixed (Table 6.7). Polymer contamination from closures and labels can render the recyclate unusable or downgrade its quality so that it cannot be reused for high value or closed loop applications.

Polyolefins, such as HDPE, LDPE and PP are incompatible with polyesters such as PET, PVC and PS. They can be separated during the washing process at recycling facilities, which takes advantage of the differing polymer densities (see Table 6.3). However, the separation of polymers with a similar density, such as PET, PVC and PS, is more difficult. This can cause problems during reprocessing. For example, when PET is recycled a float/sink separation process is used. PET and PVC have very similar densities (see Table 6.3), and if PVC labels or cap liners are used on PET containers they will become a contaminant in the PET stream. The two polymers are not compatible in the extrusion or moulding process because they react, reducing the melt viscosity of the PET [20, p. 125]. The PVC contaminants, although tiny, also degrade at the processing temperatures used for PET, and the PET bottle will end up with visible discolouration and black specks [20, p. 127]. Other contaminants such as food residues should be removed during washing; otherwise they can cause problems during recycling. Degradable thermoplastics, including polylactic acid (PLA), can also contaminate conventional thermoplastic recycling streams.

In Australia, the recycling rate for all plastics is 18%, and for plastics packaging it is 36%. The higher rates for plastics packaging reflects its common usage in single-use products. The recycling rates for individual polymers vary from 43% (PET) to 3% (other polymers, consisting of acrylics, acetals, cellulosics, polyethylene oxide, polyisobutylene and other propylene and styrene polymers') [21].

> Refer to Table 2.14 for a compilation of all packaging material recycling rates

Practical Limits to Close Loop Recycling?

Legislation controls the use of recyclate in food contact applications because of concerns about contaminants migrating into food stuffs (see Sect. 4.2.5); that is, 'assimilation of adjuvant/additives in the recycled plastic not approved for food-contact use' [22, p 31]. This limits opportunities for closed loop recycling. One food contact application that often uses post-consumer recyclate is PET bottles. The PET post-consumer recyclate is purified using complex proprietary technology before reuse in bottle production.

> ⇒ See Case study 2.13 on use of recycled PET packaging by Marks & Spencer

Consider Degradability

The degradation behaviour of thermoplastics is an important design consideration. Degradable thermoplastics have the ability to 'break down, by bacterial (biodegradable), thermal (oxidative) or ultraviolet (photodegradable) action' [23, p 11].

A biodegradable polymer is capable of being broken down by micro-organisms in the presence of oxygen (aerobic process) to carbon dioxide, water, biomass and mineral salts or any other elements that are present; or in the absence of oxygen (anaerobic process) to carbon dioxide, methane and biomass [24].

The degradation of non-renewable thermoplastics is generally limited. However, there are some non-renewable thermoplastics that readily degrade in different environments. 'Oxodegradable' materials combine a conventional polymer such as HDPE with an additive to enhance degradation via heat, light or stressors acting as catalysts. These materials are not currently certified as meeting relevant composting standards (refer to Sect. 6.7), and the value of these materials in landfill or in the litter stream is not clear. Applications of oxodegradable polymers include shopping bags, bubble wrap and compost bags. Polyester polymers can degrade via a reaction with water (hydrolysis). The rate of degradation ranges from weeks for aliphatic polyesters such as polyhydroxyalkanoates to decades for aromatic polyesters such as PET. Photodegradable polymers degrade through the action of ultraviolet light, which breaks the chemical bonds in the polymer chains. This process can be assisted by the presence of UV-sensitive additives in the polymer.

6.7 Renewable Thermoplastics

The use of renewable thermoplastics, such as those derived from crops, in packaging applications is increasing. Types of renewable thermoplastics include:
- starch and thermoplastic starch (TPS) based polymers
- cellulose-based polymers
- polylactic acid (PLA or polylactide).

Material Properties and Packaging Applications

Starch-Based Polymers

Starch-based polymers can be clear to translucent. These are blended 'to achieve the necessary performance properties for different applications' [25, p. 16]. Starch-based polymers can be used in injection moulding and film applications, including loose fill, cutlery, cups, foamed trays, shopping bags, mulch film, compost bags and film. The density of pure TPS is 1.2–1.25 g/cm^3 (Photo 6.9).

Cellulose-Based Polymers

Cellulose-based polymers are transparent and permeable to water and oxygen. They are strong and stiff with moderate impact resistance, and are easy to extrude and injection-mould. Applications include film wrap for bakery products, confectionary and fresh produce, twist wrap and glued bags. The density of cellulose-based polymers is approximately 0.9 g/cm^3.

Photo 6.9 Example of a packahe made from a biodegradable starch-based polymer (Photo: Garden Express)

Polylactic Acid

Polylactic acid (PLA) is transparent, has good gas-barrier properties and is resistant to oils. It is stiff, with moderate heat and impact resistance and can be fabricated by injection moulding, sheet extrusion, blow-moulding, thermoforming and film-forming. Applications include extruded and thermoformed food containers and bottles. The density of PLA is 1.24–1.30 g/cm^3 (Photo 6.10).

Photo 6.10 Biodegradable Polylactic acid bottles (Photo: Nature Works)

Life Cycle

Figure 6.8 presents a simplified life cycle flow chart for a generic renewable thermoplastic. Post-consumer recovery and recycling of renewable thermoplastics is currently limited due to technical difficulties in sorting renewable thermoplastics from recyclable non-renewable plastics, as well as commercial limitations due to the low volumes and poor economies of scale [26]. Recovery of renewable thermoplastics is likely to become more viable in the future.

In the cradle-to-cradle model, renewable thermoplastics belong in the biological metabolism, that is, in organic recycling systems such as composting.

Starch-Based Polymers

Starch can be derived from potato, corn, wheat or cassava (tapioca) [2]. It can be used as a polymer, but it is modified in most applications to counteract property losses due to its fast biodegradation. Starch can be blended with non-renewable polymers to improve some properties such as strength at the expense of others such as biodegradability (Photo 6.11).

TPS is produced by the gelatinisation of starch, typically with a high amylose content (typically greater than 70%). Non-renewable polymers such as LDPE and polyesters, plasticisers such as glycerol, compatabilisers and processing aids can be added to TPS to tailor the polymer for specific properties, including processability, mechanical properties and degradability.

Cellulose-Based Polymers

Cellulose can be derived from a variety of plants including trees and cotton. Cellulose is modified with organic or inorganic acids to produce cellulosic polymers including cellulose acetate, cellulose acetate propionate and cellulose acetate butyrate.

6 Packaging Materials

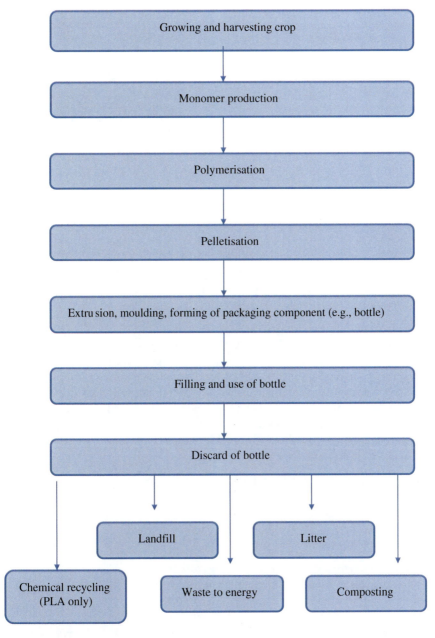

Fig. 6.8 Life cycle of a generic renewable thermoplastic

Photo 6.11 Limes nets. Sainsbury's use biodegradable plastics for some of their organic produce. These nets are made from a corn starch resin. (Photo: Sainsbury's)

Polylactic Acid

PLA is produced by using micro-organisms to convert corn starch (dextrose) into lactic acid via fermentation. The lactic acid molecules link to form rings called lactide monomer. The process of polymerisation involves the lactide ring opening and linking together to form a long chain of PLA. The polymer is then formed into pellets to be converted into different packaging applications.

Environmental Impacts

Products manufactured from renewable raw materials have different environmental impacts to those derived from non-renewable sources [2].

The renewable feedstock requires agricultural resources including land, water and energy for agricultural machinery and fertiliser. Hence environmental impacts include:
- land transformation and occupation for the growing and harvesting of crops
- application of fertilisers and other chemicals and their subsequent run-off into waterways and/or air emissions
- consumption of fossil fuels by agricultural machinery and equipment and the release of air emissions.

As some of these polymers are designed to degrade in specific environments, such as water or industrial composting, they may generate additional environmental impacts if not disposed appropriately.

Life Cycle Assessment

An important consideration when assessing the environmental impacts of renewable thermoplastics is their density. Many renewable thermoplastics have a higher density than non-renewable thermoplastics, so a general comparison between them may be misleading. Rather, environmental impacts should be assessed according to the material's specific function (for example, minimum thickness to provide a particular strength) over the full life cycle.

> Refer to Table 5.4 for a description of the functional unit

Recovery and Disposal

Renewable thermoplastics can be disposed via a number of streams, including:
- composting
- landfill
- incineration
- material recycling.

Match Degradability with End-of-Life Process

Contrary to popular belief, not all renewable thermoplastics are designed to degrade. Their degradation behaviour in landfill and composting systems depends on a number of factors, including:
- type of renewable thermoplastic and type and content of processing aids, plasticisers and compatabilisers
- degradation environment, including temperature, moisture content, pH and microbial type, activity and content
- nature of the polymer bonds, including molecular weight, chain flexibility and crystallanity
- water uptake of the polymers (hydrophilicity)
- size and shape of the polymer. Polymers with a higher surface to volume ratio, such as shredded sheet, will degrade faster than those with a lower surface to volume ratio, such as. non-shredded thick sheet.

These degradation factors not only influence the rate and amount of degradation but also the type of degradation (anaerobic and/or aerobic), which results in different types of gaseous emissions.

Many renewable thermoplastic materials are compostable, although facilities for the recovery of compostable products are limited at the present time. Another option for the recovery of biodegradable packaging is a home composting system. However, not all biodegradable polymers will break down in a home composting system because of the different conditions involved, such as lower temperatures. If packaging is to be promoted to consumers as compostable in a home composting system, it needs to be certified as meeting a relevant and recognised standard (these are listed in Sect. 3.5.3).

These standards specify the requirements that a polymer needs to meet to ensure that it 'biodegrades' and that the breakdown products and the speed of degradation are compatible with a commercial composting process.

Although there is potential for biodegradable packaging to be collected through municipal organic waste collections, a number of issues need to be resolved first, including the compatibility of plastics with the organics recovery process (for example, whether or not it will break down in the required time without contaminating the end product), and labelling to advise consumers about correct disposal.

The recovery of renewable thermoplastics is currently limited because there are insufficient quantities to make it economically viable and it is difficult to separate

Table 6.8 Environmental impact drivers of materials

Packaging material	Environmental impact categories				
	Climate change	Energy	Land Use	Eutrophication	Water use
Aluminium	Electricity for smelting	Electricity for smelting	Mining	Electricity for smelting	Electricity for smelting
Steel	Pig iron production	Coking coal in pig iron production	Coking coal storage	Pig iron production	Pig iron production
Glass	Electricity for melting	Natural gas in melting	Secondary and tertiary packaging	Electricity for melting	Electricity for melting
Paper and board	Electricity for pulping	Growing and harvesting of the trees	Growing and harvesting of trees	Electricity for pulping	Pulping/electricity into pulping
Non-renewable thermoplastics	Polymerisation	Raw material extraction	Electricity	Polymerisation	Electricity
Renewable thermoplastics	Polymer processing/ growing and harvesting crop	Growing and harvesting crop	Growing and harvesting crop	Growing and harvesting crop	Growing and harvesting crop

these materials from the non-renewable thermoplastic polymer recycling stream. In addition, renewable thermoplastics can contaminate the recycling streams of non-renewable polymers.

Research is ongoing to improve sorting technologies and end-of-life treatment processes, including recycling. Feedstock ('chemical') recycling of PLA is currently being commercialised. This process will use a chemical process known as hydrolysis to break it down into its primary foundation, lactic acid, which can then be converted back into PLA resin.

Some renewable polymers dissolve in water within a designated temperature range then biodegrade in contact with micro-organisms.

6.8 Conclusions

Each packaging material exhibits it own unique properties and life cycle that generate particular environmental impacts at different stages of the supply chain and for different environmental indicators (see Table 6.8). It is important to understand and regularly review these complex processes to inform the packaging design strategy (see Chap. 2). The business case for packaging sustainability within individual companies and the relative importance of specific drivers such as corporate and brand positioning, supply chain requirements, solid waste, resource efficiency and climate change (see Chap. 1), may influence the selection of materials. It's also crucial to embed an assessment of the life cycle of materials within product and packaging design (see Chaps. 5 and 8).

References

1. Marsh K, Bugusu B (2007) Food packaging—Roles, materials, and environmental issues. J Food Sci 72(3):39–55
2. Weber CJ (2000) Biobased packaging materials for the food industry. status and perspectives. The Royal Veterinary and Agricultural University, Denmark
3. McDonough W, Braungart M (2002) Cradle to cradle: remaking the way we make things. North Point Press, New York
4. Grant T, James K, Lundie S, Sonneveld K (2001) Stage 2 Report for life cycle assessment for paper and packaging waste management scenarios in Victoria, for EcoRecycle Victoria. Centre for Design at RMIT University, Centre for Packaging Transportation and Storage at Victoria University and the Centre for Water and Waste Technology at University of New South Wales
5. Lewis H (2008) National Packaging Covenant Mid-term Review. Report to the National Packaging Covenant Council. Helen Lewis Research, Melbourne
6. WSA (2008) Fact sheet, Energy. Steel and energy. http://www.worldsteel.org/climatechange/files/6/Fact%20sheet_Energy.pdf (cited 27 April 2011)
7. Dhir RK, Limbachiya MC, Dyer TD (2001) Recycling and reuse of glass cullet. Thomas Telford Publishing, London
8. DEWHA (2010) National Waste Report 2010, Department of the Environment, Heritage and the Arts, Commonwealth of Australia, Canberra

9. WRAP (2008) Market situation report—September 2008. Realising the value of recovered glass: An update. Waste and Resources Action Programme, Banbury, Oxon
10. Owens-Illionois Inc. (2010) Lean and Green Brochure. http://www.o-i.com/uploadedFiles/O-I/NorthAmerica/News/NA_LeanAndGreen_Brochure.pdf (cited 28 January 2011)
11. Lewis H, Gertsakis J, Grant T, Morelli N, Sweatman A (2001) Design + environment: a global guide to designing greener goods. Greenleaf Publishing, Sheffield, UK
12. Hischier R (2004) Life cycle inventories of packagings and graphical papers. ecoinvent-report no. 11. Part III paper and board. 2004, Swiss Centre for Life Cycle Inventories, Dübendorf
13. Mander T (2000) PAC-IT: an introduction to packaging in New Zealand, teacher guide. Minister for the Environment and the Packaging Council of New Zealand, Wellington
14. Taylor PL (2005) In the market but not of it: fair trade coffee and forest stewardship council certification as market-based social change. World Dev 33(1):129–147
15. Pattberg P (2005) What role for private rule-making in global environmental governance? Analysing the Forest Stewardship Council (FSC). Int Environ Agreem Politics, Law Econ 5(2):175–189
16. The Paper Task Force (1995) Paper Task Force recommendations for purchasing and using environmentally preferable paper. Environmental Defence Fund, New York
17. Pedroso MC, Bastros AF (2006) Recycling of aseptic carton packages in Brazil: a case study of sustainable supply chain. In: 4th Global Conference on Sustainable Product Development and Life Cycle Engineering. Sao Carlos, Brazil
18. ISO (2008) ISO 15270: Plastics—guidelines for the recovery and recycling of plastics waste. International Standards Organisation (ISO), Geneva
19. Visy Recycling (2011) Plastics automated sort screen. http://www.visy.com.au/recycling/?id=271 (cited 18 March 2011)
20. Scheirs J (1998) Polymer recycling. Wiley, New York
21. PACIA (2009) 2009 National plastics recycling survey. Available from: http://www.pacia.org.au/Content/LearningRecyclingSurvey.aspx (cited 2 February 2011).
22. Imam S, Glenn G, Chiou BS, Shey J, Narayan R, Orts W (2008) Types, production and assessment of biobased food packaging materials. In: Chiellini E (ed.) Environmentally compatible food packaging. Woodhead Publishing Limited, Cambridge, England. p 29–62
23. ExcelPlas Australia, Centre for Design at RMIT University, and Nolan-ITU (2004), The impacts of degradable plastic bags in Australia. ExcelPlas Australia, Centre for Design at RMIT University and Nolan-ITU, Melbourne
24. Standards Australia (2006) AS 4736-2006, Biodegradable plastics—biodegradable plastics suitable for composting and other microbial treatment. Standards Australia, Sydney
25. PACIA (2007) Using degradable plastics in Australia. A product stewardship guide and commitment. Department of the Environment and Water Resources, Australian Government, Canberra
26. Cornell D (2007) Biopolymers in the existing postconsumer plastics recycling stream. J Environ Polym Degr 15(4):295–299
27. EPA U (2009) Municipal solid waste generation, recycling, and disposal in the United States detailed Tables and Figures for 2008. U.S. Environmental Protection Agency
28. EC (2010) Packaging waste, Data 2008 (updated 10 December 2010). Available from http://epp.eurostat.ec.europa.eu/portal/page/portal/waste/data/wastestreams/packaging_waste eurostat: (cited 28 January 2011)
29. Envirowise (2004) Sustainable design of electrical and electronic products to control costs and comply with legislation—GG427. Didcot, Oxfordshire

Chapter 7
Selecting and Applying Tools

Karli Verghese and Simon Lockrey

Abstract Many tools to support design for sustainability are now available. This chapter provides a general overview of tools commonly used in packaging design and examples of their application.

Contents

7.1	Introduction	253
7.2	Life Cycle Assessment Tools	253
	7.2.1 Life Cycle Map	262
	7.2.2 Sustainability Impact Matrix	262
	7.2.3 Life Cycle Assessment Software	265
	7.2.4 Packaging-Specific LCA-Based Design Tools	265
7.3	Packaging-Specific Rating And Ranking Tools	273
	7.3.1 The Eco-Design Indicator Tool	273
	7.3.2 Walmart Package Modelling (Including Sustainable Packaging Scorecard)	273
7.4	Packaging-Specific Design Guidelines	274
	7.4.1 Australian Sustainable Packaging Guidelines	275
	7.4.2 Sustainable Packaging Coalition's Design Guidelines for Sustainable Packaging	276
	7.4.3 The Guide to Evolving Packaging Design by WRAP	277
	7.4.4 Envirowise Guides	278

K. Verghese (✉) · S. Lockrey
Centre for Design, RMIT University, GPO Box 2476, Melbourne, VIC 3001, Australia
e-mail: Karli.Verghese@rmit.edu.au

S. Lockrey
e-mail: Simon.Lockrey@rmit.edu.au

7.5 Case Study Databases 279
    7.5.1 The International Packaging Study Database 279
    7.5.2 Sustainable Packaging Coalition Design Library 279
7.6 Conclusions 281
References 282

Figures

Figure 7.1	Key steps/questions to pose when constructing a life cycle map	263
Figure 7.2	Screenshots of characterisation graph from SimaPro	266
Figure 7.3	Screenshots of graphs generated by GaBi	267
Figure 7.4	Comparison of packaging system formats in bar graph output of PIQET	269
Figure 7.5	Comparison of packaging system formats in spider graph output of PIQET	269
Figure 7.6	Screenshot of COMPASS component impact contribution analysis	270
Figure 7.7	Screenshot of COMPASS comparative LCA	271
Figure 7.8	Screenshot of COMPASS packaging attributes and material health	271
Figure 7.9	Screenshot of Toyota's EPIC illustrating the system boundary	272
Figure 7.10	Toyota's EPIC illustrating data distillation framework	272
Figure 7.11	The 12 strategies in the Australian Sustainable Packaging Guidelines	275
Figure 7.12	Screenshot of biscuit case study on International Packaging Study website	280
Figure 7.13	Screenshot of a case study SoilWrap - on the SPC Design Library website	281

Tables

Table 7.1	Packaging sustainability tools	254
Table 7.2	Example applications of packaging sustainability tools	257
Table 7.3	Examples of metrics reported in various tools	260
Table 7.4	Databases/sources used in life cycle tools	262
Table 7.5	Impact assessment matrix for the production of a polystyrene cup in Australia	264
Table 7.6	Impact assessment methods used in LCA software programs	266
Table 7.7	Life cycle assessment application in packaging-specific tools	268
Table 7.8	Case study: using the Australian Sustainable Packaging Guidelines	276
Table 7.9	Sustainable Packaging Coalition's design objectives and strategies for sustainable packaging	277

7 Selecting and Applying Tools

7.1 Introduction

Many tools to support design for sustainability are now available. This chapter provides a general overview of tools commonly used in packaging design and examples of their application. This is not a complete list but a snapshot in time, as the availability, scope, nature and format of tools is constantly changing.

These tools vary in a number of ways, including their:
- format: ranging from design guidelines and checklists through to rating tools and streamlined LCA tools (see Table 7.1)
- application: for example strategy development, innovation or product and packaging design (see Table 7.2)
- breadth and depth of indicators and metrics (see Table 7.3)
- scope and use of life cycle assessment (LCA): functional unit, system boundary, tool design and data sources and so on (see Table 7.4)
- mode of delivery: internal documents, interactive web tools and so on.

7.2 Life Cycle Assessment Tools

LCA tools are used in a range of formats from simple-to-use life cycle maps (see Sect. 7.2.1) to comprehensive software for the experienced LCA practitioner. Increasingly, streamlined tools are used to enable life cycle thinking to be applied practically by non-LCA practitioners such as packaging designers and meet the requirements of the day-to-day business environment.

> See Chap. 5 to learn more about life cycle assessment

Life cycle tools rely on access to LCA data that is normally housed within LCA software programs such as SimaPro and GaBi. These databases and individual data entries can be very detailed and have been compiled from various data sources such as LCA reports, published articles and direct calculations.

Packaging-specific LCA design tools such as the Packaging Impact Quick Evaluation Tool (PIQET) and the Comparative Packaging Assessment (COMPASS) draw upon aggregated datasets from life cycle inventories and some primary data. These design tools rely on calculations of material life cycle impacts undertaken in SimaPro and then aggregated impact assessment characterisation values imported into the databases of PIQET and COMPASS. Table 7.4 summarises the databases within SimaPro, GaBi, PIQET and COMPASS.

Table 7.1 Packaging sustainability tools

Format	Example	Data required	Data disseminated	Access	Region
Life cycle assessment tools	COMPASS	Data for a closed-loop system that takes into account raw resource, conversion, and end-of-life treatment and fate of the materials after use	Results include component contribution, comparative life cycle assessment, and packaging attributes that are important to the designer.	Online software application (pay per licence)	US
	GaBi	All primary elements of packaging life cycle, plus any secondary data supported by in-built databases.	Detailed and flexible quantitative life cycle environmental impacts of packaging systems, full comparative and sensitivity capability	Stand-alone software package (pay per licence)	Global
	Life cycle map	Details of materials, processes, transport, and end-of-life of package.	'High level' life cycle process tree of packaging system	In-house method, cross-functional team	Global
	PIQET	Data from all stages of the life cycle for a complete packaging system and all its components, default data included	Rapid streamlined environmental impact assessment, including comparative, tabulated, graph- and inventory-based reporting.	Online software application (pay per licence)	Global
	SimaPro	All primary elements of packaging life cycle, plus any secondary data supported by in-built databases.	Detailed and flexible quantitative life cycle environmental impacts of packaging systems, full comparative and sensitivity capability	Stand-alone software package (pay per licence)	Global
	Toyota Environmental Packaging Impact Calculator (EPIC)	Incorporates field-based input area where users enter relevant information for each package; i.e., weight, direct or standard shipping, shipping mode, material type, etc.	Toyota packaging environmental indicator comparison, life cycle cost comparison, life cycle stage analysis and life cycle inventory	In-house customised software package	Global

(continued)

7 Selecting and Applying Tools

Table 7.1 (continued)

Format	Example	Data required	Data disseminated	Access	Region
Packaging-specific rating and ranking tools	The Eco-Design Indicator Tool (EDIT)	Packaging materials (and amount), geographical location of material sourcing, transport modes, and volume of product	Generates graphical and tabular results for a range of environmental indicators for up to six different designs comparatively	Web portal	UK
	UK Packaging Benchmark Tool	Packaging product category, material, and type (to search database)	Provides information on the weight of packaging used for food and drink products found on the UK supermarket shelf, for benchmarking	Web portal	UK
	Walmart Package Modelling	Packaging material type, weight, transport distance and efficiency.	Life cycle-based environmental impact metrics, including comparative analysis and report exporting	Stand-alone software package (pay per licence)	Global
Packaging-specific codes of practice and design guidelines	Australian Sustainable Packaging Guidelines	Answers to questions on design, marketing, materials, processes, supply chain, end-of-life, and function of package	Assessment of the environmental impact and credentials of packaging designs against criteria	Paper-based (electronic if desired), online through web	Australia
	Design guidelines for sustainable packaging	Answers to criteria questions on design, marketing, materials, processes, supply chain, end-of-life, and function of package	Designing for sustainability background and design strategies based on packaging environmental impact credentials and criteria	Paper-based (electronic if desired), online through web	USA
	Envirowise PackGuide: a guide to packaging eco-design	Packaging specifications and requirements for comparison with data in guidelines	Design-focused information on materials, regulatory data, and links to further resources	Online PDF (with registration)	UK/Europe
	Envirowise guide—Packaging design for the environment: reducing costs and quantities	Packaging specifications and requirements for comparison with data in guidelines	General packaging 'how to' information, regulatory data, materials figures, and further resources links	Online PDF (with registration)	UK/Europe

(continued)

Table 7.1 (continued)

Format	Example	Data required	Data disseminated	Access	Region
	The guide to evolving packaging design	Packaging specifications and requirements for comparison with data in guidelines	Designing for sustainability data on the consumer's perspective, legal issues, brand, innovation, tools, techniques, material considerations, and further links (and glossary)	Online resource with text, video, links and PDFs	UK/Europe
Case study databases	Sustainable Packaging Coalition design library	General interest for benchmarking and best practice	Case studies and images of packaging designed for a lower environmental impact	Web portal	Global
	The International Packaging Study image bank	General interest for benchmarking and best practice	Case studies and images of packaging designed for a lower environmental impact	Web portal	Global

All tools are listed alphabetically within each separate tool format

Table 7.2 Example applications of packaging sustainability tools

Business process	Business activity	Packaging decision context	Life cycle assessment tools	Packaging specific LCA based tools	Packaging specific rating and ranking tools	Packaging specific codes of practice and design guidelines	Case study databases
Corresponding section in this chapter			7.2	7.2.4	7.3	7.4	7.5
Developing strategy	Strategic and operational planning	Identify, confirm and review:					
		• competitive position of products and packaging from a sustainable development perspective	√	√	√	√	√
		• corporate, brand and product positioning strategies	√	√	√	√	√
		• packaging's role in meeting sustainability goals	√	√	√	√	√
		• packaging design sustainability goals and targets	√	√	√	√	√
Designing packaging	Product and packaging innovation	Identify new and improved:					
		• product-packaging systems	√	√	√	√	√
		• packaging systems.					
	Product and packaging improvement	Develop, assess and optimise:					
		• product-packaging systems	√	√		√	
		• packaging systems.					
	Regulatory compliance	Comply with packaging-specific regulatory requirements		√		√	
	Supply chain compliance	Comply with customer specific sustainable development requirements			√	√	
	Label design	Inform the selection and use of logos and claims	√			√	

(continued)

Table 7.2 (continued)

Business process	Business activity	Packaging decision context	Life cycle assessment tools	Packaging specific LCA based tools	Packaging specific rating and ranking tools	Packaging specific codes of practice and design guidelines	Case study databases
Marketing and communicating sustainability	Brand management	Scientifically justify, defend or challenge environmental claims	✓				
		Promote environmental credentials of the company, its products and packaging	✓				
	Corporate reporting	Report packaging sustainability goals, challenges and achievements		✓			
Implementing strategy	Capacity building	Develop organisational understanding and commitment to sustainable development	✓	✓			
	Packaging design	Embed LCA into packaging design	✓	✓	✓	✓	
		Engage all business units in packaging design for sustainability		✓	✓	✓ ✓	✓
		Communicate sustainability impacts of packaging		✓	✓		
		Scientifically justify, defend or challenge packaging design decisions	✓	✓			
		Incorporate sustainability requirements into packaging specifications	✓	✓	✓	✓	

(continued)

7 Selecting and Applying Tools

Table 7.2 (continued)

| Business process | Business activity | Packaging decision context | Tool category |||||
			Life cycle assessment tools	Packaging specific LCA based tools	Packaging specific rating and ranking tools	Packaging specific codes of practice and design guidelines	Case study databases
	Procurement	Incorporate sustainability requirements into packaging specifications	√	√	√	√	
		Incorporate sustainability requirements into: • purchasing contracts • capital equipment approval processes.	√	√	√	√	

(√) indicates that the sub-tool category can be used for the identified packaging decision context

Table 7.3 Examples of metrics reported in various tools

Metric	Metric type	SimaPro	GaBi	PIQET	Compass	EDIT	Walmart package modelling
Climate change	LCIA	√	√	√	√	√[a]	√
Cumulative energy demand	LCIA	√	√	√	√		
Minerals and fuels	LCIA	√	√	√			
Photochemical oxidation	LCIA	√	√		√		
Eutrophication	LCIA	√	√	√	√		
Land use	LCIA	√	√	√	√		
Water use	LCIA	√	√	√	√	√[b]	
Solid waste	LCIA	√	√	√	√		
Eco-indicator 99	LCIA m	√	√		√		
Solid waste recovered	LCIA	√	√		√		√
Fossil fuel use	LCIA	√	√	√	√		
Biotic resources	LCIA	√	√		√		
Mineral use	LCIA	√	√		√		
Clean production—human Impact	LCIA	√	√		√		
Clean production—aquatic toxicity	LCIA	√	√		√		
Virgin material content	SM	√[c]	√[c]				
Recycled content	SM	√[c]	√[c]	√	√		√
Material sourcing	SM						
Material health (potential hazard identification)	SM	√[c]	√[c]				
Packaging weight per unit of product	SM	√[c]	√[c]	√	√	√	
Packaging volume per unit of product	SM	√[c]	√[c]				
Product/packaging weight ratio	SM	√[c]	√[c]				
Percentage of product remaining in packaging	SM						
Packaging to landfill	SM	√[c]	√[c]	√	√	√	
Packaging to recycling	SM	√[c]	√[c]		√	√	

(continued)

7 Selecting and Applying Tools   261

Table 7.3 (continued)

Metric	Metric type	SimaPro	GaBi	PIQET	Compass	EDIT	Walmart package modelling
Packaging to composting	SM	✓[c]	✓[c]	✓			
Packaging to incineration	SM	✓[c]	✓[c]	✓		✓	
Packaging to waste–to-energy	SM	✓[c]	✓[c]	✓			
Logistics efficiency	SM	✓[c]	✓[c]				✓
Environmentally-friendly materials	SM						✓
Transportation impacts	SM	✓[c]	✓[c]				✓
Recovery value	SM						✓
Renewable energy used	SM	✓[c]	✓[c]				✓
Sustainable innovation	SM						✓

This table is indicative only as metric definitions and calculation methodologies vary and tools are updated from time to time
LCIA = life cycle impact assessment; LCIA m = life cycle impact assessment method; SM = sustainability metric
[a] Reported as total carbon embedded CO_2e in packaging per product per use (kg)
[b] Reported as water embodied in packaging materials (per use)
[c] Can be determined by practitioner, but not a native metric

Table 7.4 Databases/sources used in life cycle tools

Databases/sources	LCA software		Packaging-specific LCA design tools	
	SimaPro	Gabi	PIQET	COMPASS
Ecoinvent v2	√	√	√	√
European life cycle database (ELCD)		√		
Danish input output database	√			
Dutch input output database	√			
Australian LCA database			√	
US LCI database	√	√		√
US input output database	√			
Japanese input output database	√			
IVAM database	√		√	
IDEAMAT			√	
ETH-ESU 96			√	
BUWAL 250			√	
LCA food database	√			
Industry data	√	√		

Source:
SimaPro: http://www.pre.nl/simapro/inventory_databases.htm
GaBi: http://www.gabi-software.com/australia/databases
PIQET: http://www.sustainablepack.org
COMPASS: http://www.design-compass.org

7.2.1 Life Cycle Map

Life cycle maps, introduced in Chap. 5, provide a visual representation of the steps required to source and produce the product-packaging system. They are used to:
- inform the sustainability strategy, goals and targets (see Chap. 1)
- confirm the role of packaging in achieving sustainability goals
- confirm the scope and goal of an LCA
- identify 'hotspots' and priority areas in the supply chain.

There are a number of ways a life cycle map can be compiled, and Fig. 7.1 presents a suggested process by which to construct one. The map should be constructed by a cross-functional and knowledgeable team.

7.2.2 Sustainability Impact Matrix

A sustainability impact matrix provides a visual snapshot of the sustainability profile of a product and/or its packaging. It can be used to inform business strategies, including marketing and communication, product and packaging development and procurement.

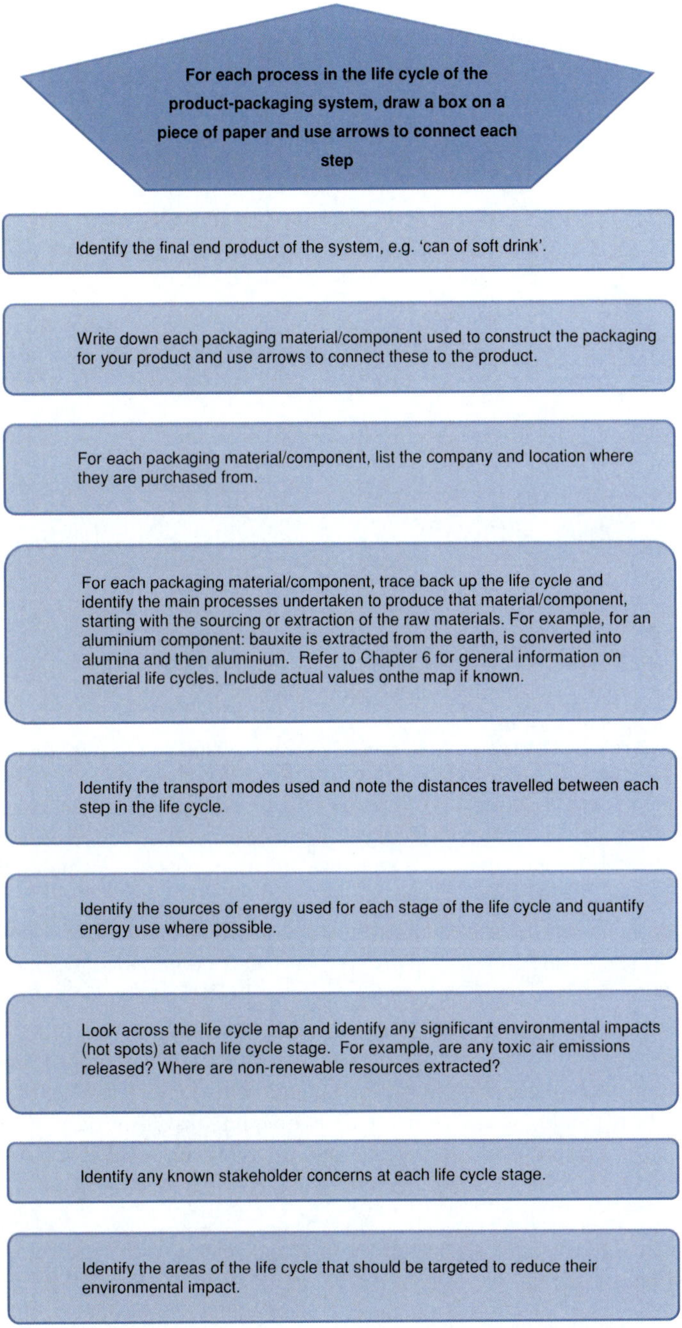

7 Selecting and Applying Tools

> **For each process in the life cycle of the product-packaging system, draw a box on a piece of paper and use arrows to connect each step**

- Identify the final end product of the system, e.g. 'can of soft drink'.

- Write down each packaging material/component used to construct the packaging for your product and use arrows to connect these to the product.

- For each packaging material/component, list the company and location where they are purchased from.

- For each packaging material/component, trace back up the life cycle and identify the main processes undertaken to produce that material/component, starting with the sourcing or extraction of the raw materials. For example, for an aluminium component: bauxite is extracted from the earth, is converted into alumina and then aluminium. Refer to Chapter 6 for general information on material life cycles. Include actual values on the map if known.

- Identify the transport modes used and note the distances travelled between each step in the life cycle.

- Identify the sources of energy used for each stage of the life cycle and quantify energy use where possible.

- Look across the life cycle map and identify any significant environmental impacts (hot spots) at each life cycle stage. For example, are any toxic air emissions released? Where are non-renewable resources extracted?

- Identify any known stakeholder concerns at each life cycle stage.

- Identify the areas of the life cycle that should be targeted to reduce their environmental impact.

Fig. 7.1 Key steps/questions to pose when constructing a life cycle map

Table 7.5 Impact assessment matrix for the production of a polystyrene cup in Australia

	Resource depletion	Global warming	Smog	Acidification	Eutrophication	Toxic waste	Biodiversity reduction
Production of basic materials	3[a]	3[b]	2[c]	1[d]	2[e]	2	2[f]
Manufacturing	0	0	2[g]	1	1	1	0
Distribution	1	1	1	1	0	0	0
Product use	-1[h]	-1[h]	0	0	0	0	0
End of product life	0	0	0	0	0	0	1[i]
TOTAL	3	3	5	3	3	3	3

Source: Lewis et al. [1, p. 51]
[a] Oil resources limited in the long term
[b] Methane emissions from venting in oil production
[c] Non-methane volatile organic compounds from venting and flaring
[d] Low sulphur oils used in Australia
[e] Nitrous oxide emissions from energy use
[f] Oil pollution
[g] Emission of propene when moulding the cup
[h] Possible benefits in use (relative to other cups with less insulating capacity)
[i] Impact of landfill and littering

The matrix has two-dimensions:
- environmental or social concern along one axis
- life cycle stage along the other.

Each element of the matrix (for example energy use in manufacture) is assessed for its relative impact on each area of social or environmental concern. Relative impacts may be scored numerically (see Table 7.5), qualitatively (see Sect. 3.4.4) or visually in different colours. The more informed the assessment, the more appropriate it is that numerical data be used.

An example of an environmental impact assessment matrix for a polystyrene coffee cup is shown in Table 7.5. This type of matrix could be adapted to meet the needs of the user; for example, by adapting the environmental indicators and life cycle stages. Impacts are ranked from 0 (no impact) to 4 (serious impact), and a negative number indicates a potential positive impact [1]. The matrix for the cup indicates that the highest impacts are associated with the production of the basic raw material, expanded polystyrene. This is manufactured from a non-renewable resource (oil), and the production process emits methane, which is a greenhouse gas. Other pollutants at this stage include volatile organic compounds and nitrous oxides.

The impact matrix can also be applied to social sustainability indicators such as health and equity [2, pp. 58–60]. In the example shown in Chap. 3 (Table 3.10), life cycle stages are listed along the horizontal axis and socio-ecological indicators on the vertical axis. Colour coding is used to indicate areas of 'low impact', 'medium impact' and 'high impact'.

7 Selecting and Applying Tools

An impact matrix can be developed using information from:
- a qualitative life cycle mapping exercise (Sect. 7.2.1)
- a quantitative LCA (Sects. 7.2.3 and 7.2.4)
- a combination of quantified LCA data and qualitative research.

A ranking exercise using either numbers or colours relies on the knowledge and experience of the person or team undertaking it, so it is important to involve people with different expertise [1].

7.2.3 Life Cycle Assessment Software

LCA software programs are specialised programs designed for experienced LCA practitioners and should be used when complex and detailed modelling and/or accreditation to the ISO 14040 series is required; for example, to support specific environmental claims.

⇒ More information on LCA methodology see Chap. 5

For the experienced user, these programs provide the maximum flexibility to control all aspects of the LCA including functional unit, system boundary, datasets and impact assessment methods. They also provide comprehensive LCA metrics and sensitivity analysis capability.

The time taken to conduct assessments can be weeks, months or years depending on the scope of the project. These programs are often used to create background data for other tools for designing for sustainability, including streamlined packaging-specific LCA design tools such as PIQET.

There are a number of programs available; SimaPro (see http://www.pre.nl/simapro) and GaBi (http://www.gabi-software.com) are the most widely used. Typical life cycle inventory databases and inventory impact assessment methods accessed with these programs are summarised in Tables 7.4 and 7.6. Users are also able to incorporate their own data.

Data entry is required for each stage of the product life cycle based on identification and quantification of the various input and output material flows. The programs provide interactive screens for the user to assist in the modelling process, and results are generally displayed graphically (see Fig. 7.2—SimaPro, and Fig. 7.3—GaBi) or tabulated, and are exportable (for example, csv files to Excel, jpg image files).

7.2.4 Packaging-Specific LCA-Based Design Tools

Packaging-specific LCA-based design tools have emerged to enable life cycle thinking to be integrated into the packaging design process. These overcome two of the main obstacles to using LCA software programs; that is, the time involved and the fact that non-LCA practitioners find them too complex.

Table 7.6 Impact assessment methods used in LCA software programs

Impact assessment method	SimaPro	GaBi
ReCiPe	✓	
Eco-indicator 95	✓	✓
Eco-indicator 99	✓	✓
Ecopoints 97	✓	
CML 92	✓	✓
CML 1996		
CML 2 (2000)	✓	
CML 2001		✓
CML 2007		✓
EDIP/UMIP	✓	
EPS 2000	✓	
Impact 2002+	✓	✓
TRACI	✓	✓
EPD method	✓	
EDIP		✓
Cumulative energy demand	✓	✓
IPCC greenhouse gas emissions	✓	✓
Ecological scarcity method (UBP)		✓
USEtox		✓
Custom		✓

Source:
SimaPro: http://www.pre.nl/simapro/impact_assessment_methods.htm
GaBi: http://www.gabi-software.com/australia/software/gabi-4/functionalities/results-and-interpretation

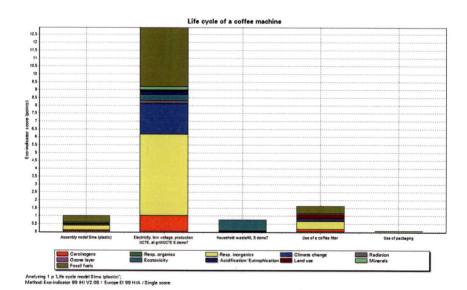

Fig. 7.2 Screenshots of characterisation graph from SimaPro. *Source*: PRe Consultants [3]

7 Selecting and Applying Tools

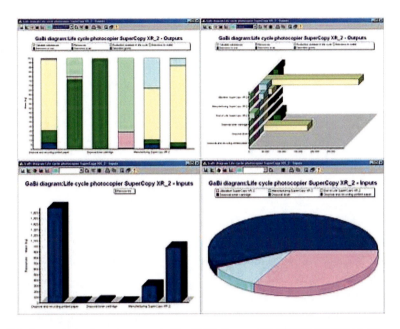

Fig. 7.3 Screenshots of graphs generated by GaBi. *Source*: PE INTERNATIONAL [4]

To enable a streamlined application of LCA, these tools require standardisation of approaches and assumptions that must be informed by comprehensive LCA modelling in LCA software programs. The tools generally include:
- other packaging-specific sustainability metrics that complement the life cycle approach or align with specific reporting requirements
- the ability to report design decisions against corporate sustainable development goals.

Their use can be integrated into all stages of the product development process to align design strategies with sustainable development goals and facilitate innovation. They provide the ability to model the sourcing, production, manufacturing and use of materials and processes across a range of different geographical regions. The datasets are region-specific (Australian, United States, Canada and so on) for the energy grid, mix of materials and processes and end-of-life material recovery, such as recycling, because these are different in each region. This enables users to model different scenarios for their product-packaging systems and maximise efficiency in material design (see Chap. 2) aligned with regulatory requirements (see Chap. 4).

Examples of packaging-specific LCA-based design tools are the PIQET (see http://www.sustainablepack.org) and the COMPASS tool (see http://www.design-compass.org). Both of these are web-accessed, project-based (different design scenarios are evaluated for each project), and use graphics and tables to report the results that can be used to inform the designer about strategies to improve their packaging designs. All results are 'multi-metric' instead of a single score to allow the

Table 7.7 Life cycle assessment application in packaging-specific tools

	PIQET	COMPASS
System boundary	Cradle-to-grave	Cradle-to-grave
Life cycle stages included:		
• raw material acquisition	√	√
• packaging material production	√	√
• packaging material conversion	√	√
• transport of packaging to filler	√	
• filling operations	√	
• transport of packed product to retailer	√	Planned for 2010
• end-of-life waste management	√	√
– landfill	√	√
– recycling	√	√
– composting	√	
– incineration	√	
– waste-to-energy	√	
Functional unit	kg of impact per kg of product on pallet	User defines the base unit and capacity of the packaging component
Accounting for recycling	Adjustable allocations by the user	Predefined recycling rates
Number of formats compared	3	4

user to establish and track designs against metrics that are relevant to their business's sustainable development goals. There are differences, however, in their application of LCA (see Table 7.7) and metrics (see Table 7.3), data sources (see Table 7.4) and reporting. Proprietary tools of this type are also in use in some businesses; for example, the Toyota Environmental Packaging Impact Calculator (EPIC).

Packaging Impact Quick Evaluation Tool

PIQET models the complete packaging system required to deliver a product from the filling location to the retailer or downstream customer. It can be used to inform the sustainability strategy and targets and highlight life cycle hot spots [5]. Once a packaging system has been defined within PIQET it can be easily replicated to allow any changes in the design or life cycle to be evaluated, providing immediate feedback for the design strategy and/or data collection requirements.

Graphical and tabular results are generated in different reporting formats; firstly to aid the designer/packaging technologist to understand the relationship between the design and its impact on specific metrics (see Fig. 7.4), and secondly to assist the designer/packaging technologist to communicate their recommendations to others involved in the decision-making process (see Fig. 7.5). PIQET relies on an extensive embedded life cycle inventory database of which users can override all, with the exception of the materials database.

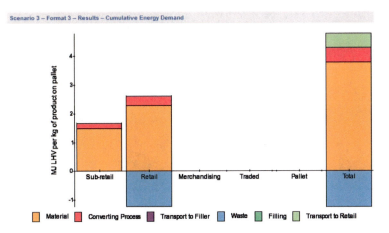

Fig. 7.4 Comparison of packaging system formats in bar graph output of PIQET. *Source*: Verghese et al. [5]

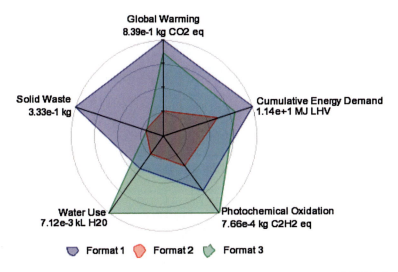

Fig. 7.5 Comparison of packaging system formats in spider graph output of PIQET. *Source*: Verghese et al. [5]

Comparative Packaging Assessment (COMPASS)

COMPASS enables the comparison of different packaging designs and is a guidance tool intended to facilitate optimal material selection [6]. For each project, the user specifies the base packaging unit and capacity that determines the functional unit. Environmental profiles of primary and secondary packages can be assessed

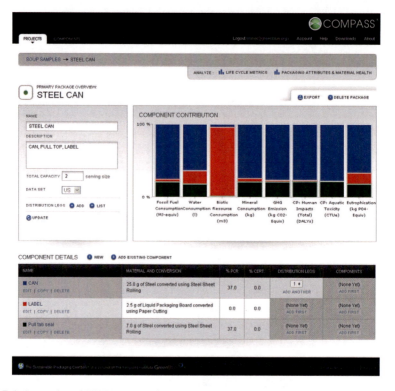

Fig. 7.6 Screenshot of COMPASS component impact contribution analysis. *Source*: Green Blue [6]

independently or in combination, as comparative assessments can be performed for up to four scenarios. Three types of analysis are undertaken and reported:

- the relationship or contribution of each component to the package on a 100% scale for each metric (see Fig. 7.6)
- a comparison of the life cycle indicators for up to four packaging scenarios (see Fig. 7.7)
- a comparison of packaging sustainability metrics (see Fig. 7.8) including material health (identifying if the material is a carcinogen; a reproductive toxicant; or a persistent, bioaccumulative or toxic substance).

Toyota Motor Sales' Environmental Packaging Impact Calculator

Toyota and its suppliers use Toyota's EPIC to reduce the environmental impacts and costs of packaging used for accessories and parts [7, 8]. Environmental metrics align with specific problems that Toyota aims to improve: air pollution, global warming, human health and toxicity, and resource depletion.

GaBi was used to construct the inventories and develop the tool, and the life cycle model is based on Toyota's specific packaging system life cycle (see Fig. 7.9). The tool requires data entry to define the life cycle, reports a number of environmental and costing metrics (see Fig. 7.10), and allows easy

7 Selecting and Applying Tools

Fig. 7.7 Screenshot of COMPASS comparative LCA. *Source*: Green Blue [6]

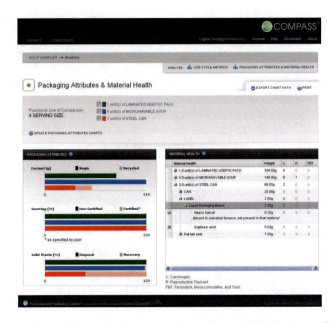

Fig. 7.8 Screenshot of COMPASS packaging attributes and material health. *Source*: Green Blue [6]

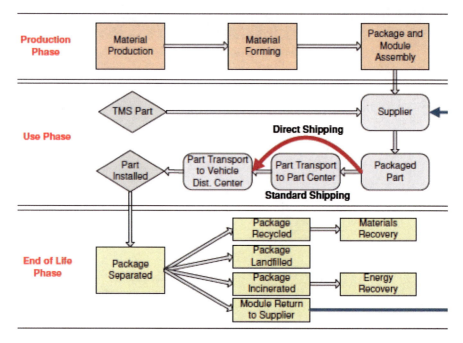

Fig. 7.9 Screenshot of Toyota's EPIC illustrating the system boundary. *Source*: Early et al. [7]

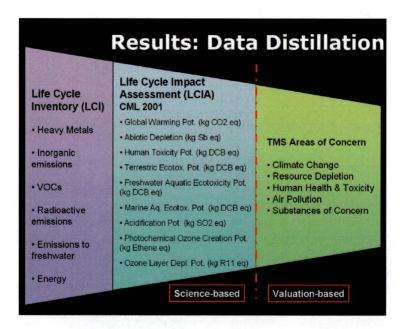

Fig. 7.10 Toyota's EPIC illustrating data distillation framework. *Source*: Early et al. [8]

identification of the impact of material choice, size/volume, logistics options, and packaging reuse. Embedded data for particular elements of the packaging life cycle are constant (for example, shipping module weight, material types recycled at a facility) and functions of inputs (for example, percentage of non-recycled waste that is incinerated, average transport costs).

7.3 Packaging-Specific Rating And Ranking Tools

7.3.1 The Eco-Design Indicator Tool

The Eco-Design Indicator Tool (EDIT) is a web-based eco-design tool that reports the environmental profile of a product and its packaging against seven metrics (indicators: see Table 7.3). (See http://www.envirowise.gov.uk/uk/Our-Services/Tools/EDIT-The-Eco-Design-Indicator-Tool.html to view EDIT) [9]. The predecessor to EDIT was the Pack-In tool, which only considered packaging.

Users need to input information on the types and weights of materials in their packaging components, the recycled content of the materials, the location of the material supplier, and the likely end-of-life destination for the packaging components. The tool includes drop-down menu selection for material sourcing, distribution and material converting as well as product distribution. One of five end-of-life waste management options can be selected: 'recycled', 'landfill', 'energy from waste', 'reuse' and 'compost'.

Using the input data and default background data embedded in the tool, graphical and tabular reports are generated for up to six different packaging designs. Automated general feedback is also provided on design for recyclability, recycled content and carbon footprint. There is no information accompanying EDIT that describes data sources or calculation methodologies. It is also not clear where the life cycle calculations begin for materials; for example, at the extraction or material conversion stage.

Indicator tools like EDIT provide the user with a basic introduction to the concept of life cycle impacts. Learning from these tools should then be supplemented by the more rigorous outputs of packaging-specific LCA-based design tools like PIQET and COMPASS (see Sect. 7.2.4) that model the packaging life cycle across different geographical regions and in combination with packaging-specific indicators. If the aim is to make marketing claims, then investment in full LCA modelling in LCA-specific software (see Sect. 7.2.3) is required.

7.3.2 Walmart Package Modelling (Including Sustainable Packaging Scorecard)

Walmart initially developed and released an online Packaging Scorecard in 2008 to gather information on suppliers' packaging and rank it using a number of weighted environmental metrics [10–12]. As a key part of its sustainability strategy, Walmart subsequently released Package Modelling as a web portal and stand-alone software package to enable suppliers to model different packaging designs and identify environmental improvement opportunities (see http://www.scorecardmodeling.com). The tool supports Walmart's goal to reduce packaging by 5% by 2013 (from the 2008 baseline). It is also available in a range of languages including French and Spanish.

The Packaging Scorecard focuses on input data associated with material selection (type), use (weight) and transport efficiency (the distance the packaging component has travelled and efficient use of space in transport vehicles). The inputs are used to calculate nine environmental indicators that each contribute to an overall rating of the packaging system (see Table 7.3). The input metrics and data have been reviewed by a range of organisations including the US Environmental Protection Agency, GreenBlue, California Integrated Waste Management Board, Rochester Institute of Technology, Michigan State and Clemson University, The Fibre Box Association, and the American Chemistry Council.

This is a very streamlined process compared to a full LCA, and the range of impact categories is therefore limited. Results from the modelling can be uploaded into Walmart's Sustainability Scorecard to enable comparisons with other options, suppliers and packaging formats. They can be printed as reports or exported to Microsoft Excel, and the capability also exists for importing data from Excel.

7.4 Packaging-Specific Design Guidelines

Many packaging industry associations and regulatory agencies have developed guidelines and checklists to inform packaging design. These provide information about specific issues to consider such as material source reduction, communicating with the consumer, and end-of-life waste management. They describe ways to pursue design strategies, such as those discussed in detail in Chap. 2. These can be adapted by individual organisations to align with the business's sustainable development goals and relevant business processes such as procurement and compliance reporting.

> ⇒ More information on regulations can be found in Chap. 4

Checklists and guidelines are relatively simple and easy to use. They often ask questions that prompt the user to capture qualitative and quantitative information relating to the design decision. Responses are typically qualitative and subjective, which can limit their role in encouraging and supporting innovation, strategy development, benchmarking or continuous improvement. Some examples are provided in the following sections.

> ⇒ More information on packaging for sustainability strategies go to Chap. 2

7 Selecting and Applying Tools

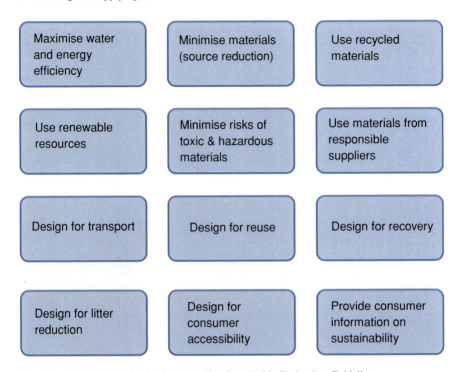

Fig. 7.11 The 12 strategies in the Australian Sustainable Packaging Guidelines

7.4.1 Australian Sustainable Packaging Guidelines

The Australian Sustainable Packaging Guidelines form part of the Australian Packaging Covenant [13] and contain 12 design strategies to consider when designing a new packaging system or updating an existing one (see Fig. 7.11). For each strategy there is a brief explanation of its meaning and purpose, and questions are provided to guide its application. In a similar manner to the packaging sustainability framework described in Chap. 2, the guidelines are based on four design principles: fit-for-purpose, resource efficiency, low-impact materials and resource recovery. A data collection template is provided to support the use of the guidelines.

The guidelines should be used early in the design process to encourage exploration of different options for materials and packaging system design.

Signatories to the Australian Packaging Covenant are required to use the guidelines to review all new and existing consumer packaging (primary/retail packaging and its associated distribution packaging). Table 7.8 provides an example of how they could be used.

Table 7.8 Case study: using the Australian Sustainable Packaging Guidelines

Example questions	Example responses
Maximise water and energy efficiency	
Have you considered using renewable energy for manufacturing, for example, by purchasing a percentage of GreenPower (Australian government accreditation program for renewable energy)?	• *We have investigated the possibility of purchasing GreenPower and are confirming contracts with energy company to supply 45% GreenPower to our manufacturing sites.*
Minimise materials (source reduction)	
Does the design of the package allow the product to be completely dispensed, i.e., to avoid product wastage?	• *Approximately 3% of the product cannot be dispensed from our current bottle design. The design team is currently reviewing the shape of the bottle neck to improve dispensing and reduce product wastage.*
Use renewable materials	
Are the renewable raw materials grown and harvested using sustainable farming or forestry practices?	• *The paper fibres for the corrugated boxboard are sourced from Forest Stewardship Council certified forests.* ⇒ Refer to Chap. 6 for information on the life cycle impacts of renewable and non-renewable materials
Minimise risks associated with potentially toxic and hazardous materials	
Have you applied conventional and conservative risk management principles in the selection of substances for packaging applications (for example, any inks, pigments, stabilisers and adhesives)?	• *Yes, we have included non-toxic inks in our packaging specifications. We are also investigating alternatives to Bisphenol A (BPA) in the epoxy lining of our metal cans.* ⇒ Refer to Chap. 2 for information on designing to minimise health and safety risks
Design for recovery	
How many materials are being used in this package? If more than one material is used, are the different materials compatible in the recycling process?	• *We are currently using a PET bottle with a PVC shrink-wrap label. We are investigating alternative film materials as both materials have similar densities they will not separate effectively in the recycling process.* ⇒ Refer to Chap. 2 for information on designing for recyclability

Source: Example questions from APCC [13]

7.4.2 Sustainable Packaging Coalition's Design Guidelines for Sustainable Packaging

In 2006, the Sustainable Packaging Coalition released their Design Guidelines for Sustainable Packaging [14] (see http://www.sustainablepackaging.org/). These provide a good general introduction to sustainability and the cradle-to-cradle design philosophy, and a comprehensive overview of possible design strategies. There is a section on quality that encourages new thinking about conventional design objectives such as cost, technical performance, regulatory compliance and appearance.

7 Selecting and Applying Tools

Table 7.9 Sustainable Packaging Coalition's design objectives and strategies for sustainable packaging

Design objectives	Strategies	Example of how to do it
Optimise resources Does the design optimise materials and energy?	Practice source reduction	Use materials with lower embodied energy
	Use recycled content	Consider the availability of recycled material in your area and technical feasibility
	Design for transport	Consider truck and container dimensions for most efficient packing
Materials health Are all materials healthy for people and the environment?	Know the chemistry of your package	Ask suppliers for information on the composition of each material you use
Responsible sourcing Has the material been produced and delivered responsibly?	Design with environmental best practice	Set goals for continuous improvement beyond compliance
	Design with fair labour and trade practices	Ask suppliers about their labour and trade practices
	Design with renewable virgin materials from sustainably managed sources	Require certification for sustainable management
Resource recovery Where will the materials go after use?	Design for reuse	Confirm whether it will work for your package requirements
	Design for recycling	Use single materials or design for disassembly
	Design for composting	Confirm that all components, additives and inks are compatible qwith the composting process

Sources: Sustainable Packaging Coalition [14, p. 22], Jedlicka [15, pp. 193–196]

Like the packaging sustainability framework presented in Chap. 2, four design objectives and ten associated design strategies are proposed that integrate sustainable development and expand conventional design objectives (see Table 7.9). Each strategy is supported by practical information outlining why the strategy is important, which life cycle stage it will influence and guidance on how to implement the strategy at the level of process or packaging.

The guidelines should be used early in the design process to encourage exploration of different options for materials and packaging design.

7.4.3 The Guide to Evolving Packaging Design by WRAP

Developed by the UK-based Waste and Resources Action Programme (WRAP), The Guide to Evolving Packaging Design [16] is a web-based resource to support rethinking about how packaging can be produced with less environmental impact (see http://www.wrap.org.uk/retail/the_guide_to_evolving_packaging_design/index.html). The web resource can be used to obtain more information about:

- consumer attitudes to packaging (also see Chap. 3)
- packaging regulations and voluntary agreements (also see Chap. 4 and Appendix C)
- design strategies, principles and decisions to be made in the new product development process (see also Chap. 2)
- general background information and case studies on common packaging materials (see also Chap. 6)
- tools and techniques to reduce the environmental impacts of packaging including the waste hierarchy, and the International Packaging Study Database (see Sect. 7.5.1).

The website and content should be used early in decision-making when the design team is exploring different options for materials and packaging designs. It may also be useful in developing and informing the sustainability strategy.

7.4.4 Envirowise Guides

Envirowise in the UK has published two useful guides to packaging design:
- *Packaging design for the environment: reducing costs and quantities*
- *PackGuide: a guide to packaging eco-design.*

These are briefly introduced below.

Packaging Design for the Environment: Reducing Costs and Quantities

This publication aims to encourage eco-efficiency in packaging design by demonstrating how to simultaneously reduce costs and environmental impacts of packaging (see http://envirowise.wrap.org.uk/uk/Our-Services/Publications/GG360R-Packaging-design-for-the-environment-Reducing-costs-and-quantities-Revised-in-February-2008.html) [17]. It outlines strategies such as resource minimisation, recycling and reduction in the use of hazardous substances, and assesses tools and techniques, including LCA.

The guide, which is similar to the Guide for Evolving Packaging Design by WRAP, contains an extensive list of resources and links including:
- packaging legislation (see also Chap. 4 and Appendix C)
- description of various tools and metrics that can be used in the design process (see Tables 7.1–7.3)
- information on packaging materials (see also Chap. 6)
- checklists that can be integrated into the packaging design process (see Sect. 7.2).

PackGuide: A Guide To Packaging Eco-Design

Envirowise also developed this guide to encourage innovation for sustainability by demonstrating how eco-design can simultaneously improve packaging functionality (for example, protect the product) and make a positive contribution to sustainable production, distribution and consumption (see http://envirowise.wrap.org.uk/uk/Our-Services/Publications/GG908-PackGuide-a-guide-to-packaging-eco-design.html). It focuses on reducing costs and quantities, as does *Packaging design for the environment* (described above). Like that guide, the main drawback is that the majority of techniques are 'rules of thumb' or descriptions of tools, although links to metrics-based tools are included for the designers to draw on.

7.5 Case Study Databases

There are two web-based case study databases that provide examples of more sustainable packaging. These are useful tools to use when trying to identify a case study for a specific product category, geographic location or innovation platform and often provide images.

7.5.1 The International Packaging Study Database

The International Packaging Study Database compiled in 2005 contains case studies from around the world of 'successful retail packaging formats, product dispensing and distribution systems, merchandising approaches and product designs' [18] (see http://www.wrap.org.uk/retail_supply_chain/research_tools/tools/international_packaging_study/index.html). The database can be searched via product category, innovation platform, country of origin or keyword [19]. For each case study, a product description is given identifying the packaging materials used, the retail outlet where the product was purchased, the product category, country of origin, date the record was added to the database, and the innovation platform (see Fig. 7.12). A section on added value and an image bank complement this case study information.

7.5.2 Sustainable Packaging Coalition Design Library

The Sustainable Packaging Coalition design library [21] (see http://www.spcdesignlibrary.org) is a database established to promote innovative packaging that addresses the SPC's eight sustainability criteria (refer to Case Study 1.6 for details). It includes case studies of award-winning designs, evaluated against the above criteria.

Fig. 7.12 Screenshot of biscuit case study on International Packaging Study website. *Source*: WRAP [20]

The case studies are presented in the following categories:
- additive
- facility/process
- cleaning product
- composite
- consumer goods
- cosmetics/personal care
- display
- electronics
- flexible
- food and beverage
- general packaging
- paper and board
- polymer (bio-based)
- polymer (petroleum-based)

7 Selecting and Applying Tools   281

Fig. 7.13 Screenshot of a case study—SoilWrap—on the SPC design library website. *Source*: SPC [21]

- rigid
- transport
- machinery
- PAC Award winner
- DuPont Award winner
- Greener Package Award winner.

For each case study, an image and brief overview is provided (see Fig. 7.13). For some case studies, details on cost effectiveness, performance, sourcing, clean production, effective recovery, community benefit, resource and energy optimisation, and material health are provided.

7.6 Conclusions

The growing range of qualitative and quantitative environmental assessment tools made available to the packaging supply chain is increasing the need for careful selection of the appropriate tool for the nominated task. Tool selection will also be dependent on where the organisation sits within the sustainability curve (Chap. 1 and Chap. 8), as it is unwise to embark upon full LCAs if the organisation is still only complying with regulations as a minimum. The sophistication of data sources and collection, and the degree of collaboration along the supply chain, will also help determine which tools to use.

References

1. Lewis H, Gertsakis J, Grant T, Morelli N, Sweatman A (2001) Design + environment: a global guide to designing greener goods. Greenleaf Publishing, Sheffield
2. Belz FM, Peattie K (2009) Sustainability marketing: a global perspective. Wiley, Chichester
3. PRe Consultants (2010) SimaPro LCA software. Available from: http://www.pre.nl/simapro/ (cited 29 September 2010)
4. PE INTERNATIONAL (2010) GaBi Software. Available from: http://www.gabi-software.com (cited 29 September 2010).
5. Verghese K, Horne R, Carre A (2010) PIQET: the design and development of an online 'streamlined' LCA tool for sustainable packaging design decision support. Int J Life Cycle Assess 15(6):608–620
6. Green Blue (2008) COMPASS—comparative packaging assessment. Available from: http://www.design-compass.org (cited 29 September 2010)
7. Early C, Kidman T, Menvielle M, Geyer R, McMullan R, (2009) Informing packaging design decisions at Toyota Motor Sales using life cycle assessment and costing. J Ind Ecolo 13(4)
8. Early C, T Kidman, M Menvielle, R Geyer, R McMullan (2008) Informing packaging design decisions at Toyota Motor Sales using life cycle assessment, Group Project Brief. University of California, Santa Barbera
9. Envirowise (2010) EDIT—The eco-design Indicator Tool. Available from: http://www.envirowise.gov.uk/uk/Our-Services/Tools/EDIT-The-Eco-Design-Indicator-Tool.html (cited 30 April 2010)
10. Sterling S (2007) Field guide to sustainable packaging. Summit Publishing Company, Chicago
11. WalMart (2008) Sustainability scorecard FAQs. Available from: http://www.sustainability-education.com/faq.aspx (cited 29 September 2010)
12. ECRM (2010) Package Modeling. Available from: http://www.ithumbprint.com/presentations/Wal-Mart_Stores_Web_Cast_files/frame.htm (cited 30 November 2010)
13. APCC (2009) Australian Packaging Covenant. A commitment by governments and industry to the sustainable design, use and recovery of packaging. Australian Packaging Covenant Council (APCC): Melbourne
14. Sustainable Packaging Coalition (2006) Design guidelines for sustainable packaging, Version 1. GreenBlue Institute, Charlottesville
15. Jedlicka W (2009) Packaging sustainability. Tools, systems and strategies for innovative package design. Wiley, Hoboken
16. WRAP (2010) The Guide to Evolving Packaging Design. Available from: http://www.wrap.org.uk/retail/the_guide_to_evolving_packaging_design/index.html (cited 11 March 2010)
17. Envirowise (2008) Packaging design for the environment: reducing costs and quantities. Oxfordshire
18. WRAP (2005) IPS—International packaging study 2005. Available from: http://www.wrap.org.uk/retail_supply_chain/research_tools/tools/international_packaging_study/index.html (cited 23 May 2011)
19. WRAP (2009) International Packaging Study—search for packaging examples. Available from: http://www.wrap.org.uk/retail_supply_chain/research_tools/tools/international_packaging_study/index.html (cited 23 May 2011)

20. WRAP (2011) Biscuits. Product description: duchy originals organic oaten biscuits. Available from: http://www.wrap.org.uk/templates/packaging_photo_library_details.rm?id=17170 (cited 6 June 2011)
21. SPC (2010) Sustainable Packaging Coalition design library. Available from: http://www.spcdesignlibrary.org (cited 23 May 2011)

Chapter 8
Implementing the Strategy

Leanne Fitzpatrick, Helen Lewis and Karli Verghese

Abstract As a general rule, sustainability can be seamlessly integrated into existing business processes. However, new knowledge, skills, and tools, as well as refinements to existing processes, policies and procedures, are required. The business processes most significantly affected are: strategic and operational planning; marketing, communications and sales activities; product and packaging development; procurement and supply chain management; and process and environmental improvement. How evolved a business is, in sustainable development terms, determines its corporate sustainability goals and, in turn, how easy or difficult it is to accommodate and make changes. (For example, is it focused only on compliance, practicing 'strategic pro-activity' or aiming to be a 'sustaining corporation'?) In this final chapter we outline how these business processes are involved in packaging for sustainability decisions and provide a framework for a packaging for sustainability action plan.

Contents

8.1	Introduction	287
8.2	Strategic and Operational Planning	292
	8.2.1 Establish Sustainable Development Goals	292
	8.2.2 Commit to and Invest in Innovation	293
	8.2.3 Understand the Life Cycles of Your Products and Packaging	295

L. Fitzpatrick (✉)
Birubi Innovation, 5 Brooklyn Avenue, Dandenong, VIC 3175, Australia
e-mail: leanne@birubi.com.au

H. Lewis · K. Verghese
Centre for Design, RMIT University, GPO Box 2476, Melbourne, VIC 3001, Australia
e-mail: lewis.helen@bigpond.com

K. Verghese
e-mail: Karli.Verghese@rmit.edu.au

8.2.4	Benchmark Current Sustainability Performance	295
8.2.5	Embed Packaging Sustainability into Business Plans	297
8.2.6	Establish a Packaging Sustainability Team	304
8.2.7	Understand Current Packaging	305
8.2.8	Identify Packaging-Specific Sustainable Development Goals and Metrics	307
8.2.9	Understand and Engage with Stakeholders	311
8.2.10	Build or Leverage Like Minded Supply Chains	312
8.2.11	Develop a Packaging for Sustainability Action Plan	313
8.3	Marketing, Communication and Sales Activities	319
8.4	Developing Products and Packaging	319
8.4.1	Embed Sustainability into the Development Process	320
8.4.2	Conduct Formal Ideas Generation Sessions	321
8.4.3	Document Packaging Specifications	322
8.5	Procurement and Supply Chain Management	324
8.6	Process and Environmental Improvement	326
8.7	Conclusion	327
References		327

Figures

Figure 8.1	Generalised organisational structure showing business functions impacted by packaging sustainability	288
Figure 8.2	The Innovation Diamond	295
Figure 8.3	Businesses involved in the packaging supply chain	311
Figure 8.4	Example of a new product development process with three 'gates'	320

Tables

Table 8.1	Examples of business functions involved in packaging sustainability	289
Table 8.2	Managing innovation for sustainable development	296
Table 8.3	Characteristics of companies at different stages of sustainability evolution	298
Table 8.4	Example activities to develop sustainable development capability	301
Table 8.5	Examples of business metrics	308
Table 8.6	Packaging sustainability metrics	309
Table 8.7	Packaging metrics under development by the Consumer Goods Forum (draft 2011)	310
Table 8.8	Engaging with external stakeholders	312
Table 8.9	Example of a packaging for sustainability action plan	314
Table 8.10	Writing the packaging specification	323
Table 8.11	Roles of each sector in the supply chain	325

8 Implementing the Strategy

Case Studies

Case Study 8.1	Sustainability strategy at VIP Packaging	293
Case Study 8.2	Evaluating regulatory compliance	303
Case Study 8.3	Undertaking a packaging review	306
Case Study 8.4	Sara Lee Australia Packaging Reviews	307
Case Study 8.5	Running an ideas generation session	322
Case Study 8.6	Incorporating sustainability in the packaging specification	324
Case Study 8.7	Sustainable procurement at Hewlett Packard	326

8.1 Introduction

The packaging for sustainability strategy is informed by:
- internal business functions including marketing, research and development, procurement and supply chain management, operations, and environmental management who are also responsible for its implementation
- the external environment and other stakeholders including customers, suppliers, consumers, consumer groups and regulators.

For a particular business, the goals and metrics of the strategy should reflect:
- an understanding of the life cycle impacts of its products and packaging (see Chaps. 1 and 5)
- corporate, brand and product positioning (see Chap. 3)
- regulatory requirements (see Chap. 4).

Sustainable development goals are achieved, however, through operational plans and business processes, which is where priorities are set, resources allocated and sustainability considered alongside other goals and metrics relating to sales, market share, financial performance and so on.

New Knowledge, Skills and Tools are Required

As a general rule, sustainable development can be seamlessly integrated into existing business processes and work flows. However, this requires the development and application of new knowledge, skills, and tools as well as refinements to existing processes (including policies and procedures) and work flows.

How easy or difficult it is to accommodate, and make these changes, will depend on:

- how evolved a business is, in sustainable development terms (see Chap. 1)
- the specific focus of the corporate sustainability goals (for example, internal resource efficiency versus step change product innovation).

⇒ Learn more about organisational sustainability evolution in Sect. 1.3.1

Many Business Units and Processes are Involved

A generalised organisational structure highlighting the range of business units and their degree of involvement in packaging-related sustainable development decisions is shown in Fig. 8.1.

Although every business has its unique structure and processes, and businesses differ widely in their scale, position in the supply chain and business model, it is possible to identify common operational activities involved in setting and achieving packaging-related sustainable development goals (see Table 8.1).

For all businesses, the processes most significantly affected are:

- strategic and operational planning
- marketing, communication and sales activities (corporate, brand and product)
- product and packaging development
- procurement (packaging components, technologies, equipment) and supply chain management
- process and environmental improvement.

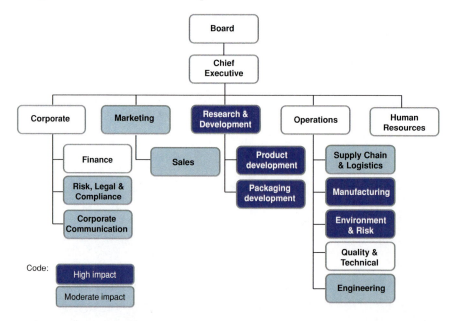

Fig. 8.1 Generalised organisational structure showing business functions impacted by packaging sustainability

Table 8.1 Examples of business processes and functions involved in packaging sustainability

Packaging decision context	Indicative business function involvement							
	Board and executive	Corporate functions	Marketing	Sales and business development	Research and development	Procurement and supply chain	Environment and sustainability	Engineering	Manufacturing	Human Resources
Strategic, Operational Planning and Capacity Building										
Set corporate sustainability goals and targets	✓✓✓	✓✓	✓✓✓	✓	✓✓	✓	✓✓✓		✓✓	
Determine corporate, brand and product sustainability positioning	✓✓✓	✓✓	✓✓✓	✓	✓✓	✓	✓✓✓		✓✓	
Assess competitive position of products and packaging from a sustainable development perspective			✓✓✓		✓✓✓		✓✓✓			
Confirm packaging's role in meeting sustainable development goals and set packaging sustainability goals and targets		✓	✓✓	✓	✓✓✓	✓	✓✓	✓	✓✓	
Develop organisational understanding and commitment to sustainable development		✓			✓✓✓	✓	✓✓			✓✓
Marketing, communications and sales activities										
Report sustainable development goals, challenges and achievements in packaging	✓	✓✓			✓		✓✓✓			
Promote environmental credentials of the business, its products and packaging	✓✓	✓✓	✓✓✓	✓✓✓	✓	✓	✓✓			
Determine and use environmental claims, logos and on-pack labelling to support achievement of sustainable development goals and ensure they are consistently and accurately applied		✓✓	✓✓		✓✓✓	✓	✓✓			

(continued)

Table 8.1 (continued)

Packaging decision context	Board and executive	Corporate functions	Marketing	Sales and business development	Research and development	Procurement and supply chain	Environment and sustainability	Engineering	Manufacturing	Human Resources
Developing products and packaging										
Identify, assess and develop new and improved packaging formats to meet marketing AND sustainable development goals			✓✓✓		✓✓✓	✓✓	✓	✓	✓✓	
Comply with packaging-specific environmental regulatory requirements.					✓✓✓		✓✓✓	✓	✓	
Procurement and supply chain management										
Set or ensure compliance with customer-specific sustainable development requirements.		✓	✓	✓	✓	✓✓✓	✓✓	✓	✓✓	
Incorporate sustainable development requirements into purchasing contracts					✓✓	✓✓✓	✓✓✓	✓✓	✓✓	
Incorporate sustainable development requirements into packaging specifications					✓✓✓	✓✓✓				
Incorporate sustainable development requirements into capital planning, procurement and investment decision-making	✓✓	✓✓			✓✓	✓✓✓	✓✓	✓✓	✓✓	

(continued)

Table 8.1 (continued)

Packaging decision context	Indicative business function involvement							
	Board and executive	Corporate functions	Marketing	Sales and business development	Research and development	Procurement and supply chain	Environment and sustainability	Engineering	Manufacturing	Human Resources

Process improvement and environmental management

Packaging decision context	Board	Corporate	Marketing	Sales	R&D	Procurement	Environment	Engineering	Manufacturing	HR
Improve resource (materials, energy, water) efficiency of manufacturing processes and unit operations involved in producing, using, recovering and reprocessing packaging					✓	✓	✓	✓✓	✓✓✓	
Reduce waste and emissions from manufacturing processes involved in the production of packaging					✓	✓	✓	✓✓	✓✓✓	
Develop new and improved processes to manufacture and use packaging more sustainably e.g. increased recycled content, improved end-of-life recovery					✓✓✓			✓✓	✓	

Notes: Corporate functions may include corporate risk or legal management, finance, communications
✓ = May be involved; ✓✓ = Significant involvement; ✓✓✓ = Highly involved

Leadership from Brand Owners and Retailers

Brand owners and retailers are leading sustainable development initiatives and progressively adopting a product stewardship approach based on an understanding of the life cycle impacts of their products. Their expectations and demands require immediate and upstream suppliers to reshape and position themselves to support current initiatives and create better options for the future.

⇒ Learn more about retailer initiatives in Sect. 1.5.2

No Longer Business as Usual for Suppliers

For material and packaging component or equipment suppliers, significant changes in sales, manufacturing and engineering are likely. As well as maintaining a focus on resource efficiency in strategic planning and product development, these businesses must develop a better understanding of the sustainability performance of their product-packaging systems from a life cycle perspective and consider the sustainable development goals of their supply chains.

'Sustainability' Not Environmental Management

Environmental management is significantly reshaped as organisations evolve in sustainable development. It broadens from an inward focus emphasising regulatory compliance, waste reduction and resource efficiency to an outward focus applying life cycle assessment (LCA). This requires new knowledge and skills.

8.2 Strategic and Operational Planning

8.2.1 Establish Sustainable Development Goals

The corporate strategy must include sustainability metrics, targets and baseline data that confirm the business's sustainable development goals and allow performance to be monitored. These will include goals that affect packaging decisions indirectly as well as directly (see Marks and Spencer Case Study 1.1 and VIP Packaging Case Study 8.1).

⇒ Learn more about corporate sustainability in Chap. 1

High-level goals and metrics in corporate strategies vary widely. Examples include [1, p 11]:
- Walmart— reduce amount of packaging in the supply chain by 5% by 2013
- Marks and Spencer—reduce weight of non-glass packaging by 25% by 2012
- Cadbury—reduce absolute carbon emissions by 50% by 2020

- Coca-Cola Enterprises—reduce overall carbon footprint of business operations by 15% (compared to 2007 baseline) and recover the equivalent of 100% of packaging by 2020.

> **Case Study 8.1 Sustainability Strategy at VIP Packaging**
>
> VIP Packaging has linked its business model and sustainability strategy to an ambitious packaging recovery target.
>
> **Sustainability Strategy**
> VIP Packaging has a sustainability strategy called 'Sustainable Choices Program'. This is supported by a roadmap, which outlines the company's objectives for 'people, profit and planet'. The four objectives for the planet are:
> - 'Consider the lifecycle impacts of our packaging during the design and development process
> - Develop innovative recycling and collection solutions for our packaging to avoid landfill disposal
> - Use natural resources responsibly and endeavour to use renewable alternatives
> - Implement an environmental management system to minimise the environmental impact of our activities, products and services, including greenhouse gas emissions and water consumption'.
>
> **Packaging Target**
> The company has a target to ensure that 100% of its products are able to be recycled, reconditioned or composted at their end-of-life by 2015.
> *Source*: VIP Packaging [2]

8.2.2 Commit to and Invest in Innovation

Innovation is the process by which a business can meet or exceed its goals by doing things differently. It is more than the generation of ideas—it is the capacity to source or create ideas, develop the best of these and deliver triple bottom line value from them.

Packaging has undergone a long history of innovation, and continued innovation is essential to achieve business goals such as:
- growing market share
- creating and entering new markets
- reducing product and supply chain costs and waste
- meeting sustainable development goals.

⇒ Read more about the business case for sustainability in Chap. 1

All Business Units Must Commit to Innovation

As sustainable development goals become integral to corporate strategies, sustainability becomes innovation's new frontier [3] in all business units of companies involved in or, servicing, the packaging supply chain.

The willingness and capacity to innovate, in particular the step-changes required for eco-effectiveness, must therefore be developed within and across business units and supply chains. For example:
- The business must learn to manage a portfolio containing a greater number of potentially higher risk initiatives.
- Marketers need to be willing to rethink the product or service they are providing and potentially change their sales and distribution models.
- Supply chain and procurement must reposition the business to be more sustainable from a life cycle perspective.
- New development partnerships need to be created with current supply chain partners (customers and suppliers) and new supply chain businesses (recyclers, composters).

Apply a Strategic Approach to Innovation

Innovation is a process, although far from linear, that can be strategically developed and managed and should include investment in building organisational capacity. The 'Innovation Diamond' applied by Proctor and Gamble for new product development [5] is an example of a strategic approach to innovation. This has been adapted in Table 8.2 and Fig. 8.2 to highlight some factors to consider when developing or repositioning innovation for sustainable development and packaging's role in it (Fig. 8.2).

Fig. 8.2 The innovation diamond. *Source*: Adapted from Cooper and Mills [5, p 9]

8.2.3 Understand the Life Cycles of Your Products and Packaging

Strategic planning for sustainability requires a good understanding of:
- the life cycle impacts of products and their packaging
- the role of packaging in achieving the corporate sustainability goals.

⇒ Learn more about life cycle assessment in Chap. 5 and decision-support tools and their application in Chap. 7

This understanding can be obtained in different ways ranging from the use of simplified tools, such as life cycle maps, the sustainability impact matrix and packaging-specific assessment tools, to the use of comprehensive LCA data. The method chosen should be matched with the specific objectives of the life cycle review, the resources available (time, money and internal capacity) and how evolved the business is in sustainable development terms. Regardless of the method chosen and stage of business development, annual updates of strategies and operational plans should ensure continuous improvement in sustainable development outcomes.

8.2.4 Benchmark Current Sustainability Performance

One of the difficulties faced when trying to embed sustainable development into a business is not knowing where to start. It is often the greatest barrier to change—the lack of clarity on where to focus and why, or what is the 'right' thing to do. A sense of being overwhelmed by the issues and challenges often becomes the reason for not doing anything.

Table 8.2 Managing innovation for sustainable development

Innovation framework element	Factors to consider
Strategy	What are the short (1–2 years), medium (2–5 years) and long (>5 years) term sustainable development goals of the business and its customers?
	What are the short, medium and long term goals for each product category? Are current products sustainable?
	How will sustainable development and product category goals be achieved simultaneously?
	What new product categories should be developed?
	What barriers does the current corporate model place on the scope, targets for and achievement of medium and long term goals?
	What are the marketing and sales barriers to making products more sustainable?
	What are the technological and manufacturing barriers to achieving medium and long term goals?
	What are the technological developments and trends in packaging materials, processing, end-of-life recovery and downstream processing? How can these be leveraged and where do they align with short, medium and long term goals?
Portfolio	What level of resources will be allocated to each product category?
	What level of resources will be allocated to eco-efficiency versus eco-effectiveness projects?
	What level of resources will be invested in new products and packaging versus improvements to current products and packaging?
Capacity	What level of training in sustainable development is required? Who needs to be trained? How will this be done?
	What are the roles of different business units in meeting sustainable development goals? What new skills and knowledge do they need?
	What level of LCA capability is required in-house?
	What partnerships need to be developed to complement current strengths and fill strategic knowledge or skill gaps?
	What decision-support tools are required? How will they be used and by whom?
Process	What information is required to inform the generation of new ideas (ideation processes)?
	How will sustainable development goals be incorporated into the product and packaging development processes?
	How will sustainability be accounted for in the evaluation and approval of investments (capital, joint ventures, distribution strategies)?
	What information needs to be handed over to procurement and manufacturing in the commercialisation phase of new products, packaging and processes?

Get Started or Refine the Strategy

But actually getting started may be more important than where to start. Once a conscious decision has been made to address sustainability issues, the pathway evolves and there is increased clarity about what to focus on and why. Achievements

8 Implementing the Strategy

ultimately speak for themselves, and experiences provide the essential feedback loop to refine and improve strategies. This book itself has been a journey that has progressively evolved from the authors' involvement in packaging and environment issues in early 2001 to championing packaging for sustainability in 2011.

For readers who have already commenced or are well-advanced on their journey, our approach is hopefully accessible and provides new insights to assist in reviewing and refining their own approach.

Identify Strengths, Weaknesses and Priorities for Action

Benchmarking current sustainability performance helps to identify strengths and weaknesses and priorities for action.

Based on the sustainability phase model (see Sect. 1.3.1):
- Table 8.3 highlights behaviours and characteristics of businesses at different stages of sustainability evolution with respect to packaging
- Table 8.4 highlights examples of the types of activities businesses could undertake at each stage to improve their sustainable development performance.

Assessing current performance will help to identify priorities for action. A business may not fit into one specific category ('rejection', 'compliance', 'strategic pro-activity' or 'sustaining corporation'—see Sect. 1.3.1). Rather, it may have strengths in some areas, which should be leveraged, and gaps and weaknesses in others, which require targeted initiatives to address.

Use External Benchmarking Data

Benchmarking sustainable development can also be informed by reviewing a range of external sources including:
- independent sustainability assessments; for example, the Dow Jones Sustainability Index (see Sect. 1.5.1)
- corporate sustainability reports of leading companies and competitors
- LCAs (see Sect. 5.3.2).

8.2.5 Embed Packaging Sustainability into Business Plans

Business goals and priorities change over time. Most organisations have a business plan that outlines their short and medium term goals and priorities. Some of these may be relevant to packaging, for example:
- Cost reduction or productivity targets could be supported by packaging efficiency improvements
- Market development targets may provide an opportunity to explore new products or brand positioning strategies based on sustainability.

Table 8.3 Characteristics of companies at different stages of sustainability evolution

	Actions and characteristics of companies in different stages of sustainability evolution			
	Rejection	Compliance	Strategic pro-activity	Sustaining corporation
Developing strategy				
Corporate sustainable development goals and targets	Do not exist	Focus on immediate compliance issues	Beyond compliance, reflecting a commitment to sustainable development	Embedded within the culture of the business and reflected in strategies and business plans across the organisation
Corporate, brand and product sustainability positioning	Not relevant	Compliance-focus	Clear strategy and positioning identified	Reflected in the values and operating activities of the business
Competitive position of products and packaging from a sustainable development perspective	Not considered	May be considered where relevant and reflect internally focused metrics e.g., waste reduced, resource efficiency improvements	Embedded in product and packaging development process. Internal focus and increasing use of LCA	Critical success factor for all products based on a product-packaging life cycle approach
Packaging's role in meeting sustainable development goals	Not considered	Regulations are the reason for addressing packaging	Aim to meet regulatory requirements but start to challenge some issues due to increased understanding of packaging's role from a product-packaging system and life cycle perspective	Clearly understood targets for packaging's contribution to sustainability goals are well-defined
Implementing strategy				
Organisational understanding and commitment to sustainable development	Not relevant	Focus on environmental compliance and cost benefit activities	Investment in training and development across the business	Pre-requisite for doing any role in the company

(continued)

Table 8.3 (continued)

Implementing strategy	Rejection	Compliance	Strategic pro-activity	Sustaining corporation
Reporting of packaging sustainable development goals, challenges and achievements	None	Report against regulatory requirements	Exceed regulatory requirements and reflect product-packaging system and life cycle perspective	Reporting life cycle impacts and triple bottom line packaging-specific metrics
Promotion of environmental credentials of the business, its products and packaging	None	Focus on compliance reporting	Promote strategy and achievements across a number of channels	Clear strategy that links corporate, brand and product positioning across various channels including on-pack labelling
Development of new and improved packaging formats to meet marketing AND sustainable development goals	Never considered	Unlikely to be considered	Emerging or low level activity as part of the portfolio	Essential for new product and product improvement strategies
Packaging-specific environmental regulatory requirements	Not considered	Meet with reluctantly	Meet or exceed to position for the future	Already well ahead—and may challenge substance and approach of regulations
Setting or complying with customer-specific sustainable development requirements	Not considered	Meet with reluctantly where this can be done	Meet or exceed to position for the future	Already well ahead—and may challenge substance and approach of customers and suppliers
Sustainability requirements incorporated into purchasing contracts and/or packaging specifications	Not considered	Not considered	Emerging on specific issues	Essential to align supply chain with business's values and sustainability goals

(continued)

Table 8.3 (continued)

Actions and characteristics of companies in different stages of sustainability evolution				
	Rejection	Compliance	Strategic pro-activity	Sustaining corporation
Implementing strategy				
Sustainable development requirements considered in investment decision-making	Not considered	Not considered	Add on to existing criteria	Benefits to be clearly demonstrated
Process improvement and environmental management	Links between improvement strategies and sustainable development not made	Eco-efficiency and waste reduction focus in response to regulatory compliance priorities	Eco-efficiency embedded and eco-effectiveness considered as part of development and capital investment plans	Eco-efficiency and eco-effectiveness strategies identified and aligned

Source: Builds on the phase model developed by Dunphy et al. [10]

8 Implementing the Strategy

Table 8.4 Example activities to develop sustainable development capability

Business activities	Activities to develop sustainable development capability			
	Rejection	Compliance	Strategic pro-activity	Sustaining corporation
Capacity building	← Understand sustainable development →			
		← Understand & apply life cycle thinking →		
			← Manage innovation for sustainable development →	
Strategic and operational planning		← Confirm current regulatory requirements →		
			← Monitor and assess future regulatory requirements →	
			← Benchmark corporate & product/packaging sustainability performance →	
			← Conduct market research – consumer attitudes & understanding of sustainability issues →	
		← Conduct life cycle mapping →		
				← Conduct LCAs →
			← Report on corporate sustainability reporting →	
				← Promote environmental credentials of products →
Marketing, communications and sales activities			← Apply on-pack labelling to support sustainability goals →	
		← Identify & introduce decision-support tools →		
		← Introduce a packaging sustainability framework →		
		← Develop a packaging database →		
		← Review & benchmark current packaging →		
		← Revise 'new product development' process to embed sustainability →		

(continued)

Table 8.4 (continued)

Business activities	Activities to develop sustainable development capability			
	Rejection	Compliance	Strategic pro-activity	Sustaining corporation
Developing products and packaging		← Identify & implement eco-efficiency initiatives →		← Conduct formal ideas generation sessions to rethink the product-packaging system →
				← Identify & implement eco-effective initiatives →
			← Develop or use new technologies and processes →	
Procurement and supply chain management			← Identify & introduce decision-support tools →	
			← Revise procurement process to embed sustainable development →	
			← Review current supply chain sustainable development initiatives →	
			← Identify & implement supply chain eco-efficiency initiatives →	
				← Identify & implement supply chain eco-effective initiatives →
Process and environmental improvement		← Identify & implement waste reduction and eco-efficiency initiatives →		
			← Benchmark sustainability performance →	
				← Develop or use new technologies and processes →

Source: Builds on the model developed by Dunphy et al. [10]

To be effective, packaging sustainability goals and strategies should be embedded within, and be consistent across, the operational plans of relevant business units. The packaging for sustainability action plan outlined in Sect. 8.2.7 is one way this can be achieved and needs to be reflected in the performance goals of the individuals involved.

Specific goals and strategies should be selected to:
- support business sustainability and other goals and priorities
- reduce the life cycle impacts of products and operations
- address stakeholder expectations (Sect. 8.2.9)
- meet relevant trade practices and sustainability regulations (See Chap. 4 and Case Study 8.2).

Existing business programs or special initiatives should also be assessed to determine whether they could be supported by or potentially affect packaging sustainability strategies (for example, business improvement, lean manufacturing, capital and capacity planning).

Case Study 8.2 Evaluating Regulatory Compliance

The US Sustainable Packaging Coalition has published a series of design guidelines for sustainable packaging. These include the following questions that a business should ask itself about regulatory compliance:
- In what country or countries are you selling this product and packaging?
- Have all applicable regulatory requirements been identified?
- What regulations related to your product might also apply to packaging?
- What materials are banned or restricted at the package's final destination?
- Are there packaging fees, and could they be reduced by using different materials?
- Are there design requirements such as void space, layers, recycled content, recyclability, recoverability, and so on?
- If extensive legal labelling is required (as in pharmacy and personal products), can alternatives such as fold-out labels be used instead of making the package bigger?
- Are Certificates of Compliance with heavy metals or hazardous substances regulations and wood-treatment standards on file or available from each of your suppliers?
- Is it possible for your business to adopt a policy of using the strictest standards for all package designs?

Source: Sustainable Packaging Coalition [4, p 17]

8.2.6 Establish a Packaging Sustainability Team

Many businesses have established a cross-functional team that develops and oversees the implementation of their packaging for sustainability strategy. This helps ensure that:
- the strategy is aligned with other business goals and priorities
- goals, targets and activities are relevant and achievable
- the sustainable development capacity of the business is simultaneously developed.

The individual responsible for leading packaging sustainability must have a very good understanding of the business including:
- its goals and priorities
- what and how packaging decisions are made and who is involved (see Table 8.1)

Establishing and Managing the Packaging Sustainability Team

1. Identify business units involved in packaging-related decisions.
2. Identify potential sustainable development champions within these business units.
3. Obtain line management support for champions to be involved in the team.
4. Conduct a meeting with the sustainable development champions to develop a draft terms of reference. Include business goals to be addressed, summary of business case for the team, scope of activities, resource allocations and performance measures.
5. Present terms of reference to executive/senior management team for feedback and approval.
6. If a member of the executive/senior management team member is not in the team request appointment of a team sponsor from the executive/senior management team.
7. Develop and document the packaging for sustainability action plan. This may require background research by the team and some initial training or awareness activities. It is also useful to consider engaging an experienced facilitator to help the team through this process efficiently and effectively.
8. Conduct monthly meetings to review progress against the plan and update the action plan if required.
9. Report team progress on a monthly basis to relevant line managers and the Executive/Senior management team through the team sponsor.
10. Report achievements annually against the terms of reference and the plan—and update as appropriate.

- the life cycle impacts of products and packaging
- expectations of external stakeholders, including customers, suppliers and regulators, and the skills to interact with them.

8.2.7 Understand Current Packaging

Maintain a Packaging Database

A packaging database should be maintained that catalogues packaging materials and/or components with information such as:
- type
- amounts used/quantity sold (number of units and weight)
- supplier
- application
- recycled content and its source (post-consumer and pre-consumer)
- end-of-life recovery processes
- recovery rate in end-markets.

The database is a useful resource to:
- assist benchmarking and conducting reviews of packaging
- support the packaging development process
- support the engagement process with customers and suppliers
- provide information for reporting and measuring performance (including regulatory compliance).

Benchmark Current Packaging

Current materials and/or packaging components or systems should be benchmarked to help identify current strengths, priority areas for action and new business opportunities (new products, cost reductions, risk reduction).

The benchmarking process should consider:
- life cycle impacts of the product-packaging system, packaging materials and components
- alignment or otherwise with corporate sustainability goals and metrics
- compliance with packaging-specific regulatory requirements
- performance versus 'best practice' (not necessarily best in category) and competitive products.

Tools and resources outlined in this book that can be used for benchmarking include:
- life cycle tools (Chap. 7)
- packaging material life cycles (Chap. 5)
- packaging sustainability framework (Chap. 2).

⇒ Read more about packaging evaluation tools in Chap. 7

In the absence of access to life cycle data and other decision-support tools, material weight can be used as a guide. The 'best in class' database published by WRAP in the United Kingdom [8] provides weights for the heaviest and lightest packs for specific products in the UK market. This is a useful benchmarking tool and highlights the significant potential for improvement in most categories. However, users need to remember that their packaging may not necessarily be in the same 'class' as the lightest pack in the WRAP survey, for example it may have different distribution channels or markets [8]. It is therefore important to be clear on the purpose of the benchmarking exercise and to use the most appropriate functional unit and packaging system definition.

Conduct Packaging Reviews

More detailed reviews of current packaging can be used to identify opportunities for improvement and select the most appropriate design strategies from those outlined in Chap. 2.

Start by grouping current packaging into logical categories (see Case Study 8.4). The packaging in each of these categories should be similar, and there may also be other criteria that would make the review process efficient and productive. For example, it might be useful to group packaging from the same supplier, manufacturing site or business group. The aim is to review each packaging category within a realistic timeframe (such as 2–3 years) and build the opportunities identified into future upgrades or redesigns.

The packaging review can be undertaken in a number of ways. For example, it could use one of the LCA tools outlined in Sect. 7.2. If these are not accessible due to resource constraints (skills, time or money), life cycle maps, guidelines and checklists can be used. Prepare a set of guidelines or a checklist that will ensure that relevant issues are considered for each of the packaging groups. The codes of practice and design guidelines introduced in Sect. 7.4 provide a good starting point and can be adapted to meet specific regulatory requirements, stakeholder expectations and corporate goals. Case Study 8.3 provides an example of the steps that can be undertaken at a packaging review session.

Case Study 8.3 Undertaking a Packaging Review

With the involvement of internal and external stakeholders including suppliers, review the sustainability impacts of each category by:
- drawing up a life cycle map
- identifying important sustainability impacts and 'hot spots' (priorities for action)
- recording these impacts in a sustainability matrix
- going through the guidelines/checklist to identify specific opportunities for improvement
- documenting the outcomes of the review and taking action to investigate the opportunities further.

> **Case Study 8.4 Sara Lee Australia Packaging Reviews**
>
> Signatories to the Australian Packaging Covenant, which commenced in mid-2010, are required to review all their existing packaging against the Sustainable Packaging Guidelines within a reasonable timeframe [6]. The guidelines suggest that packaging should first be grouped into logical categories; for example, by product or packaging material. The purpose of the review is to identify opportunities for improvement.
>
> Ten companies were asked to pilot the grouping and review process and to publish the results on the covenant website. One of the case study businesses was Sara Lee Australia, a wholly owned subsidiary of global food and beverage company Sara Lee. The company decided initially to group its packaging into three business categories—Bakery, Retail Tea and Coffee, and Foodservice Tea and Coffee—to ensure integration in their current business structure. A member of each Marketing and Sales Unit was asked to complete a matrix of products and their packaging, to inform a more detailed grouping process. The result of the process was a timetable to review the existing packaging in the following categories:
> - Bakery—11 product groups based on packaging types and 4 individual Stock Keeping Units
> - Retail Tea and Coffee—4 packaging groups (glass jars, beans and bricks in flexible film, tins, sachets)
> - Foodservice Tea and Coffee—large number of categories, to be prioritised on the basis of highest percentage of sales and greatest opportunity for improvement.
>
> *Source*: Adaptation Environmental Services [7]

8.2.8 Identify Packaging-Specific Sustainable Development Goals and Metrics

The contribution of packaging to the achievement of sustainable development goals will be primarily delivered through three business processes:
- product (including packaging) development
- packaging design
- supply chain management (includes procurement, logistics).

In all cases, sustainability will need to be considered alongside many other factors, such as cost, consumer acceptance, function, capital availability and risk, that are already considered in these processes and are generally well-characterised and quantified (see Table 8.5).

Table 8.5 Examples of business metrics

Performance criteria	Metrics
Financial	Product price point
	Product margin
	Return on investment
	Cost of capital
	EBIT (earnings before interest and tax) contribution
Consumer acceptance	Purchase intent
	Market share
	Sensory evaluation scores
Packaging function	Product testing (microbial, chemical and/or physical)
	Consumer acceptance
	Product shelf life
	Damaged goods (returns)

Accordingly, metrics should be identified to:
- set packaging-specific sustainable development goals, targets and benchmarks
- inform packaging related decision-making
- measure packaging sustainability achievements.

Different types of metrics need to be considered:
- LCA metrics such as waste production, global warming potential, water consumption and land use occupation (see Table 5.7)
- packaging sustainability metrics that provide information about the functions and performance of the product-packaging system but do not address the full life cycle, such as product-packaging ratio, cube utilisation and percentage of post-consumer recycled content
- other relevant economic and social metrics.

Table 8.6 provides a list of LCA and packaging-specific metrics that could be used to measure performance against the packaging sustainability framework outlined in Chap. 2. A list of metrics being developed by The Consumer Goods Forum is also provided in Table 8.7.

Goals and targets within the corporate sustainability strategy should guide the setting of packaging specific goals and metrics to ensure they are consistent and mutually supporting. For example:
- A carbon reduction target can be supported by design strategies such as lightweighting, increased recycled content or increased recyclability.
- Supply chain programs to improve the sustainability of product raw materials (such as palm oil, timber and so on) can be extended to packaging.

8 Implementing the Strategy

Table 8.6 Packaging sustainability metrics

Principles	Metrics
Effective: fit for purpose ★	Functionality of each component of the packaging system (list)
	Social and economic benefits of the packaging system as a whole (list)
	Whether the package can be opened by consumers with limited strength or movement, e.g. arthritis sufferers (yes/no)
Efficient: minimal use of materials and energy $	Packaging weight and minimisation*
	Packaging-to-product weight ratio*
	Cube utilisation*
	Percentage of product that becomes waste before it reaches the consumer (e.g. is damaged in transit)
	Percentage of product remaining in retail unit packaging (once consumer has dispensed product)
	Material waste*
	Cumulative energy demand*
	Fresh water consumption*
	Number of truck movements before and after packaging redesign
Cyclic: renewable and recyclable materials ◯	Renewable material content*
	Percentage of stationary energy use from a renewable source
	Percentage of transport energy use from a renewable source
	Packaging reuse rate*
	Whether the packaging is recyclable (yes/no)
	Whether the packaging is compostable (yes/no)
	Packaging recovery rate*
	Recycled content*
	Packaging types as a percentage of items collected in the litter stream (e.g. from national litter statistics if available)
	Recycling information and advice on recyclable and compostable packaging (yes/no)
	An anti-litter message and/or logo for products consumed away from home (yes/no)
Safe: non-polluting and non-toxic ✚	Use of heavy metal-based additives (list) and concentration (ppm)
	Compliance with heavy metal limits (yes/no)
	VOCs generated in manufacturing processes
	Actions taken to minimise migration into food (list)
	Percentage of paper fibre from ECF or TCF processes
	Greenhouse gas emissions (global warming potential)*
	An EMS is in place (yes/no)*
	Policies are in place to promote ecological stewardship (yes/no)
	Chain of custody*
	Number of suppliers with policies and procedures in place to promote ecological stewardship

Notes: KPIs marked with * are from the Consumer Goods Forum [14]

Table 8.7 Packaging metrics under development by the Consumer Goods Forum (draft 2011)

Category	Metrics	
Environmental	Packaging weight and minimisation	Toxicants concentration
	Packaging to product weight ratio	Water used from stressed or scarce sources
	Material waste	Packaging reuse rate
	Recycled content	Packaging recovery rate
	Renewable content	Cube utilisation
	Chain of custody	
	Overall performance checklist	
	Environmental management system (EMS) use	Energy audit
Economic	Total cost of packaging	Packaged product wastage
Social	Packaged product shelf life	Community investment
	Overall performance checklist	
	Child labour	Occupational health
	Excessive working hours	Discrimination
	Responsible workplace practices	Safety performance
	Forced or compulsory labour	On pack end of life communication
	Remuneration	Product safety
	Freedom of association	
Life cycle indicators	*Inventory indicators*[a]	
	Cumulative energy demand	Land use
	Fresh water consumption	
	Impact category indicators[b]	
	Global warming potential	Photochemical ozone creation potential
	Ozone depletion	Acidification potential
	Toxicity, cancer	Aquatic eutrophication
	Toxicity, non cancer	Freshwater eco-toxicity potential
	Particulate respiratory effects	Non-renewable resource depletion
	Ionising radiation (human)	

Source: Consumer Goods Forum ([14, p 21, 15])

[a] A life cycle inventory lists the quantity of resources (materials, fuels and energy) and wastes and emissions (air, land, water) associated with the packaging life cycle

[b] The numbers in the life cycle inventory are classified into different environmental impact categories, such as climate change. The life cycle inventory results are converted to common units, e.g. carbon dioxide equivalents (CO_2-e) for greenhouse gas emissions, and then aggregated for each category. The outcome of the calculation is a numerical indicator result known as an impact category indicator

8.2.9 Understand and Engage with Stakeholders

Although corporate models vary, generally speaking the packaging supply chain consists of a complex web of material producers, packaging component manufacturers (converters), packaging equipment suppliers, brand owners (users of packaging), retailers and waste recovery facilities/reprocessors (see Fig. 8.3). Packaging sustainability requires all parts of the supply chain to better understand their product's role in the life cycle of a packaging-product system.

⇒ Learn more about understanding and engaging with stakeholders in Sect. 3.4.2

Businesses have many other stakeholders with different roles and influences on the business and different expectations about sustainability. A stakeholder analysis can be useful to help identify organisations that could support or hinder the business's progress towards sustainability, the issues of concern to each group, and engagement strategies. The packaging for sustainability action plan should identify stakeholders and how interactions with them will be managed. Table 8.8 highlights some questions that a packaging sustainability champion might ask key stakeholders about a particular type of packaging. The stakeholders who need to be consulted will depend on the business's location in the supply chain.

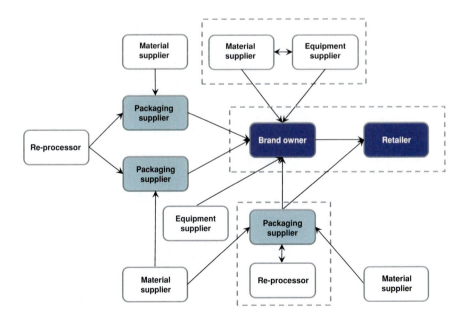

Fig. 8.3 Businesses involved in the packaging supply chain

8.2.10 Build or Leverage Like Minded Supply Chains

In their report *Sustainable Packaging: Threat or Opportunity?*, PriceWaterhouseCoopers argue that to become more proactive in addressing sustainability businesses need to take the following actions [1, p. 3]:

- Review their customer base to understand which ones have made public announcements on their commitment to sustainability and begin talking to them about what this means in practice for their packaging needs

Table 8.8 Engaging with external stakeholders

Questions for external stakeholders	
Raw material suppliers	What are the environmental impacts of material extraction and processing?
	Has there been any innovation in material or processing technologies to reduce its environmental impact?
	Does the material contain any heavy metals? If so, what is the concentration? Can the heavy metal content be eliminated or reduced?
	Does the material contain any other substances that may be potentially toxic during use or disposal?
	Is there any recycled content in the material? Can the level of recycled content be increased without compromising functionality or cost?
Packaging manufacturers	What new technologies, materials or formats are available?
	How can the functionality and/or environmental performance of the packaging be improved?
	How can the cost of the packaging be reduced (including costs in distribution and handling)?
Brand owner	What are the requirements for the packaging - cost, functionality, labelling, distribution, shelf display, environmental performance, etc.?
	How effective is the current packaging?
	How can it be improved?
Retailers	What are the requirements for the packaging - distribution, shelf display, environmental performance, etc.?
	Has the business received any complaints from consumers about the packaging?
	How effective is the current packaging?
	How can it be improved?
Regulators	What are the legal requirements for the packaging—labelling, environmental, void fill, etc.?
Reprocessors	Is the current packaging technically recyclable?
	Is there any design change required to improve its recyclability?
	Is the packaging actually collected and reprocessed?
Environment and consumer groups	Does the group have any concerns about this type of packaging?
	Could they provide suggestions on potential design strategies?
Local government associations	Does local government have any concerns about the impacts of this type of packaging in the disposal or recovery stream?
	Are there any new technologies or systems being introduced or considered which will make this type of packaging more or less recoverable?

8   Implementing the Strategy   313

- Work with their most significant customers to build a common understanding of their sustainability requirements and re-align their product offering
- Based on this common understanding, agree with customers which metrics (carbon footprint, energy usage, waste and so on) to monitor and report to demonstrate ongoing improvements in the sustainability of their packaging
- Investigate other market segments where it could be legitimately argued that the packaging is more sustainable than the competition's, and target 'sustainability aware' customers in these new markets.

8.2.11 Develop a Packaging for Sustainability Action Plan

A packaging for sustainability action plan converts an intention to contribute to sustainable development through packaging into action. This ultimately embeds packaging for sustainability into strategic and operational planning.

The plan should reflect the business's current stage of evolution, leveraging current strengths and putting in place initiatives to address gaps and weaknesses. The stage of evolution in sustainable development is critical. For example, if a business currently sees no business case for sustainable development, there is little likelihood of obtaining resources to conduct LCAs on products. However, there may be an opportunity to build packaging efficiency goals and process improvement activities into product development because these can also reduce business costs. On the other hand, if a business is well-advanced in sustainable development the priority might be to increase the number of eco-effective projects that require more innovation and risk to achieve.

Development of the plan should consider:
- how packaging (each business unit and packaging decision) will contribute to achievement of the corporate sustainable development goals
- the impact of other sustainability goals on packaging's role and environmental impact
- what packaging-related decisions are made and who is involved
- how packaging for sustainability targets, challenges and achievements will be communicated internally and externally
- which stakeholders to involve and how to engage them.

One approach to documenting a packaging for sustainability plan is provided in Table 8.9. The plan should be developed for at least a 3-year period and reviewed at least annually as part of the strategic and corporate planning process.

Table 8.9 Example of a packaging for sustainability action plan

Business activity	Objective	Action required	Timing	Responsibility
Capacity building	Obtain board and senior executive commitment to sustainable development	Conduct sustainable development workshop and strategic planning session	By end Q1 in 1st year of plan	Sustainability Manager
		Conduct annual update on sustainable development issues and their impact on the business, including packaging for sustainability	Annual strategy planning session	
	Embed life cycle thinking into the culture of the organisation	Develop training program for all levels and functions within the organisation	By end of 1st year of the plan	Sustainability Manager and Innovation/R&D Manager
		Roll out the training program and incorporate it into new employee induction programs	Ongoing	Human Resources Manager
		Include sustainable development goals and performance metrics in performance evaluation processes	By end Q2 in 1st year of plan	
Strategic and operational planning	Set sustainable development goals and targets for product and packaging innovation	Conduct market research to understand and track views on and relevance of sustainable development issues (target markets for current products), including perceptions of packaging	1 product per Qtr on ongoing basis	Brand Managers

(continued)

8 Implementing the Strategy

Table 8.9 (continued)

Business activity	Objective	Action required	Timing	Responsibility
		Review current and future regulatory requirements, including regulations about packaging	Annual strategy planning session	Sustainability Manager
		Benchmark current products and packaging, also against competitors	1 product per Qtr on ongoing basis	Product and packaging development managers
		Develop a packaging database	By end Q2 in 1st year of plan	Packaging development managers and Procurement Managers
		Review literature on relevant LCAs and develop a life cycle map of current products	By end Q2 in 1st year of plan	Product development managers
		Conduct streamlined LCAs of products and packaging	2nd year, ongoing as part of NPD process	Product and packaging development managers
Marketing, communications and sales activities	Establish current positioning from a sustainable development perspective	Review corporate social responsibility benchmarks and other reports to benchmark current performance against competitors	By end Q1 in 1st year of plan	Marketing Manager
	Promote environmental credentials of products and packaging	Confirm positioning and key messages (dependent upon LCA data)	From Q3 in 1st year of plan	Brand Managers and Sales representatives
		Confirm role of on-pack labelling and logos to be used—document and communicate policy	From Q3 in 1st year of plan	Brand Managers and packaging development managers
		Ensure compliance of on-pack labelling and logos with corporate policy and relevant legislation	From Q3 in 1st year of plan	Packaging development managers

(continued)

Table 8.9 (continued)

Business activity	Objective	Action required	Timing	Responsibility
Developing products and packaging	Embed sustainable development into product and packaging development processes	Develop new policies, guidelines and process changes	By end Q2 in 1st year of plan	Innovation/R&D Manager
		Train development teams and executive team in process changes	By end Q3 in 1st year of plan	
		Review performance annually	Annual strategy planning session	
		Include routine updating of the packaging database in the development process	From Q3 in 1st year of plan	
		Identify decision-support tools	By end of year 1 of the plan	Product and packaging development managers
		Implement decision-support tools including training	Year 2 of plan	
	Meet or exceed annual sustainability goals and targets for product and packaging development	Identify and prioritise next year's projects	Annual strategy planning session	Innovation/R&D Manager, development teams and Executive Team
		Review prior year's performance and assess lessons learned	Annual strategy planning session	Innovation/R&D Manager and development teams
	Maintain an active pipeline of projects, balance eco-efficiency and eco-effectiveness	Conduct ideation workshop based around life cycle map, competitive analysis and benchmarking of current products	Start of Year 2 of plan	Innovation/R&D Manager
		Review portfolio of current projects versus achievement of short and long term sustainable development goals	Annual strategy planning session	Innovation/R&D Manager, Marketing Manager, Sustainability Manager and Executive Team

(continued)

8 Implementing the Strategy

Table 8.9 (continued)

Business activity	Objective	Action required	Timing	Responsibility
Procurement and supply chain management	Embed sustainable development into procurement and supply chain processes	Develop new policies, guidelines and process changes	By end Q2 in 1st year of plan	Supply Chain Manager and Procurement Managers
	Reduce carbon emissions of product supply chains from X to Y	Identify, prioritise and implement strategies (dependent upon LCA data)	From Q3 in 1st year of plan	
	Increase use of renewable energy from x% to y%	Identify strategies and develop the business case	By end of Year 1 of the plan	
		Obtain approval for implementation and initiate changes	By end Q2 in 2nd year of plan	
	Reduce energy and water consumption associated with the production or use of packaging throughout the supply chain to X and Y/kg of packaging	Identify, prioritise and implement strategies (dependent upon LCA data)	From Q3 in 1st year of plan	
	Source all paper and board packaging from FSC certified suppliers	Survey current suppliers to confirm current certification and proposed plans	By end Q2 in 1st year of plan	
		Develop and document policy and transition plan and advise suppliers of intentions to change	By end Q3 in 1st year of plan	
		Update vendor approval process and roll out changes	By end Q4 in 1st year of plan	

(continued)

Table 8.9 (continued)

Business activity	Objective	Action required	Timing	Responsibility
Process and environmental improvement	Reduce carbon emissions of operations from X to Y	Quantify current emissions and sources	By end Q3 in 1st year of plan	Sustainability Manager, Operations/Plant Manager and Process improvement managers
		Conduct brainstorm session to identify short, medium and long term strategies	Q4 in 1st year of plan	
		Assess feasibility of ideas and implement	From Q1 in 2nd year of plan	
	Inform and participate in packaging environmental regulations setting	Become a member of the sustainability committee of relevant industry/professional associations	By end Q1 in 1st year of plan	Sustainability Manager
	Embed packaging sustainability into the sustainability program	Update policy documents to specifically address packaging for sustainability goals and values	Month 1 of plan	
	Meet or exceed best practice in package waste generation from the manufacturing process	Identify and benchmark current waste management practices	By end Q1 in 1st year of plan	Operations/Plant Manager and Process improvement managers
		Identify and prioritise improvement initiatives	From Q2 in 1st year of plan	
	Increase recycling rates of packaging produced or used from x% to y%	Confirm current recycling processes and rates (which requires visits to MRFs in 3 locations) and identify barriers to recycling	By end Q3 in 1st year of plan	Sustainability Manager and Packaging/Technical development managers
		Identify, prioritise and implement strategies to address barriers to recycling	From Q4 in 1st year of plan	

8.3 Marketing, Communication and Sales Activities

Commitment from marketing and sales business units to sustainable development—and their willingness to lead it—will have a significant impact on the business's ability to set and achieve corporate sustainable development goals. Obtaining this commitment requires:

⇒ Learn more about marketing and communicating sustainability in Chap. 3

- confirmation of the business opportunities and risks for embracing or ignoring sustainable development within the marketing and sales strategies, informed by market research, business case analysis, awareness of regulatory impacts and training
- identifying the appropriate strategies for marketing and communicating sustainability (see Chap. 3)
- developing the marketing and sales teams' understanding of the life cycle impacts of the products and packaging (see Chap. 5) and how to communicate these to customers, consumers and other stakeholders (see Chap. 3).

8.4 Developing Products and Packaging

Most businesses have an established process for product and packaging development that supports product improvements or renovations as well as completely new products. Increasingly, this process is formalised and documented and, referred to as the new product development (NPD) process.

The NPD process typically involves a number of 'stages and gates' reflecting various phases of the innovation process such as ideas generation, feasibility, development and commercialisation (see Fig. 8.4). At each 'gate', decisions are made to approve, reject or review a particular development initiative. Gate decisions are made by representatives from various business units after assessing the submission of prescribed information. Often the gate decisions apply scorecards to assess the initiative against specific criteria such as strategic alignment, business case and technical and commercial risk. Results from the scoring process highlight areas of weakness to feedback to the development team and also assists in prioritising competing projects.

Fig. 8.4 Example of a new product development process with three 'gates'

8.4.1 Embed Sustainability into the Development Process

The development process should be modified to:
- inform ideas generation about the benefits and impacts of sustainability so that sustainability is inherently built into the design of new products and packaging
- present information at each gate to assess the sustainability benefits and impacts, and check how the new product-packaging system aligns with and contributes to the corporate sustainable development goals
- establish policies for approving or rejecting an initiative at each gate based on its sustainability benefits and impacts, particularly when these may be at odds with other corporate goals, such as consumer appeal or financial return.

There are many decision-support tools available to assist integration of sustainable development into the NPD process including:
- the packaging sustainability framework outlined in Chap. 2
- industry-specific packaging guidelines or checklists that can be revised to incorporate sustainability criteria
- LCA tools outlined in Chap. 7.

These or similar tools should be used as early as possible and practicable in the design process.

8.4.2 Conduct Formal Ideas Generation Sessions

The ideas generation stage of the NPD process provides new ideas to consider, and when assessed on an ongoing basis these maintain a pipeline of development initiatives to meet short, medium and long term business goals. Ideas are generated from many sources and range widely in their complexity, application and investment requirements.

Formal ideation and brainstorming processes are an important part of this phase particularly to:
- identify bigger picture ideas to meet stretch goals or specific strategic outcomes
- obtain commitment across business units to the same ideas
- develop the innovation capacity of the business.

From a sustainable development perspective, the formal ideation process is therefore best suited for new product and packaging ideas generation.

Prepare, Conduct, Assess, Feedback

A formal ideation process is:
- shaped by its preparation
- remembered for its atmosphere and spirit, and
- marketed on the basis of its success.

Preparation requires obtaining the commitment and interest of participants and developing an interesting and focused agenda that includes:
- the business case for sustainable development (Chap. 1) including regulatory requirements (Chap. 4)
- values, attitudes and expectations of consumers (Chap. 3) and other stakeholders
- environmental life cycles of existing or competing packaging systems (Chaps. 5 and 6).

Success of the process (which will involve multiple sessions) requires a good facilitator to ensure:
- full participation and consideration of a range of business units and all business perspectives
- a good balance between listening and learning versus doing and contributing
- initial assessments of ideas in the first session based on available facts
- follow-up processes and sessions to complete and review assessments and convert some ideas into projects.

With a focus on sustainability, Case Study 8.5 outlines an agenda and process for an ideas generation process, which can be used to inform any stage of the packaging development process.

Case Study 8.5 Running an Ideas Generation Session

Select a team to include representatives from a range of functional areas within the business such as technical, marketing and environment. It may also be useful to involve external stakeholders such as sustainability specialists or suppliers.

The agenda could include:
- introduction to the concept or market opportunity (presentation by marketing/discussion)
- review or preparation of a life cycle map (team exercise)
- review of LCA results if available (presentation/discussion)
- brainstorming ideas for the packaging, including sustainability features (team exercise)
- brainstorming ideas for new concepts for the product-packaging system
- preliminary prioritisation of ideas (sustainability benefits and feasibility) (team exercise)
- next steps (convenor).

8.4.3 Document Packaging Specifications

The sustainability principles introduced in Chap. 2 can be used to identify the design objectives relating to functionality, cost and sustainability performance (see Table 8.10). These can then be incorporated into the packaging specifications (Case Study 8.6).

8 Implementing the Strategy

Table 8.10 Writing the packaging specification

Sustainable packaging principle	Design objective	Examples of questions to be addressed in the packaging specification
Effective	Functionality Cost	What is the product being packed (e.g. description, purpose, size, weight, contents)? Where will it be sold (e.g. geographic region, distribution channel, type of store)? Who is the target consumer and what do we know about their needs/wants/expectations (e.g. for functionality, environmental performance, openability)? What are the mandatory performance requirements (e.g. physical protection, shelf life, moisture or gas barrier)? What are the mandatory transport and handling requirements (e.g. method of transport, size and height of pallet, retailer requirements)? What are the mandatory labelling requirements (e.g. ingredients, dangerous goods, nutritional information)? What is the price point for the product and the maximum cost of the packaging? Is there a requirement for a certain packaging format (e.g. to meet an industry standard)?
Efficient	Cost Environmental performance	Is every material component of the current (or comparable) packaging system essential to achieve the required level of functionality? What are the minimum packaging requirements to achieve the required level of functionality? Can any packaging component be eliminated or reduced in size/weight?
Cyclic	Environmental performance	Is there an existing recovery system for the packaging system (e.g. reuse, material recycling, composting)? Is it appropriate and desirable to design the packaging system for recovery through one or more of these channels? What are the technical requirements that need to be met for the packaging system to be recovered through the target recovery system (e.g. type of material, adhesives, inks)? How should the packaging be labelled to inform consumers about appropriate disposal or recovery (e.g. logos, directions for separation of material components, advice)?
Safe	Functionality Environmental performance	Are there any specific health or safety issues that need to be addressed during the design process (e.g. openability, tamper evidence, safe use of the product, migration of substances, safe disposal of the packaging)? Are there any packaging materials or components that should be avoided for health, safety or environmental reasons? Are there any manufacturing or printing processes that should be avoided for health, safety or environmental reasons?

8.5 Procurement and Supply Chain Management

A life cycle approach to packaging sustainability requires engagement and collaboration with suppliers to implement changes in materials, product and packaging design, distribution and recovery. For this reason, the supply chain team is critical to implementation of the packaging sustainability strategy.

From the customer's point of view (for example, the brand owner), supply chain engagement is necessary to:
- ensure that suppliers understand the business's sustainability needs
- collect information on the environmental and social impacts of the supply chain to support life cycle mapping exercises or a LCA
- ensure that chain-of-custody documentation is available where necessary (for example, for fibre-based packaging)
- gather market intelligence on new materials or technologies that could improve sustainability
- identify collaborative research and development projects to develop more-sustainable packaging.

Case Study 8.6 Incorporating Sustainability in the Packaging Specification

A typical specification might include the following material:
- Introduction: description of the product, intended markets
- Mandatory requirements: packaging material specifications, product protection, transport and handling, labelling, openability, environmental requirements
- Cost: product price, maximum packaging cost
- Sustainability considerations: sustainability issues and suggested strategies
- Aesthetic philosophy: emotional appeal, colour, shape, feel, shelf presence
- Production considerations: any limitations on materials/processing methods, design for assembly.

Supply chain engagement also has significant benefits for the supplier (for example, the packaging manufacturer). Responsiveness to their customer's sustainability expectations will help to build stronger, longer term relationships. Relationships based on collaboration and shared goals will justify increased investment in research and development.

Packaging sustainability cannot be achieved by one business or sector in the supply chain. Each sector has a role to play, and effective supply chain engagement can optimise sustainability through communication and collaboration

Table 8.11 Roles of each sector in the supply chain

Sector in supply chain	Roles
Raw material suppliers	Implement sustainability programs internally (water, energy, waste, emissions).
	Undertake R&D on environmentally improved materials/additives.
	Monitor and share information on global 'best practice' and innovation with customers.
	Provide sustainability information to customers (packaging manufacturers) and others in the supply chain.
Packaging manufacturers	Implement sustainability programs internally (water, energy, waste, emissions).
	Undertake R&D on environmentally improved packaging materials, formats & processing technologies.
	Integrate sustainability in design and procurement processes.
	Monitor and share information on global best practice and innovation with customers.
	Provide sustainability information to customers (product manufacturers) and others in the supply chain.
Product manufacturers	Implement sustainability programs internally (water, energy, waste, emissions).
	Integrate sustainability in design and procurement processes.
	Provide sustainability information to customers (retailers) and end-consumers.
Retailers	Implement sustainability programs internally (water, energy, waste, emissions).
	Integrate sustainability in procurement processes; for example, for distribution and shelf-ready packaging, retail shopping bags and so on.

(see Table 8.11). Procurement processes, including specifications, guidelines, questionnaires and regular meetings, should be amended to ensure that sustainability is considered in all packaging decisions.

Procurement Sustainability Strategies

The following strategies can be used by the supply chain team to support the achievement of packaging sustainability goals:
- a sustainability review for all packaging procurement, for example, through a supplier questionnaire or checklist (and documented in procurement procedures)
- inclusion of sustainability requirements in all packaging specifications, including labelling
- an agenda item to discuss packaging sustainability opportunities and innovation at all supplier meetings
- a commitment to buy products made from recycled packaging to help 'close the loop' in the general procurement policy.

Case Study 8.7 presents an example of Hewlett-Packard's approach to supply chain engagement in order to achieve packaging sustainability.

8.6 Process and Environmental Improvement

Build Packaging Sustainability into Environmental Management Systems

Packaging sustainability should be built into existing systems for process and environmental improvement. Relevant initiatives include:
- a commitment to packaging sustainability in the business's environment or sustainability policy
- objectives, metrics and targets for packaging in the environmental or sustainability strategy
- Information about packaging sustainability impacts and review processes included in the business's environmental management system.

Case Study 8.7 Sustainable Procurement at Hewlett Packard

Businesses in the Information and Communications Technology (ICT) sector are required to meet a range of environmental regulations, and many individual businesses are also implementing their own sustainability strategies that place new requirements on suppliers.

Hewlett Packard includes detailed packaging requirements in its 'General specification for the environment' [11]. These cover:
- restricted materials
- ozone-depleting substances in packaging materials
- heavy metals in packaging materials
- polyvinyl chloride (PVC)
- recyclable materials
- material coding
- packaging and pallets made of wood
- secondary packaging restrictions
- wood, paper, and other plant-based packaging restrictions.

Hewlett Packard has also collaborated with other ICT sector businesses to develop a sustainability self-assessment tool for suppliers [12]. This is designed to:
- raise supplier awareness about the importance of sustainability principles
- clarify ICT customer expectations about suppliers' sustainability practices
- support customer assessments of supplier characteristics and potential risks
- enable suppliers to evaluate, improve and communicate their performance
- reduce the burden on suppliers of responding to multiple questionnaires.

Sources: Hewlett Packard [11], GeSI and EICC [12]

Sustainable development goals such as water, energy and material efficiency should also be built into business improvement and capital expenditure processes. This can be done, for example, by including the impact on quantitative metrics (such as relevant indicators from Table 8.6) and other business benefits, such as improved corporate reputation, in proposals for new plant or equipment.

8.7 Conclusion

Packaging is required to supply products that create social, economic and environmental value and is one of many factors to address in the complexity of sustainable development challenges. By learning to understand and respond to environmental concerns about packaging, the global packaging supply chain is emerging as a case study for leadership in sustainable development.

However, businesses operating within the packaging supply chain are at various stages of evolution in sustainable development. Emerging leaders are addressing the sustainability of their product-packaging systems using life cycle thinking. They are adopting product stewardship, thereby not only addressing their immediate environmental impacts but taking ownership of the environmental consequences of their decisions and progressively evolving the terms of business and their product mix.

For less developed businesses, packaging sustainability provides a platform from which to create a sustainable development capacity. They need to respond to pressure from the supply chain as well as regulatory pressure for change. In this endeavour, they are able to leverage the significant amount of global research and practical projects undertaken on packaging sustainability over the last decade. This book collates much of this work. The aim of the book has been to turn this information into useful resources for businesses so that they can benchmark their current performance and gain tools and strategies to inform packaging for sustainability strategies.

References

1. PricewaterhouseCoopers LLP (2010) Sustainable Packaging: Threat or Opportunity? PricewaterhouseCoopers LLP, London
2. VIP Packaging (2011) Sustainability, packaging and VIP packaging. http://www.vippackaging.com.au/?id=407 (cited 11 March 2011)
3. Nidumolu R, Prahalad CK, Rangaswami MR (2009) Why sustainability is now the key driver of innovation. Harv Bus Rev 87(9):56–64
4. Sustainable Packaging Coalition (SPC) (2006) Design guidelines for sustainable packaging, version 1.0. Sustainable Packaging Coalition Charlottesville, Virginia
5. Cooper RG, Mills MS (2005) Succeeding at New Product Development the P&G way: a key element is using the 'Innovation Diamond'. PDMA Vis XXIX(4): 9–13
6. Australian Packaging Covenant (2010) A commitment by governments and industry to the sustainable design, use and recovery of packaging. APC Secretariat, Sydney

7. Adaptation Environmental Services, Australian Packaging Covenant (2010) Case Study: industry implementation of the Sustainable Packaging Guidelines. Australian Packaging Covenant, Sydney
8. WRAP (2006) UK Packaging Benchmark. http://www.wrap.org.uk/retail/tools_for_change/uk_packaging_benchmark/ (cited 11 January 2010)
9. Envirowise (2008) Packguide: a guide to packaging eco-design. Oxfordshire
10. Dunphy D, Griffiths A, Benn S (2003) Organisational change for corporate sustainability. Routledge, London
11. Hewlett Packard (2010) HP standard 011 general specification for the environment
12. Global e-Sustainability Initiative (GeSI), Electronic Industry Code of Conduct (EICC) (2007) The information and communications technology (ICT) supplier self-assessment questionnaire. Global e-Sustainability Initiative (GeSI) Supply Chain Working Group and Electronic Industry Code of Conduct (EICC) Implementation group
13. Raja S, Hagedorn R (2010) Global packaging project: pilot findings, in Presentation to Global Packaging Project, Paris (14 October 2010)
14. The Consumer Goods Forum (2011) Global protocol on packaging sustainability 2.0: draft for consultation
15. The Consumer Goods Forum (2010) Packaging sustainabiltiy indicators and metrics framework 1.0

Editors Biography

Dr. Karli Verghese Karli is Program Director, Sustainable Products and Packaging at the Centre for Design at RMIT University, Melbourne, Australia. She has been involved in research specialising in life cycle assessment (LCA), packaging sustainability and eco-design projects since 1996. Her research interests also include developing qualitative and quantitative tools to assist designers, packaging technologists and other professionals in environmental decision-making. She has been a key researcher in the Sustainable Packaging Alliance (SPA) and was the Product (Research) Development Manager for PIQET—Packaging Impact Quick Evaluation Tool, an online life cycle-based decision support tool for the packaging supply chain. Karli is a Fellow of the Australian Institute of Packaging. She has an active publishing profile of which includes her other book (with Ralph Horne and Tim Grant) 'Life Cycle Assessment: Principles, Practice and Prospects' that was published in 2009 (CSIRO Publishing).

Dr. Helen Lewis Helen is an environmental consultant specialising in sustainable packaging design, product stewardship and environmental communication. Her involvement with the packaging industry began in the early 1990s as Environment Manager for the Plastics and Chemicals Industries Association. She moved to the Centre for Design at RMIT University, becoming Director in 2001. This role included the design and management of research projects on the environmental impacts and ecodesign of buildings, products and packaging. As Director of the Centre she helped to establish the Australian Sustainable Packaging Alliance in collaboration with representatives from Birubi Innovation and Victoria University's Packaging and Polymer Research Unit. Helen is a Fellow of the Australian Institute of Packaging and Adjunct Professor at RMIT University. Her other publications include (with John Gertsakis) 'Design + Environment: a Global Guide to Designing Greener Goods' (Greenleaf 2001).

Dr. Leanne Fitzpatrick Leanne is currently the Executive Director of the Australian based Sustainable Packaging Group of companies and General Manager of Birubi Innovation and Birubi Foods. She began her career in the food industry in 1987 as a

Process Development Engineer with Kraft Foods and her career has involved a broad range of food innovation, research, and education/technology transfer roles across many science and technology platforms. She has planned, designed, managed and delivered R&D/innovation, consulting, training, mentoring and technology transfer programs in private, industry and public sector organisations including Warrnambool Cheese and Butter, Dairy Australia, Dairy Innovation Australia, the Cooperative Research Centres for Innovative Dairy Products and, Food and Packaging, Bonlac Foods, Monash University, and RMIT University. She has extensive experience managing multi-disciplined/multi-organisational teams/projects. In 2002, in collaboration with representatives from RMITs Centre for Design and Victoria University's Packaging and Polymer Research Unit, Leanne initiated the establishment of the Sustainable Packaging Alliance as a focal point to facilitate the development of packaging sustainability by Australian businesses.

Appendix A
Application of a Sustainable Packaging Framework

Retailers around the world are trying to respond to consumer concerns about the environmental impact of single-use plastic bags, but which option is best? For example, should they be replaced with paper bags, or should consumers be encouraged to buy a sturdy reusable polypropylene (PP) bag?

Create a Life Cycle Map

Simplified life cycle maps of two bag options are illustrated in Figs. A.1 and A.2. In practice, a more detailed life cycle map is required to evaluate the sustainability impacts.

Apply the Principles

Assess each principle (effective, efficient, cyclic, safe) in turn and then together in order to identify factors to be considered in choosing a particular bag format. As each principle is discussed, also identify opportunities for improvement—as optimisation of a particular option may be better than changing formats.

Table A.1 shows how the framework might be used to assess these retail bag options.

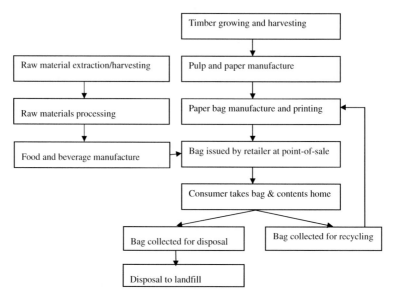

Fig. A.1 Life cycle of a paper bag

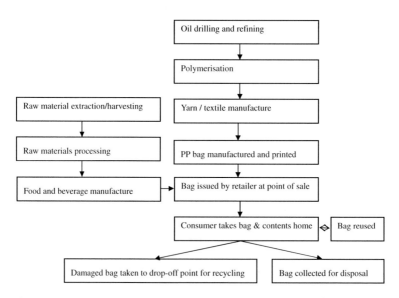

Fig. A.2 Life cycle of a reusable PP bag

Appendix A: Application of a Sustainable Packaging Framework

Table A.1 Using the packaging sustainability framework to evaluate impacts and identify opportunities

Principles	Single use paper bag impacts and opportunities	Reusable PP bag impacts and opportunities
Effective: sustainable value The packaging system achieves its functional requirements with minimal social and environmental impact.	The bag is functional for many products but can break if used for wet products (e.g. frozen/refrigerated). Can only be used once. *Opportunities* • Redesign to strengthen the bag, e.g. reinforce the base or handles, or use a biodegradable coating.	The bag is functional for all products. The bag is very durable and can be reused many times for shopping and other tasks. LCA's assume a usage rate of a least 100 [e.g. 1] By encouraging a change in consumer habits, reusable bags might help to promote an ethic of environmental responsibility. Many retailers use income from the sale of bags to fund community or environmental initiatives. *Opportunities* • Design for ease of use by consumers and retailers. • Promote environmental benefits. • Encourage maximum reuse.
Efficient: minimal use of materials, energy and water The packaging system is designed to use materials and energy efficiently throughout the product life cycle. Efficiency can be defined through reference to world's best practice at each stage of the packaging life cycle.	The bag weighs 47 g (total material consumption over a 2 year period is approx. 24 kg). Pulp and paper manufacture uses a relatively large amount of water and energy. *Opportunities* • Lightweight as much as possible without compromising functionality. • Design pulp mill to be energy- and water-efficient.	The bag weighs 116 g and can be reused many times. Consumption of water and energy is significantly less than for the paper bag if reused approx. 100 times. Some water and energy will be used if the bag is washed by the consumer. *Opportunities* • Lightweight as much as possible without compromising functionality. • Design textile mill to be energy- and water-efficient.

(continued)

Table A.1 (continued)

Principles	Single use paper bag impacts and opportunities	Reusable PP bag impacts and opportunities
Cyclic: minimising waste Packaging materials used in the system are cycled continuously through natural or industrial systems, with minimal material degradation. Recovery rates should be optimised to ensure that they achieve energy and greenhouse gas savings.	Paper is a renewable resource. Paper bags are highly recyclable and can be manufactured back into bags. Some fibre is lost in the recycling process. *Opportunities* • Use 100% recycled content. • Design for recycling. • Promote recyclability. • Use renewable energy. • Recycle water from the pulp mill.	The bag minimises waste by replacing approx. 100 single use paper or plastic bags. PP is recyclable, but collection facilities for the bags are limited at present due to the small volume of material available. The overall impact on solid waste is roughly the same for both bags. *Opportunities* • Use some recycled content. • Design for recycling. • Promote reusability. • Establish a recycling program. • Use renewable energy. • Recycle water from the textile mill.
Safe: non-polluting and non-toxic Packaging components used in the system, including materials, finishes, inks, pigments and other additives do not pose any risks to humans or ecosystems. When in doubt the precautionary principle applies.	Paper bags have a much higher impact on global warming because of the larger mass of material used over a 2 year period. Paper bags have a higher impact on eutrophication (nutrients to waterways) and land degradation due to forestry operations. *Opportunities:* • Fibre certified from sustainably managed forests. • 100% recycled content. • Minimal printing. • Vegetable-based inks.	Green pigments can be toxic. If overloaded, the bags can be an occupational health and safety risk for retail staff. *Opportunities:* • Verify that the pigment is non-toxic. • Redesign checkouts to avoid lifting. • Reduce bag size.

Source: Based on information from Lewis H, Verghese K, Fitzpatrick L (2010) Evaluating the sustainability impacts of packaging: the plastic carry bag dilemma. Packaging Technol Sci 23(3):145–160

Note The table compares the sustainability impacts and benefits of shopping bags over a 2 year period, assuming one shopping trip per week

Appendix B
Labels and Logos

This table explains and illustrates some of the many labels and logos used to promote sustainability performance on packaging. The list is not comprehensive.

Sustainability strategy	Label	Purpose and details	Further information
Material recycling to reduce waste	*Mobius loop* (example only)	*Purpose*: To indicate that a product or package is recyclable *Used*: Worldwide *Owned*: Public domain *Legal status*: Voluntary	ISO 14021: 1999 ISO 7000, symbol no. 1135
	'Green Dot'	*Purpose*: To indicate that a financial contribution has been paid for that package to a qualified national packaging recovery organisation that has been set up in accordance with the principles defined in European Packaging and Packaging Waste Directive 94/62 and the respective national law *Used*: European Union *Owned*: Duales System Deutschland (DSD), licensed to PRO EUROPE *Legal status*: Voluntary	DSD http://www.gruener-punkt.de PRO EUROPE http://www.pro-e.org
	On-pack recycling label (example only)	*Purpose*: To provide more specific information to consumers on the availability of recycling services *Used*: United Kingdom *Owned*: British Retail Consortium *Legal status*: Voluntary	OPRL http://www.onpackrecyclinglabel.org.uk

(continued)

Appendix B: Labels and Logos

Sustainability strategy	Label	Purpose and details	Further information
	Resin identification codes	Purpose: To identify the type of plastic used to make packaging in order to facilitate source separation of recyclable packaging by consumers and recyclers Used: Worldwide Owned: Public domain (originally developed by US Society of the Plastics Industry) Legal status: Introduced as a voluntary code but regulated in parts of the United States (see Chap. 4)	US Society of the Plastics Industry (SPI) http://www.plasticsindustry.org/AboutPlastics/content.cfm?ItemNumber=825&navItemNumber=1124 British Plastics Federation www.bpf.co.uk/Sustainability/Plastics_Recycling_Markings.aspx
Organic recycling to reduce waste	Compostable packaging	Purpose: To show that finished products or packaging (including components and residual product) are compostable and tested and certified as meeting the requirements of DIN EN 13432 in connection with ASTM D6400, DIN EN 14995 and ISO 17088 Used: Europe, Australasia Owned: registered trademark of European Bioplastics Legal status: voluntary	European Bioplastics http://www.european-bioplastics.org (For information on certification in Australasia to AS 4736, contact the Australasian Bioplastics Association—http://www.bioplastics.org.au)

(continued)

338 Appendix B: Labels and Logos

(continued)

Sustainability strategy	Label	Purpose and details	Further information
Recycled content of packaging to reduce waste and improve efficiency	*Mobius loop* (example only)	*Purpose*: To indicate that a product or package is made from recycled material (number indicates percentage of recycled material) *Used*: Worldwide *Owned*: Public domain *Legal status*: Voluntary	ISO 14021: 1999
	Recycled cartonboard	*Purpose*: To indicate that the package contains recycled material, is recyclable, and made in Australia *Used*: Australia *Owned*: Organisation of the Australian Recycled Cartonboard Campaign *Legal status*: Voluntary	Organisation of the Australian Recycled Cartonboard Campaign http://www.arc.org.au
Encourage consumers not to litter	*Tidyman*	*Purpose*: To encourage consumers to dispose of the package in a rubbish or recycling bin *Used*: Worldwide *Owned*: Public domain *Legal status*: Voluntary	Victorian Litter Action Alliance http://www.litter.vic.gov.au/www/html/1320-tidyman-logo.asp
Paper manufactured with a chlorine free bleaching process	*Totally chlorine free* (virgin paper)	*Purpose*: To indicate that the paper meets environmental criteria including no bleaching with chlorine and no fibre from old growth forests *Used*: Worldwide *Owned*: Chlorine Free Products Association *Legal status*: Voluntary	Chlorine Free Products Association http://www.chlorinefreeproducts.org

(continued)

Appendix B: Labels and Logos

(continued)

Sustainability strategy	Label	Purpose and details	Further information
	Process chlorine free (recycled paper)	*Purpose*: To indicate that the paper meets environmental criteria including at least 30% post-consumer recycled fibre, no bleaching or re-bleaching with chlorine and no fibre from old growth forests. *Used*: Worldwide *Owned*: Chlorine Free Products Association *Legal status*: Voluntary	Chlorine Free Products Association www.chlorinefreeproducts.org
Forest stewardship—paperboard manufactured with fibre sourced from sustainably managed forests	*Forest Stewardship Council*	*Purpose*: To certify that a product or package meets the Forest Stewardship Council's minimum standards for sustainable resource extraction and processing *Used*: Worldwide *Owned*: Forest Stewardship Council *Legal status*: Voluntary	Forest Stewardship Council http://www.fsc.org
Carbon emissions being reduced to address climate change	*The Carbon Reduction Label* (example only)	*Purpose*: Shows which products have had their footprints certified and committed to a carbon emission reduction. The footprint is based on the whole life cycle of the product including raw materials and packaging, through to manufacture, transportation, use and disposal. *Used*: Worldwide *Owned*: Carbon Trust Footprinting Company, a subsidiary of The Carbon Trust *Legal status*: Voluntary	The Carbon Reduction Label http://www.carbon-label.com The Carbon Trust Footprinting Certification Company http://www.carbontrustcertification.com The Carbon Trust http://www.carbontrust.co.uk Planet Ark (Australia) http://carbonreductionlabel.com.au

(continued)

(continued)

Sustainability strategy	Label	Purpose and details	Further information
Use of renewable energy to reduce carbon emissions	Green-e Marketplace	*Purpose*: To certify that a business has purchased (or generated) a qualifying amount of renewable energy to offset its electricity use *Used*: United States *Owned*: Center for Resource Solutions *Legal status*: Voluntary	Center for Resource Solutions http://www.green-e.org/getcert_bus_what.shtml

Appendix C
Matrix of International Regulations, Policies and Standards

This builds on Lewis, H. (2009), *Packaging and product stewardship: International regulations and policies*, Sustainable Packaging Alliance, Melbourne.

Appendix C: Matrix of International Regulations, Policies and Standards

Australasia

Country	Legislation/Policy	Description	Policy measure	Link
Australia	Australian Packaging Covenant (2010-)	The Covenant is the voluntary component of a co-regulatory policy framework for packaging. It's an agreement between governments (federal, state and local) and the packaging supply chain to reduce the environmental impacts of consumer packaging. The APC replaced the previous National Packaging Covenant (*Mark I 1999–2004, Mark II 2005–2010*)	Voluntary agreement Code of practice	http://www.packagingcovenant.org.au
	National Environment Protection Measure (NEPM) for Used Packaging Materials	The regulatory 'safety net' for the Covenant. It does not apply to compliant Covenant signatories or those who can satisfy governments that they are producing 'equivalent outcomes' to those achieved through the Covenant. The NEPM is implemented by state and territory governments.	Extended producer responsibility	http://www.ephc.gov.au/taxonomy/term/48
	Container Deposit Legislation (South Australia)	South Australia has had a deposit and return system for beverage containers since 1975.	Container deposit legislation	http://www.epa.sa.gov.au/households/waste_and_recycling/10c_refund
	Container Deposit Legislation (Northern Territory)	Legislation to introduce a deposit scheme for beverage containers was passed in February 2011.	Container deposit legislation	http://www.cashforcontainers.nt.gov.au
	Ban on plastic checkout bags (South Australia)	South Australia banned the use of single-use, non-biodegradable plastic checkout bags from 1 May 2009.	Ban	http://www.byobags.com.au

(continued)

Appendix C: Matrix of International Regulations, Policies and Standards

(continued)

Australasia

Country	Legislation/Policy	Description	Policy measure	Link
	Ban on plastic checkout bags (Northern Territory)	Lightweight single-use, non-biodegradable plastic checkout bags were banned from 1 September 2011.	Ban	http://www.nt.gov.au/nreta/environment/plasticbagban/
	Australian Consumer Law	'Green marketing and the Australian Consumer Law' (2011) provides advice from the Australian Competition and Consumer Commission on environmental claims and labels.	Trade practices legislation	http://www.accc.gov.au/content/index.phtml/itemId/815763
New Zealand	Product Stewardship Scheme	In 2010 the Packaging Council of New Zealand launched a voluntary Packaging Product Stewardship Scheme which builds on the successes of the previous Packaging Accord and meets the requirements of the Waste Minimisation Act 2008. Its objectives are: • improved packaging design to reduce packaging waste • improved systems to reduce packaging waste • increased reuse of packaging • increased recycled content in packaging to replace virgin material • increased consumer awareness and understanding of sustainable packaging.	Voluntary agreement	http://www.packaging.org.nz/packaging_stewardship/packaging_stewardship.php
	Code of Practice for Packaging Design, Education and Procurement 2010	The Packaging Council's Code of Practice was updated in 2010 and is an integral component of the Product Stewardship Scheme.	Code of practice	http://www.packaging.org.nz/packaging_info/packaging_code.php

(continued)

Appendix C: Matrix of International Regulations, Policies and Standards

Americas	Legislation/Policy	Description	Policy measure	Link
Brazil	Pesticide containers	Brazil introduced take-back regulations for pesticide containers in 2002.	Extended producer responsibility	http://ec.europa.eu/environment/ppps/pdf/container_management.pdf
	Brasília Recycling Law (No 3651)	The federal district of Brasília enacted a law requiring manufacturers and distributors of plastic bottles to collect and reuse a percentage of their products.	Extended producer responsibility	
Canada	Canada-wide Strategy for Sustainable Packaging (2009)	Commits all jurisdictions to work towards the establishment of EPR programs for packaging (among other things) within 6 years, and sets out general principles and guidance for provincial/territorial regulators.	Extended producer responsibility	http://www.ccme.ca/assets/pdf/sp_strategy.pdf
	Beverage Container Recycling Regulation (Alberta)	Deposits cover a broad range of beverage containers, including plastic bottles, sealed cups, liquid paperboard cartons, plastic pouches, aluminium cans, bag-in-a-box packs, glass bottles and steel cans.	Container deposit legislation	http://www.bottlebill.org/legislation/canada/alberta.htm
	Beverage Container Stewardship Program Regulation (1997) (British Columbia)	In 1997 this replaced the mandatory deposit-refund system for soft drink and beer containers. It requires all brand owners of ready-to-drink beverages (with the exception of milk, milk substitutes, liquid meal replacements and infant formula) to establish a province-wide collection program under a deposit-refund system.	Extended producer responsibility	http://www.bottlebill.org/legislation/canada/britishcolumbia.htm
	Waste Reduction and Prevention Act (Manitoba)	Beverage producers have been given a choice of either introducing a deposit-return system or paying a levy per container. Beer manufacturers have chosen to introduce a deposit scheme, but all others are paying the levy. The levy is used to pay for 80% of the costs of kerbside recycling in the province.	Levy	http://www.ec.gc.ca/epr/default.asp?lang=En&n=3374501D-1

(continued)

Appendix C: Matrix of International Regulations, Policies and Standards

(continued)

Americas

Legislation/Policy	Description	Policy measure	Link
Beverage Containers Act (1992) (New Brunswick)	Deposits are charged on all beverage containers except those for milk. Only half of the deposit is returned for non-refillable containers—the remainder is retained as an 'environmental fee'.	Container deposit legislation	http://www.bottlebill.org/legislation/canada/newbrunswick.htm
Waste Management regulations (2003) (Newfoundland)	Deposits are charged on all beverage containers except those for milk. Only part of the deposit is returned for non-refillable containers; the remainder is retained as an 'environmental fee'.	Container deposit legislation	http://www.bottlebill.org/legislation/canada/newfoundland.htm
Beverage Container regulations (2005) (Northwest Territories)	Deposits are charged on all beverage containers except those for milk, wine and spirits.	Container deposit legislation	http://www.bottlebill.org/legislation/canada/northwest.htm
Solid Waste-Resource Management Regulations (2000) (Nova Scotia)	Deposits are charged on all beverage containers except those for milk. Only half of the deposit is returned for non-refillable containers; the remainder is retained as an 'environmental fee'.	Container deposit legislation	http://www.bottlebill.org/legislation/canada/novascotia.htm
Waste Diversion Act (2002) (Ontario)	Industry and municipal representatives established the Blue Box recycling program in the province in the mid-1980s on a voluntary basis. The Waste Diversion Act now requires all businesses that introduce packaging or paper into the Ontario market to fund 50% of the Blue Box collection program.	Extended producer responsibility	http://www.stewardshipontario.ca
Ontario Deposit Return System (Ontario)	Ontario has mandatory deposits on alcoholic beverage containers. Other beverage containers are collected through the Blue Box program.	Container deposit legislation	http://www.bottlebill.org/legislation/canada/ontario.htm

(continued)

(continued)

	Legislation/Policy	Description	Policy measure	Link
Americas	Litter Control Regulations (amended 2008) (Prince Edward Island)	Deposits are charged on all beverage containers except those for milk. Non-refillable containers for beer and soft drink were originally banned (since 1977), but the ban was repealed in 2008. Beverages may not be sold if they are connected by plastic rings or any other connecting device which is not biodegradable or photodegradable.	Container deposit legislation	http://www.bottlebill.org/legislation/canada/princeedward.htm http://www.beveragecontainers.pe.ca/index.php?lang=E
	Litter Control Act (Saskatchewan)	Mandatory deposits are charged on beverage containers excluding milk. (Milk is covered by a voluntary deposit scheme.)	Container deposit legislation	http://www.bottlebill.org/legislation/canada/saskatchewan.htm
	Act respecting the sale and distribution of beer and soft drinks in non-returnable containers (Quebec)	Quebec has had a deposit on soft drink and beer containers since 1984. Each beer manufacturer has to provide no more than 37.5% of their production (as measured by the number of containers produced) in non-refillable containers.	Container deposit legislation	http://www.bottlebill.org/legislation/canada/quebec.htm
United States	Resource Conservation and Recovery Act (1976)	The Act covers treatment, storage and transport of hazardous waste; establishes requirements for state solid waste management plans; and directs federal agencies to buy recycled-content items.		

(continued)

Appendix C: Matrix of International Regulations, Policies and Standards 347

(continued)

Americas

Legislation/Policy	Description	Policy measure	Link
Federal Trade Commission Act	'Part 260—Guides for the use of environmental marketing claims' provides advice from the Federal Trade Commission on legal requirements for environmental marketing and advertising. Proposed amendments to the 'Green Guides' were published in 2010.	Trade practices legislation	http://www.ftc.gov/bcp/grnrule/guides980427.htm http://www.ftc.gov/os/fedreg/2010/October/101006greenguidesfrn.pdf
Toxics in Packaging Bill	In 1990 a model Toxics in Packaging Bill was introduced by the Coalition of Northeastern Governors. The Bill calls for: • a ban on the intentional use of lead, cadmium, mercury and hexavalent chromium in packaging • a limit on the sum of the concentration of incidentally introduced lead, cadmium, mercury and hexavalent chromium to 600 ppm 2 years after the law is introduced, 250 ppm after 3 years and 100 ppm after 4 years. There are a number of exemptions including packaging made from recycled materials. As at 2010 the Bill has been adopted by 19 states: California, Connecticut, Florida, Georgia, Illinois, Iowa, Maryland, Maine, Minnesota, Missouri, New Hampshire, New Jersey, New York, Pennsylvania, Rhode Island, Vermont, Virginia, Washington and Wisconsin.	Design requirements	http://www.toxicsinpackaging.org

(continued)

(continued)

Americas

Legislation/Policy	Description	Policy measure	Link
Resin Identification Codes	The Society of the Plastics Industry introduced the plastics identification code in 1988 to assist in the identification of plastic containers for recycling. As at 2009 its use is mandatory for plastic bottles and rigid containers in 39 states: Alaska, Arizona, Arkansas, California, Colorado, Connecticut, Delaware, Florida, Georgia, Hawaii, Illinois, Indiana, Iowa, Kansas, Kentucky, Louisiana, Maine, Maryland, Massachusetts, Michigan, Minnesota, Mississippi, Missouri, Nebraska, Nevada, New Jersey, North Carolina, North Dakota, Ohio, Oklahoma, Oregon, Rhode Island, South Carolina, South Dakota, Tennessee, Texas, Virginia, Washington and Wisconsin. The association is investigating a potential expansion of the coding system to include a wider range of resins. (There are currently codes for the six most commonly used resins and one for 'other'.)	Labelling	http://www.plasticsindustry.org/AboutPlastics/content.cfm?ItemNumber=823

(continued)

Appendix C: Matrix of International Regulations, Policies and Standards 349

(continued)

Americas

Legislation/Policy	Description	Policy measure	Link
Restrictions on the use of expanded polystyrene (EPS) packaging	San Francisco (*Food Service Waste reduction Ordinance—400121*, June 2007) has banned EPS food service packaging and requires all food service packaging to be biodegradable, compostable or recyclable (subject to an affordability clause). Santa Monica has banned polystyrene (EPS and clear polystyrene) takeaway food packaging and requires all plastic takeaway food packaging to be compatible with the city's recycling infrastructure (subject to an affordability clause). Other cities and counties with a ban on EPS takeaway packaging include Portland, Oregon; Suffolk County, New York; Freeport, Maine; Galena and Kotlik, Alaska. Portland appears to have the most extensive ban on EPS. Exemptions are available if the ban is likely to result in 'undue hardship'.	Ban	http://www.sfgov2.org/index.aspx?page=886 http://www.portlandonline.com/auditor/index.cfm?c=ciij#cid_215460
Restrictions on the use of Bisphenol A (BPA)	Connecticut has passed legislation, effective in October 2010, to ban the use of BPA in containers for infant formula and baby food and for all reusable food and beverage containers. Minnesota, Chicago and Suffolk County have banned the use of BPA in baby bottles and 'sippy' cups. On 15 January 2010 the Food and Drug Administration (FDA) shifted its position on BPA, stating that it had concerns about its safety and would undertake further research.	Ban	http://www.greenbiz.com/blog/2010/01/25/what-does-fdas-bpa-decision-mean-companies

(continued)

(continued)

Americas

Legislation/Policy	Description	Policy measure	Link
Rigid plastic packaging container (RPPC) statute 1995 (California)	California introduced the RPPC statute in January 1995. It required industry to maintain an overall 25% aggregate recycling rate for rigid plastic containers or individual brand owners would face a variety of alternative requirements. In September 2004 Senate Bill 1729 (Chesbro, Chap. 561, Statutes of 2004) repealed the requirement for the California Integrated Waste Management Board to publish the 'all container' recycling rate for RPPCs. As of 1 January 2005 the minimum recycling rate is no longer available as a compliance option for regulated businesses. Product manufacturers (brand owners) must now demonstrate that their containers fulfil at least one of the following requirements: • made from at least 25% post-consumer resin • source-reduced (lightweighted) by 10% • reused or refilled at least five times • have a recycling rate of 45% if it is a brand-specific or particular type of RPPC.	Design requirements Recycling requirements	http://www.ciwmb.ca.gov/plastic/rppc/ http://www.ciwmb.ca.gov/regulations/title14/ch4a3a.htm#top
Plastic bags—Anacostia River Cleanup and Protection Act (2009) (District of Columbia)	Legislation has been introduced to require Retail Food Establishment license holders in the District of Columbia to charge consumers for plastic and paper carry bags. The regulation was implemented in January 2010.	Levy	http://www.examiner.com/x-16393-DC-Environmental-News-Examiner~y2009m7d9-District-of-Columbia-bag-tax-to-go-in-effect-January-2010

(continued)

Appendix C: Matrix of International Regulations, Policies and Standards 351

(continued)

Americas

Legislation/Policy	Description	Policy measure	Link
Beverage Container Recycling and Litter Reduction Act (1986) (California)	There is a deposit and return system for all non-refillable beverage containers (except milk).	Container deposit legislation	http://www.bottlebill.org/legislation/usa/california.htm
Plastic bag regulations (California)	The City of San Francisco (Plastic Bag Reduction Ordinance) has banned supermarkets and pharmacists from using non-recyclable plastic checkout bags made from petroleum products. Retailers are permitted to use compostable plastic bags (certified to the ASTM standard), recyclable paper bags and reusable bags of any kind. In November 2010 the City of Los Angeles banned plastic checkout bags and imposed a 10% levy on paper checkout bags.	Ban	http://sf311.org/index.aspx?page=552 http://plasticbaglaws.org/
Beverage Container Deposit and Redemption Law (1978, amended 2009) (Connecticut)	There is a deposit and return system for certain beverage containers including those for beer and carbonated soft drinks. Containers for non-carbonated soft drinks and HDPE bottles are excluded. Bottled water was added on 1 April 2009. Metal containers with removable tabs and containers with non-biodegradable holders (plastic 6-pack rings) are banned from being sold in the state.	Container deposit legislation Ban	http://www.bottlebill.org/legislation/usa/connecticut.htm

(continued)

(continued)

Americas

Legislation/Policy	Description	Policy measure	Link
Plastic bags – House Bill 15 (2009) (Delaware)	Retailers and chain stores that give out plastic bags to consumers are required to provide collection bins for recycling plastic bags.	Recycling requirements	http://whyy.org/cms/news/government-politics/2009/08/17/delaware-moves-to-limit-plastic-grocery-bag-use/15098
Beverage Container Regulation (1982) (Delaware)	There is a deposit and return system for beer, malt, ale, soft drinks, mineral water and soda water.	Container deposit legislation	http://www.bottlebill.org/legislation/usa/delaware.htm
Solid Waste Management; Deposit Beverage Container Law (2002) (Hawaii)	Hawaii introduced a bottle bill in 2002 imposing a refundable deposit on beer, mixed spirits, mixed wine, coffee and teas, carbonated soft drinks and water (excludes dairy, wine and liquor).	Container deposit legislation	http://www.bottlebill.org/legislation/usa/hawaii.htm
Beverage Container Deposit Law (1981) (Iowa)	There is a deposit and return system for beer, soft drinks, wine and liquor. Deposit containers were banned from landfill in 1990.	Container deposit legislation	http://www.bottlebill.org/legislation/usa/iowa.htm
Maine Returnable Beverage Container Law (1976) (Maine)	There is a deposit and return system for beverage containers except dairy products and non-alcoholic cider.	Container deposit legislation	http://www.bottlebill.org/legislation/usa/maine.htm
Beverage Container Recovery Law (1981) (Massachusetts)	There is a deposit and return system for beer, malt, carbonated soft drinks and mineral water.	Container deposit legislation	http://www.bottlebill.org/legislation/usa/massachusetts.htm
Michigan Beverage Container Act (1976) (Michigan)	There is a deposit and return system for beer, soft drinks, carbonated and mineral water, wine coolers and canned cocktails.	Container deposit legislation	www.bottlebill.org/legislation/usa/michigan.htm
New York State Returnable Container Law (1982) (New York)	There is a deposit and return system for beer, malt, carbonated soft drinks, mineral water and wine coolers.	Container deposit legislation	http://www.bottlebill.org/legislation/usa/newyork.htm

(continued)

Appendix C: Matrix of International Regulations, Policies and Standards 353

Asia	Legislation/Policy	Description	Policy measure	Link
China	Excessive Packaging Law and National Standard (2009)	The law sets mandatory legal standards and controls for packaging. These cover permitted packaging layers and free space and require that packaging cost be no more than 15% of the sale value of the product. A National Standard for food and cosmetics packaging has also been developed.	Design requirements	http://www.researchandmarkets.com/reportinfo.asp?report_id=655287
	Solid Waste Act (1995)	Product packaging must be chosen from materials that are 'easily recovered/reused, environmentally disposed of or biodegradable'.	Design requirements	
	General Rules for Packaging Waste Disposal and Utilization (GB/T 16716-1996)	The 'General Rules for Packaging Waste Disposal and Utilization' were introduced in 1996. They are voluntary and apply to all packaging wastes. They specify that: • All packaging should be readily treated and reused. • Industry is encouraged to reduce packaging waste and avoid over-packaging. • Use of heavy metals and other toxic materials is restricted.	Design requirements	
	Municipal laws	A central government order in 1999 required all large and medium-sized cities to ban the sale of foam plastic tableware by the end of 2000. Many Chinese municipalities have adopted recycling mandates or bans for certain food packaging materials, including Beijing, Chongqing, Dalian, Harbin, Chengdu, Guangzhou, Qiqihar, Shijiazhuang, Tianjin, Wuhan and Xiamen.	Ban	
	Plastic bags	China banned lightweight plastic shopping bags from 1 June 2008. Businesses are prohibited from manufacturing, selling or using bags less than 0.025 mm thick. More durable bags are permitted as long as they are sold to consumers.	Ban	

(continued)

(continued)

Asia

Legislation/Policy	Description	Policy measure	Link
Product Eco-Responsibility (Plastic Bags) Regulation (2009) (Hong Kong)	A levy on plastic shopping bags was introduced on 7 July 2009 through the Product Eco-Responsibility (Plastic Bags) Regulation. This is the first scheme to be introduced under the Product Eco-Responsibility Ordinance, which is the framework legislation that provides the legal basis for producer responsibility schemes in Hong Kong.	Levy	http://www.epd.gov.hk/epd/english/environmentinhk/waste/prob_solutions/env_levy.html
India	Plastic bags	In 2000 the Indian Government introduced a law banning the manufacture and use of plastic bags thinner than 20 microns in Bombay, Delhi and the entire states of Maharashtra and Kerala. A law has been introduced in the state of Himachal Pradesh banning the production, storage, use, sale and distribution of polyethylene bags. In March 2006 the state government of Maharashtra introduced the *Bio-degradable Garbage (Control) Ordinance* which bans plastic bags thinner than 50 microns.	Ban	http://www.reusablebags.com/facts.php?id=9 http://envis.maharashtra.gov.in/envis_data/pdf/nonbio.pdf
Bangladesh	Plastic bags	In March 2002 Bangladesh banned all polyethylene bags in the capital Dhaka after they were found to have been a major contributor to floods in 1988 and 1998. Discarded bags were blocking the drainage system. There are plans to extend the ban to the whole country.	Ban	http://www.edie.net/news/news_story.asp?id=5029
Japan	Law for the Promotion of Effective Utilisation of Resources (2001)	The law covers requirements for specified industry sectors to reduce waste in production; design products to reduce waste and improve recyclability; and to label packaging to promote source separation by consumers for recycling. A mandatory labelling scheme to identify materials for recycling applies to steel cans, aluminium cans, PET bottles, paper packaging and plastic containers.	Design requirements Labelling	http://www.meti.go.jp/policy/recycle/main/english/law/promotion.html

(continued)

Appendix C: Matrix of International Regulations, Policies and Standards

(continued)

	Legislation/Policy	Description	Policy measure	Link
Asia	Containers and Packaging Recycling Law (2000, amended 2006)	Producers are required to recycle consumer packaging. Companies pay a fee to the Japan Container and Recycling Association to cover costs. Materials covered by the law are: plastic bottles, plastic containers and other plastic packaging (including plastic bags and polystyrene foam containers), paper containers and other paper packaging, and glass containers. Retailers are required to reduce the 'excess' use of containers and packaging and to provide information to consumers to encourage them to reduce packaging. Municipalities are required to collect and separate materials for recycling. Specified container users are also required to set a target for reduction of packaging and to take measures to achieve the target of 'rational use of containers and packaging'. (Examples include promotion of reusable bags, lightweighting, reducing pre-packed items and avoiding double packaging.)	Extended producer responsibility	http://www.meti.go.jp/policy/recycle/main/data/pamphlet/pdf/handbook2008_eng.pdf
	Amendments to the Food Sanitation Law (2002)	The Minister for Health, Labour and Welfare introduced changes to the Food Sanitation Law to restrict use of the phthalate plasticizer DEHP in PVC packaging materials in contact with oily foods.	Design requirements	
Malaysia	Environment Quality Act (1974, amended 1995)	The Act was amended to include compulsory use of recycled materials in manufacturing. A packaging ordinance and 'closed substance cycle' law modelled on the German legislation is being considered by the government.	Design requirements	

(continued)

(continued)

Asia	Legislation/Policy	Description	Policy measure	Link
Singapore	Singapore Packaging Agreement (2007-2012)	The Singapore Packaging Agreement was signed by the National Environment Agency, five packaging industry associations, 19 individual companies, two non-government organisations, the Waste Management and Recycling Association and four public waste collectors. It came into effect on 1 July 2007. The aim is to promote voluntary efforts to reduce packaging waste; for example, through lightweighting, reuse and recycling.	Voluntary agreement	http://app2.nea.gov.sg/topics_packagereement.aspx
	Deposits	There is a deposit system for glass bottles filled locally. There is no glass manufacturing in Singapore. 85% of local glass drink bottles are refilled.	Container deposit legislation	
South Korea	Act on the Promotion of Saving and Recycling of Resources (1993)	This requires manufacturers, importers and retailers of recyclable products and packaging to take back and recycle their products. There are restrictions on the use of disposable cups, plates, plastic bags and paper bags in restaurants, public baths, department stores and some other businesses. In 2002 a voluntary agreement was signed between the Ministry of the Environment and 29 companies running fast-food and coffee businesses to facilitate the use of reusable containers within their shops and to introduce a deposit system on disposable takeaway containers. A measure approved in September 2002 makes it compulsory for all instant noodle containers and PS foam containers to be collected and recycled separately by their producers. Plastic bags used for packaging had to be recycled separately. PVC shrink wrap, laminated PVC and EPS for packaging are banned.	Extended producer responsibility Ban	http://eng.me.go.kr/board.do?method=view&docSeq=8098&bbsCode=law_law_law

(continued)

Appendix C: Matrix of International Regulations, Policies and Standards

(continued)

Asia

Legislation/Policy	Description	Policy measure	Link
Ordinance on the Standards of Packaging Methods and Materials (1993)	The Ordinance sets 'empty space ratio' goals for most packaged products and sets a limit on the number of layers of packaging a product should have. For example, processed foods must have an empty space ratio of 15% or less and two or less layers of packaging. The Ordinance also applies to manufacturers and importers of electronics.	Design requirements	
Extended Producer Responsibility Scheme (EPRS) (2003)	The EPRS imposes waste collecting and recycling obligations on the producers and importers of several products, including electrical products and some packaging (cardboard, metal cans, glass bottles and synthetic plastic packaging). A number of producer responsibility organisations have been established to meet industry's recycling obligations. Businesses failing to meet their take-back obligations will be subject to a recycling tax.	Extended producer responsibility	http://www.envico.or.kr/language/Eng_new/waste/extend.jsp
Taiwan			
Waste Disposal Act	Recycling fees are charged for post-consumer household packaging materials covered by the Act. Materials covered include PET, PVC, PS, steel, aluminium, glass, cardboard and foil-lined containers.	Levy	http://law.epa.gov.tw/en/laws/24567619.html
	Taiwan's EPA announced restrictions on packaging, effective from July 2006. These are similar to the South Korean legislation restricting packaging materials based on the ratio of empty space, number of packaging layers and materials used.	Design requirements	

(continued)

(continued)

Asia

Legislation/Policy	Description	Policy measure	Link
Plastics bags and fast food packaging	Targets for restricted use of plastic bags and disposable plastic tableware have been progressively implemented since 2003. In 2003 the Government banned department stores, shopping malls, supermarkets, convenience stores, fast food restaurants and regular restaurants from providing free plastic carry bags to their customers. In 2006 they decided to allow free plastic bags to be offered by food service operators because of concerns that bags used for food could pose a health risk if reused. In 2003 the Government banned the free distribution of disposable tableware in restaurants, department stores, supermarkets, convenience stores and fast-food outlets. In 2006 they completely banned cafeterias in government organisations from providing disposable utensils, bowls and plates to customers.	Ban	

(continued)

Appendix C: Matrix of International Regulations, Policies and Standards

Africa

Country	Legislation/Policy	Description	Policy measure	Link
Kenya	Plastic bags	In June 2007 the Kenyan Government banned the importation and use of lightweight plastic bags and introduced a tax on thicker bags.	Ban	http://news.bbc.co.uk/2/hi/africa/6754127.stm
Rwanda	Plastic bags	In 2005 the Rwandan Government banned the importation or use of lightweight plastic shopping bags.	Ban	http://news.bbc.co.uk/2/hi/africa/6754127.stm
South Africa	Section 24(d) of the Environment Conservation Act, 1989 (Act No. 73 of 1989)	From March 2003 plastics bags with a thickness of less than 30 micron were banned. Retailers face a fine of 100,000 rand or a 10-year jail sentence if they give out the bags.	Ban	http://www.elaw.org/node/1779
Uganda	Plastic bags	In July 2007 the Ugandan Government banned the importation and use of lightweight plastic bags and introduced a tax of 120% on thicker bags.	Ban	http://news.bbc.co.uk/2/hi/africa/6754127.stm
Tanzania	Plastic bags	In 2006 the Tanzanian Government completely banned the importation or use of plastic shopping bags.	Ban	http://www.suite101.com/blog/ethekwinigirl/east_african_ban_on_plastic_bags
Togo	Plastic bags	The importation and sale of plastic bags were banned from 1 July 2011.	Ban	http://www.suite101.com/blog/ethekwinigirl/east_african_ban_on_plastic_bags

(continued)

Europe

	Legislation/Policy	Description	Policy measure	Link
European Union	European Packaging and Packaging Waste Directive (94/62/EC; 2004/12/EC; 2005/20/EC)	The aims are to harmonise national measures to remove obstacles to trade, and to reduce the environmental impacts of packaging by reducing packaging at source, maximising recovery and eliminating harmful substances. It specifies a number of 'Essential Requirements' for packaging (see CEN Standards below) and recycling targets: • a minimum of 55% by weight of packaging to be recycled by 2008 • minimum recycling rates of 60% for paper/board, 50% for metals, 60% for glass, 15% for wood and 22.5% for plastics (only counts recycling back into plastics) by 2008. Directive 2005/20/EC set a later deadline for the 10 new Member States (the Czech Republic, Estonia, Cyprus, Latvia, Lithuania, Hungary, Malta, Poland, Slovenia, Slovakia) to meet the targets of the revised Packaging Directive. In 2006 a report published by the Commission (COM (2006) 767) recommended that the targets in Directive 2004/12/EC remain valid well beyond 2008.	Design requirements	http://ec.europa.eu/environment/waste/packaging/legis.htm

(continued)

Appendix C: Matrix of International Regulations, Policies and Standards

(continued)

	Legislation/Policy	Description	Policy measure	Link
Europe				
	CEN Standards for implementation of Essential Requirements under Annex II of the European Packaging and Packaging Waste Directive 94/62/EC	Only packaging which complies with the Essential Requirements of the Packaging and Packaging Waste Directive can be placed onto the market. The Essential Requirements can be summarised as: • Packaging weight and volume should be minimised to the amount needed for safety and acceptance of the packed product. • Noxious and other hazardous constituents of packaging should have minimum impact on the environment at end of life. • Packaging should be suitable for material recycling, energy recovery or composting or for reuse if reuse is intended.	Design requirements	
	REACH (Registration, Evaluation, Authorisation and restriction of CHemicals)	The REACH regulation came into force on 1 June 2007. It imposes obligations directly on industry and does not require the implementation of national laws. Materials used in packaging which could be affected include pulp and paper, plastics, metals, glass, coatings, packaging inks and adhesives. Unregistered substances will eventually have to be taken off the market.		http://www.europen.be/index.php?action=onderdeel&onderdeel=5&titel=News+Room&categorie=2&item=60
Austria	Packaging Ordinance 1993 (revised January 2001)	Producers, fillers, distributors and importers are solely responsible for ensuring collection of used packaging. They can either take back their own packaging and reuse or recover it, or they can transfer this responsibility to a third party.	Extended producer responsibility	http://www.raymond.com/promo_raymond-international/austria02.pdf

(continued)

(continued)

Europe	Legislation/Policy	Description	Policy measure	Link
	Order Concerning Refillable Plastic Bottles (1990)	There is a mandatory deposit on refillable PET containers. Quantitative targets have been set for refilling, recycling and energy recovery.	Container deposit legislation	http://www.bottlebill.org/legislation/world/austria.htm
Belgium	Packaging taxes	Eco-taxes apply to non-refillable beverage containers (on a per unit basis), and industrial packaging for solvents, glues, inks and pesticides (on a volume basis). Beverage manufacturers only pay the tax if they are not a member of the FOST Plus recovery organisation or if the FOST Plus recovery rates are below the prescribed targets.	Extended producer responsibility Tax	http://www.europen.be/index.php?action=onderdeel&onderdeel=5&titel=News+Room&categorie=2&item=60 http://people.engr.ncsu.edu/barlaz/pdfdoc/benelux.pdf
	Deposit on drink containers	Similar to eco-taxes, deposits will only be enforced if recycling targets are not met.	Container deposit legislation	http://people.engr.ncsu.edu/barlaz/pdfdoc/benelux.pdf
	Carbon packaging tax	The tax was introduced from 1 July 2007 on plastic carrier bags, plastic films, aluminium foil and disposable cutlery.	Tax	http://www.europen.be/index.php?action=onderdeel&onderdeel=5&titel=News+Room&categorie=2&item=60
Bulgaria	Waste legislation	The legislation includes recovery and recycling targets for 2007—2013. If the targets are not met, the law makes provision for charging a packaging fee.		http://www.europen.be/europen/files/File/Bulletin/EUROPEN%20Bulletin%20Issue%2034%20(Screen%20Version).pdf
Croatia	Waste legislation	A new Decree in 2005 introduced packaging fees. They are applied to beverage packaging by material type (weight based), product type (unit-based) and volume. Other types of packaging are subject to a weight-based management fee levied by material types. A deposit has been introduced for single-use beverage containers.	Container deposit legislation	http://www.europen.be/europen/files/File/Bulletin/EUROPEN%20Bulletin%20Issue%2032%20(Print%20Version).pdf,

(continued)

Appendix C: Matrix of International Regulations, Policies and Standards

(continued)

Europe				
	Legislation/Policy	Description	Policy measure	Link
Denmark	Statutory Order on Waste, no. 619 of 27 June 2000	Businesses are required to source-separate their wastes for recovery. Local authorities are required to implement recovery programs. Revised legislation came into force on 1 September 2005, requiring businesses to sort a wider range of packaging including wood packaging and more types of plastic and metal packaging. Local authorities must ensure that more packaging is source-separated.	Design requirements	http://www.europen.be/europen/files/File//Bulletin/EUROPEN%20Bulletin%20Issue%2033%20(Print%20Version).pdf.
	Packaging taxes	There is a volume-based tax (per unit basis) for glass and plastic packaging, metal cans and cartons for liquids (excluding milk). This has been in place since 1978 for most beverages. Since 1 January 1999 there has also been a weight-based tax on packaging for dairy products (excluding milk), packaged foodstuffs and non-food household products. There is a tax on non-reusable plastic and paper shopping bags and disposable tableware. From 1 January 1999 a weight-based tax was introduced on certain types of PVC film packaging, and from 1 July 2001 on PVC and phthalates.	Tax	http://www.europen.be/index.php?action=onderdeel&onderdeel=6&titelPublications&categorie=0&item=5&back=%3Faction%3Donderdeel%26onderdeel%3D6%26titel%3DPublications%26page%3D1 http://eea.eionet.europa.eu/Public/irc/eionet-circle/etc_waste/library?l=/country_fact_sheets/denmarkpdf/_EN_1.0_&a=d

(continued)

(continued)

Europe

Legislation/Policy	Description	Policy measure	Link
Deposit-return system for beverage containers (introduced 1970s, revised 2002 and 2005)	Prior to 2002 deposits only applied to refillable containers, and aluminium cans were effectively banned. Since 2002 they have applied to all beverage containers for beer and carbonated soft drinks. In 2005 deposits were extended to include all alcoholic sodas, ciders and energy drinks.	Container deposit legislation	http://eea.eionet.europa.eu/Public/irc/eionet-circle/etc_waste/library?l=/country_fact_sheets/denmarkpdf/_EN_1.0_&a=d
Targets	When the EU Packaging Directive was introduced, Denmark decided to focus on transport packaging to meet their recycling targets because household packaging was already being recycled at relatively high rates or incinerated with energy recovery. Denmark has voluntary agreements rather than mandatory targets.	Voluntary agreement	http://www.eiatrack.org/s/169
Estonia			
Packaging taxes	There is an eco-tax on packaging for beer, soft drinks, juices, waters, flavoured milk, wines and spirits. Packaging is exempt from the tax if the specified percentage of beverage container material is collected and reused, recycled or used as fuel.	Tax	http://www.europen.be/index.php?action=onderdeel&onderdeel=6&titel=Publications&categorie=0&item=5&back=%3Faction%3Donderdeel%26onderdeel%3D6%26titel%3DPublications%26page%3D1
Packaging Act (2005)	Mandatory deposits have been introduced for certain beverage containers: glass containers, metal cans and plastic packaging.	Container deposit legislation	http://www.riigiteataja.ee/ert/act.jsp?id=12964621
Finland			
Excise taxes (1970s)	Excise taxes were imposed on one-way beer and soft drink containers in the mid 1970s. The Waste Management Act allows for exemptions. The tax is reduced to 25% of the base level for non-refillable containers subject to an approved deposit-return system.	Container deposit legislationTax	www.bottlebill.org/legislation/world/finland.htm

(continued)

Appendix C: Matrix of International Regulations, Policies and Standards

(continued)

Europe

	Legislation/Policy	Description	Policy measure	Link
	Waste Act (1993, amended 1997)	Amendments to the Act were introduced in 1997 to formally implement the EU Packaging Directive. Prior to this, Finland had relied on a voluntary agreement with the packaging industry which came into force on 1 July 1995. They had already met the EU's recycling targets as a result of the deposit scheme and taxes on one-way containers.		http://www.merit.unimaas.nl/tep/reports/ppwd-synthesisreport.pdf
France	Decree on Household Packaging Waste (1 April 1992)	The decree requires packers, fillers and importers to join an approved recovery organisation, introduce their own approved recovery program or operate a deposit system. The decree did not include a quantitative recycling target. Under the national French recycling program, Eco-Emballages, manufacturers pay a fee with a sliding scale based on volume, weight, packaging material, and recyclability to help fund collection and recycling of packaging.	Extended producer responsibility	http://www.pro-e.org/France1.htm
	Deposits	There are voluntary deposits on refillable containers.		
	Plastic bags	The island of Corsica was the first French region to ban non-biodegradable bags in 1999, and a similar ban was introduced in Paris in 2007. The French Senate approved a ban on non-biodegradable plastic supermarket bags from 1 January 2010.	Ban	http://www.europen.be/europen/files/File/Bulletin/EUROPEN%20Bulletin%20Issue%2034%20(Screen%20Version).pdf, p. 8

(continued)

(continued)

Europe	Legislation/Policy	Description	Policy measure	Link
Germany	Ordinance on the Avoidance of Packaging Waste (1991, 1998, 2005)	The Ordinance sets mandatory recycling rates for packaging materials and requires businesses to take back and recover their own packaging from consumers. Most manufacturers and distributors meet their obligations through a collective industry initiative, the DSD (Duales System Deutschesland), which is separate to the publicly funded waste management system.	Extended producer responsibility	http://www.bmu.de/english/waste_management/acts_and_ordinances/acts_and_ordinances_in_germany/doc/20203.php http://www.bmu.de/english/waste_management/general_information/doc/35155.php
	Beverage container deposits	The Ordinance includes a quota and deposit system for refillable containers, based on the status quo for beverage containers in Germany 1991. There is a compulsory deposit on all one-way beverage packaging. This requirement is suspended as long as the percentage of refillable beverage containers does not fall below 72%.	Container deposit legislation	colname="c5">http://www.bottlebill.org/legislation/world/germany.htm http://ec.europa.eu/environment/waste/studies/packaging/050224_final_report.pdf
Hungary	Packaging taxes	There is a tax on non-reusable packaging with different rates for composites, plastics, aluminium, paper, wood, textiles, tinplate and glass. There is an exemption for packaging that meets specified rates of reuse or recycling.	Tax	http://www.europen.be/index.php?action=onderdeel&onderdeel=6&titel=Publications&categorie=0&item=5&back=%3Faction%3Donderdeel%26onderdeel%3D6%26titel%3DPublications%26page%3D1

(continued)

Appendix C: Matrix of International Regulations, Policies and Standards

(continued)

Europe	Legislation/Policy	Description	Policy measure	Link
Ireland	Waste Management (Environmental Levy) (Plastics Bag) Regulations 2001	Ireland has an environmental levy on all plastic bags except those used for fresh produce and reusable bags (introduced March 2002).	Levy	http://www.citizensinformation.ie/categories/environment/waste-management-and-recycling/plastic_bag_environmental_levy
	Waste Management Act 1996 (amended 2001 and 2003)	The Act puts a priority on waste prevention and recovery, and allows for the introduction of 'producer responsibility regulations'.		
	Waste Management (Packaging) Regulations 1997 (amended 1998 and 2003)	Introduced under the Waste Management Act, the regulations implement the European Packaging and Packaging Waste Directive.	Extended producer responsibility	http://www.pro-e.org/ireland1.htm
Italy	Legislative Decree 22/97	The Decree implements the European Packaging Directive. Packaging producers, packers, importers and end-users may fulfil their obligations by participating in CONAI, the Italian producer responsibility scheme. The collection of packaging waste is carried out by local authorities under an agreement with CONAI.	Extended producer responsibility	http://www.conai.org/hpm00.asp

(continued)

(continued)

Europe

Legislation/Policy	Description	Policy measure	Link
2007 Finance Act	The 2007 Finance Act introduced a tax on plastic bottles for mineral or table water. The revenue is used to finance projects aimed at providing universal access to water.	Tax	http://www.europen.be/europen/files/File/Bulletin/EUROPEN%20Bulletin%20Issue%2040%20(to%20print).pdf, p. 7
Plastic bags	Italy imposed a tax of 100 lira on plastic bags in 1988. Non-biodegradable bags were banned from 1 January 2011.	Ban	http://www.europen.be/europen/files/File/Bulletin/EUROPEN%20Bulletin%20Issue%2040%20(to%20print).pdf, p. 7
Malta			
Levy on plastic bags	The Maltese government has introduced a €0.14 levy on plastic bags. A €0.02 levy applies to degradable bags and biodegradable plastic bags are exempt.	Levy	http://www.europen.be/europen/files/File/Bulletin/EUROPEN%20Bulletin%20Issue%2029%20(Print%20Version).pdf, p. 7
Latvia			
Packaging taxes	There is an eco-tax on packaging for consumer goods and disposable food service packaging. Different rates apply to glass, plastics, metals and composites, paper/board, wood and others.	Tax	http://www.europen.be/index.php?action=onderdeel&onderdeel=6&titel=Publications&categorie=0&item=5&back=%3Faction%3Donderdeel%26onderdeel%3D6%26titel%3DPublications%26page%3D1
Netherlands			
Decree on the Management of Packaging, Paper and Board (March 2005)	The Decree introduces producer responsibility for all packers, fillers and importers of packaged goods. This represents a significant departure from the voluntary covenant model which had been used since 1991.	Extended producer responsibility	http://www.europen.be/europen/files/File/Bulletin/EUROPEN%20Bulletin%20Issue%2033%20(Print%20Version).pdf, p. 9
	The Decree provides for an extension of the mandatory deposit system to smaller plastic drink containers, drink cans and other glass bottles. This will only be enforced if recycling targets are not met. The packaging industry has also agreed to meet a voluntary target of 25% post-consumer recycled content in plastic bottles.	Container deposit legislation	http://internationaal.vrom.nl/docs/internationaal/engelsevertalingamvb.pdf

(continued)

Appendix C: Matrix of International Regulations, Policies and Standards

(continued)

Europe

	Legislation/Policy	Description	Policy measure	Link
	Carbon-based packaging tax	A packaging tax was introduced on 1 January 2008. The tax is based on the amount of carbon dioxide generated in the manufacture of each type of packaging material. Reusable packaging is exempt.	Tax	http://www.nl.pwc.com/extweb/Bn/Taxnews.nsf/Public/YR68388 http://www.sustainableisgood.com/blog/2007/10/netherlands-to-.html
	Landfill ban	Since 1996 it has been illegal to landfill used packaging.		
	Deposit Law on Soft Drinks and Waters 1993	The law imposes a mandatory deposit on soft drinks and water in glass and PET containers (refillable and non-refillable). It replaced an agreement on minimum deposits for beverage containers which was introduced in the 1960s.	Container deposit legislation	http://www.bottlebill.org/legislation/world/netherlands.htm
Norway	Product Control Act (1974, 1994)	There is a basic tax on non-refillable containers and refillable containers making their first trip (except milk and fruit juice). Containers reaching return rates above 95% are exempt from the tax.	Tax	http://www.bottlebill.org/legislation/world/norway.htm
Poland	Recycling law (2005)	A law implementing Directive 2004/12/EC was adopted in September 2005. Recycling and recycling targets are the same as in the Directive, apart from aluminium to steel which must each achieve a 50% recycling rate.		http://www.europen.be/europen/files/File/Bulletin/EUROPEN%20Bulletin%20Issue%2033%20(Print%20Version).pdf, p. 9
Sweden	Ordinance on Producers' Responsibility for Packaging (SFS 2006:1273).	Producers must ensure that packaging is separated from general waste and reused or recovered in an 'environmentally acceptable way'. Priority is given to reuse. A number of material recovery companies have been established to administer the collection and recovery of waste.	Extended producer responsibility	www.pro-e.org/Legal_Basis_Sweden.htm

(continued)

(continued)

Europe

	Legislation/Policy	Description	Policy measure	Link
	Deposits	Aluminium beverage cans have had a deposit since 1984 and PET beverage bottles since 1994. In 2005 deposits were extended to all plastic and metal containers for ready-to-drink beverages. It is now illegal to sell beverages in containers that are not part of an authorised Swedish container deposit system, with the exception of beverages that mainly consist of dairy products or vegetable, fruit, or berry juice.	Container deposit legislation	http://www.europen.be/europen/files/File/Bulletin/EUROPEN%20Bulletin%20Issue%2030%20(Print%20Version).pdf, p. 7 http://en.wikipedia.org/wiki/Container-deposit_legislation#Sweden
Switzerland	Ordinance on Beverage Container Packaging (1990)	Deposits are required on all refillable beverage containers. One-way beer, soft drink and mineral water containers that exceed a maximum amount (by weight) allowed in the waste stream require deposits. (All container types are currently within the maximum allowable amount.)	Container deposit legislation	http://www.bottlebill.org/legislation/world/switzerland.htm
United Kingdom	Producer Responsibility Obligations (Packaging waste) Regulations (1997, amended 2007) Amended by SI (Statutory Instrument) 1999 No 1361 and SI 1999 no 3447.	The aim is to meet the recovery and recycling targets in the European Packaging and Packaging Waste Directive by increasing the amount of packaging recycled, composted or burned with energy recovery.	Extended producer responsibility	http://www.berr.gov.uk/whatwedo/sectors/sustainability/packaging/Packaging%20in%20the%20UK/page38930.html http://www.opsi.gov.uk/si/si2007/pdf/uksi_20070871_en.pdf

(continued)

Appendix C: Matrix of International Regulations, Policies and Standards

(continued)

Europe

Legislation/Policy	Description	Policy measure	Link
The Packaging (Essential Requirements) Regulations (1998)	These mirror the wording in the European Packaging and Packaging Waste Directive to enforce the Essential Requirements.	Design requirements	http://www.hmso.gov.uk/si/si1998/19981165.htm
Packaging Strategy (2009)	The strategy aims to optimise packaging over its life cycle through increased uptake of environmental design in industry and to maximise recycling. Priority sectors will be identified for action based on the extent of 'over-packaging' and their potential for waste reduction. These sectors will be invited to negotiate voluntary agreements with the government for the period 2011-2016. Consideration will be given to carbon-based rather than weight-based targets for recovery.	Design requirements Voluntary agreement	http://www.defra.gov.uk/environment/waste/producer/packaging/strategy.htm
Climate Change (Scotland) Act (2009)	The Bill contains powers to introduce waste reduction targets for packaging, deposit and return systems and charges on carrier bags.	Container deposit legislation	http://www.scottish.parliament.uk/s3/bills/17-ClimateChange/b17bs3-aspassed.pdf

Middle East

Legislation/Policy	Description	Policy measure	Link
Plastic bag ban (UAE) from 2013	The United Arab Emirates has agreed to ban plastic bags from 2013. Alternatives such as biodegradable plastic and jute are being considered.	Ban	http://www.arabianbusiness.com/plastic-bags-set-be-banned-in-uae-from-2013-11593.html
Plastic bag ban (Ajman) 2010	Ajman introduced a ban on plastic bags from 1 July 2010. They must be replaced by cloth or jute bags, or recyclable plastic bags that meet certain specifications.	Ban	http://www.khaleejtimes.com/DisplayArticle08.asp?xfile=data/theuae/2010/January/theuae_January563.xml&section=theuae

United Arab Emirates

Glossary

Bagasse Fibrous material that remains after sugarcane stalks have been crushed to extract their juice.

Biodegradable polymer A polymer capable of being broken down by microorganisms in the presence of oxygen to carbon dioxide, water, biomass and mineral salts, or in the absence of oxygen to carbon dioxide, methane and biomass.

Biogenic carbon dioxide Carbon dioxide that is derived from biomass, but not fossilised or from fossil sources [3, p. 2].

Conversion Processes used to transform basic raw materials into finished packaging.

Cradle-to-cradle In this book reference is made to the cradle to cradle model promoted by McDonough and Braungart [4]. They promote the recovery of all products and materials in what they refer to as technical or biological metabolisms, in contrast to the linear, one-way 'cradle to grave' model that has existed since the industrial revolution.

Cradle-to-gate A term used in life cycle assessment (LCA) to mean the LCA has incorporated all the processes required to extract and transform materials from the environment and deliver a product to the 'gate' of the factory or retail outlet.

Cradle-to-grave A term used to describe the one-way, linear material flows that exist in industrialised societies. Resources are extracted, manufactured into products, used and often disposed of in a landfill or waste to energy facility. In life cycle assessment the term is used to imply that an LCA study has considered the entire product life cycle, in contrast to studies that only consider impacts from 'cradle-to-gate'.

Cullet Collected glass packaging that has been crushed ready for reprocessing.

Design for accessibility An approach to design that aims to make products, packaging and services accessible by all potential users, including those with a disability.

Design for environment An approach to design that considers and aims to reduce the environmental impacts of a product and/or packaging over its life cycle.

Design for sustainability An approach to design that considers and aims to reduce the environmental and social impacts of a product and/or packaging over its life cycle. Social issues that may be considered include child labour, workplace practices, freedom of association, discrimination and safety.

Downgauging Reducing the amount of material in a package, for example by reducing wall thickness.

Eco-design See *Design for environment*.

Eco-effectiveness An approach to design promoted by McDonough and Braungart in their book 'Cradle to cradle' [4]. It looks for more than the incremental change that can be achieved through an 'eco-efficiency' approach by completely rethinking the design of products and services based on sustainability principles.

Eco-efficiency A term originally developed by the World Business Council for Sustainable Development [5] to refer to strategies that deliver products and services with less consumption of materials, energy and water.

Eco-label Generally used to describe a third party certified label on a product or package that indicates the product has been evaluated across its entire life cycle against a range of environmental criteria.

Embodied energy Although not an impact category but rather an environmental indicator, embodied energy takes into account the energy demand per functional unit. All energy use including fossil, renewable, electrical and feedstock (e.g. energy incorporated into materials such as plastic) are considered.

Endocrine disrupter An endocrine gland produces hormones which are introduced directly into the blood stream. 'Endocrine disrupting chemicals' are believed to act like hormones and disrupt the normal operation of the endocrine system.

Feedstock recycling A recycling process for plastics in which they are converted back into a monomer or new raw materials by changing their chemical structure [6, p. 3].

Green procurement An approach to organisational procurement that evaluates all products and services against a range of environmental criteria in addition to conventional criteria such as cost and value for money. Environmental criteria and assessment processes are built into procurement policies and procedures.

Glossary

Greenwash A derogatory term to describe marketing and communication activities that imply an organisation is doing more to reduce environmental impacts than they are.

Hazardous A waste stream is considered hazardous if its characteristics pose a threat or risk to public health, safety or the environment. These include substances that display at least one of the following properties: explosive, oxidising, flammable, irritant, harmful, toxic, carcinogenic, corrosive, infectious, toxic for reproduction, mutagenic, sensitising or ecotoxic [7].

Hot spots Points in the value chain of a product where specific environmental impacts are greatest.

Indicator A proxy for an issue or characteristic that an organisation wants to measure (also see *Metric*).

Leachate The liquid generated as rainwater filters through a landfill and extracts materials from solid waste by leaching.

Life cycle All of the linked stages of a product or packaging system, from raw materials acquisition or generation of natural resources through to manufacture, use and final disposal.

Life cycle assessment According to ISO 14040, LCA involves the 'compilation and evaluation of the inputs, outputs and the potential environmental impacts of a product system throughout its life cycle' [8, p. 2] from raw material acquisition through production, use and disposal.

Life cycle management A business management approach that aims to improve the sustainability of a business's products throughout their life cycle.

Life cycle map A diagram showing the interrelated processes of the product or service system.

Life cycle thinking A thought process that considers environmental impacts over the entire life cycle of a product and not just at one point (e.g. manufacturing or recovery).

Mechanical recycling The reprocessing of waste materials back into secondary raw materials or products through mechanical processes without significantly changing the chemical structure of the material. Examples include recycling glass bottles into new packaging or filtration products, or plastics packaging into new packaging or textile products.

Materials recovery facility (MRF) Plant and equipment for sorting and preprocessing (e.g. crushing/baling) materials from the waste stream for recycling.

Metric The method used to express an indicator. Metrics are often quantitative but can also be qualitative (also see *Indicator*).

Mobius loop The 'chasing arrow' recycling symbol, which is recommended by the International Standards Organisation (ISO) for claims about recyclability or recycled content [9].

Organic recycling The reprocessing of waste materials back into secondary raw materials or products through organic processes such as composting or anaerobic digestion.

Oxodegradable polymer A conventional polymer (e.g. polyethylene) that undergoes controlled degradation through the addition of a catalyst that can trigger and accelerate the degradation process. These polymers start to break down through exposure to daylight, heat and/or mechanical stress.

Packaging supply chain The whole complex of consecutive steps necessary to manufacture a packaged product, including raw material processing, material production, container production and packaging process operation, and to transport and distribute the packaged product to the end user (also see *Supply chain*).

Packaging sustainability The extent to which packaging contributes to sustainable development.

Precautionary principle One of the principles in the *Rio Declaration on Environment and Development*, which was agreed at the 1992 United Nations Conference on Environment and Development. It states that: Where there are threats of serious or irreversible damage, lack of full scientific certainty shall not be used as a reason for postponing cost-effective measures to prevent environmental degradation [10].

Pre-consumer material Material diverted from the waste stream during a manufacturing process, but excluding scrap or regrind that is capable of being reclaimed in the same process that generated it [9, p. 14].

Post-consumer material Material generated by households or commercial, industrial and institutional facilities in their role as consumers which can no longer be used for its intended purpose [9, p. 14].

Primary packaging (also sales, consumer or retail packaging) The sales unit at the point of purchase [11].

Recovery Any operation that results in a waste serving a useful purpose by replacing other materials that would otherwise have been used to fulfil a particular function [7, p. 10].

Recyclable Defined by the International Standards Organization (ISO) as: a characteristic of a product, packaging or associated component that can be diverted from the waste stream through available processes and programs and can be collected, processed and returned to use in the form of raw materials or products [9, p. 13].

Glossary

Recyclate Recycled material that has undergone some form of processing, such as conversion of waste plastics to plastic pellets, and will be used to make new products or materials.

Recycling Any recovery operation that processes waste materials into products, materials or substances, whether for the original or another purpose. It includes reprocessing of organic material but not energy recovery [7, p. 10].

Renewable EUROPEN has proposed that renewable materials be defined as materials that meet all of the following requirements [12, p. 41]:
- Be composed of biomass, which can be continually regenerated within a finite lifetime
- Are replenished at a rate that is equal to or greater than the rate of depletion
- From sources that are manage in accordance with the principles of sustainable development
- Where a verifiable traceability system is in place.

Secondary packaging (also display or merchandising packaging) Packaging used at the point of purchase to contain or present a number of sales units; it can be removed from the product without affecting its characteristics [11].

Supply chain Normally refers to organisations supplying goods and services to a product manufacturer, including tier-one suppliers (supplying directly to the final product manufacturer), tier-two suppliers (providing goods or services to tier-one suppliers) and so on.

Sustainability A contestable concept but generally interpreted as the goal of sustainable development (see *Sustainable development*). EUROPEN refers to the common dictionary definition of sustain: 'to maintain or keep going indefinitely' [12, p. 41].

Sustainable development '[D]evelopment that meets the needs of the present without compromising the ability of future generations to meet their own needs' [13, p. 43].

Tertiary packaging Used to facilitate handling and transport of a number of sales units or grouped packages in order to prevent physical handling and transport damage; does not include road, rail, ship and airfreight containers [11].

Toxic Normally refers to a substance that, given sufficient exposure, can cause serious health effects in humans, such as poisoning, respiratory complaints or cancer (also see *Eco-toxicity* and *Human toxicity*).

Triple bottom line Refers to the three types of issues or indicators that should be considered for corporate sustainability: economic well-being (including capital and profit), social responsibility (issues such as working conditions, human rights and safety) and environmental responsibility (including legal compliance, waste management and resource efficiency).

Value chain Organisations and individuals involved at every stage of the product life cycle, including suppliers (see *Supply chain*), consumers, municipalities, waste management companies and recyclers.

Volatile organic compounds All organic compounds (substances made up of predominantly carbon and hydrogen) with boiling temperatures in the range of 50–260°C, excluding pesticides. This means that they are likely to be present as a vapour or gas in normal ambient temperatures [14].

References

1. Lewis H, Verghese K, Fitzpatrick L (2010) Evaluating the sustainability impacts of packaging: the plastic carry bag dilemma. Packaging Technol Sci 23(3):145–160
2. Kepp P (2005) Brazil's capital sets recycling targets for makers, distributors of plastics, tyres. Int Environ 28(20):709–710
3. Carbon Trust, DEFRA, BSI (2008) PAS 2050:2008 Specification for the assessment of the life cycle greenhouse gas emissions of goods and services. British Standards Institution, London
4. McDonough W, Braungart M (2002) Cradle to cradle: remaking the way we make things. North Point Press, New York
5. WBCSD. Eco-efficiency. Available from http://www.wbcsd.ch/templates/Template WBCSD1/layout.asp?type=p&MenuId=MzI4&doOpen=1&ClickMenu=LeftMenu. Cited 19 October 2005
6. ISO 15270 (2008) Plastics—guidelines for the recovery and recycling of plastics waste. International Standards Organisation (ISO), Geneva
7. Commission of the European Communities (2008) Directive 2008/98/EC of the European Parliament and of the Council of 19 November 2008 on waste and repealing certain other Directives, Brussels
8. ISO 14040 (2006) Environmental management—Life cycle assessment—Principles and framework. International Organisation for Standardisation, Geneva
9. ISO 14021 (1999) Environmental labels and declarations—Self-declared environmental claims (Type II environmental labelling) International Organisation for Standardisation, Geneva
10. United Nations General Assembly (1992) Rio Declaration of Environment and Development. Annex 1 of the report of the United Nations Conference on Environment and Development, Rio de Janeiro, 3–14 June 1992
11. Commission of the European Communities (1994) European Parliament and Council Directive 94/62/EC of 20 December 1994 on Packaging and Packaging Waste
12. ECR Europe and EUROPEN (2009) Packaging in the sustainability agenda: a guide for corporate decision makers. ECR Europe and The European Organization for Packaging and the Environment. Brussels, Belgium
13. United Nations (1987) Report of the World Commission on Environment and Development: Our common future. In: Transmitted to the General Assembly as an Annex to document A/42/427—Development and International Co-operation: Environment. UN WCED, Geneva
14. Department of Sustainability (2001) Water, Population and Communities. Air toxics and indoor air quality in Australia. Available from http://www.environment.gov.au/atmosphere/airquality/publications/sok/vocs.html. Cited 26 April 2011

Index

A
Accessibility, 50, 63, 159, 374
Acidification, 11, 191, 200, 207, 264, 310
Aluminium, 9, 54, 56, 60, 76, 80–83, 91, 96–98, 158, 162, 199–201, 213, 214–217, 225, 226, 230, 248, 344, 354, 357, 362, 364, 366, 369, 370
Amazon, 81
Aquatic toxicity, 191, 260, 310
Australian Competition and Consumer Commission (ACCC), 124, 127, 343

B
B&Q, 73, 74
Bagasse, 213, 373
Bans on packaging, 28, 157, 162, 163, 165, 166, 341–371
Biodegradable polymer, 54, 55, 72, 83–85, 90, 94, 95, 99–101, 117, 139, 141–146, 205–208, 242–249, 349, 353, 373
Biogenic carbon dioxide, 205, 373
Bisphenol A (BPA), 58, 87–89, 98–101, 163, 276, 349
Bottle Bills, 156, 157, 164, 341–371
Boxboard (folding boxboard, cartonboard, paperboard, solid board), 78–82, 84, 93, 94, 96–98, 165, 197, 224–230, 276, 338, 344 *See also* Paper and board
Bunnings, 65, 66

C
Cadbury, 59, 60, 292
Carbon footprint, 27, 35, 116, 190, 192, 273, 293, 339
Carbon reduction, 27, 56, 148, 308, 339
Carrefour, 27
Cartonboard, *See* Boxboard
Cellulose based polymers, 55, 213, 214, 242–249
CHEP, 73
Chlorine bleaching, 58, 92, 93, 98, 129, 147, 227, 338 *See also* Chlorine-free, Elemental chlorine-free bleaching, Processed chlorine-free (PCF) bleaching, Totally chlorine-free (TCF) bleaching
Chlorine-free, 137, 139, 146, 147, 227, 338, 339
Cleaner production, 57, 58, 86, 91–93
Climate change, 25, 34, 35, 248, 260, 310, 371 *See also* Global warming
COMPASS, 253, 254, 260–262, 267–270, 273
Composting, 5, 15, 30, 53–55, 83–85, 94, 99–101, 117, 129, 133, 134, 139, 142–146, 160, 227, 242, 243, 246, 247, 261, 268, 277, 293, 294, 309, 323, 337, 361, 376 *See also* Degradable, Organic recycling
Consumer Goods Forum, *See* The Consumer Goods Forum
Consumer packaging, *See* Primary packaging
Consumer research, *See* Market research
Container deposit legislation (CDL), *See* Bottle Bills
Corporate sustainability, 4, 5, 15–22, 292, 293, 295, 326 *See also* Dunphy sustainability phase model, Dow Jones Sustainability Indexes
Corrugated board (fibreboard), 9, 46, 52, 62, 68, 73, 74, 76, 78, 81, 93, 127, 129, 188, 224–230, 276 *See also* Paper and board

C (cont.)

Courtauld Commitment, 26, 27, 164
Cradle-to-cradle, 15, 24, 53, 138, 215, 221, 227, 235, 244, 276, 373
Cradle-to-gate, 183, 373
Cradle-to-grave, 183, 373
Critical review (of an LCA), 180, 184, 186, 194, 195
Cullet, 96, 220, 224, 373
Cyclic packaging, 53–56, 70–86, 139

D

Degradable, 85, 99–101, 117, 125, 144–146, 241, 242, 368 See also Biodegradable polymer, Oxodegradable polymer, Photodegradable
Dell, 52
Design for accessibility, 50, 51, 63, 374
Design for environment, 374
Design for sustainability, 63, 253, 258, 374
Display packaging, See Secondary packaging
Distribution packaging, See Tertiary packaging
Dow Jones Sustainability Indexes, 25, 26, 297
Downgauging, 374 See also Lightweighting
Dunphy sustainability phase model, 16, 17, 298, 299, 301, 302
Duracell, 50

E

Eco-design, See Design for environment
Eco-Design Indicator Tool (EDIT), 255, 260, 261, 273
Eco-effectiveness, 14, 15, 313, 374
Eco-efficiency, 14, 15, 33, 51, 278, 374
Eco-label, 13, 93, 121, 137, 229, 335–340, 374 See also Labelling
Ecological stewardship, 57, 86, 93, 94, 147, 309 See also Forest stewardship, Forest Stewardship Council (FSC), Product stewardship
Effective packaging, 48–51, 214
Efficient packaging, 51–53, 69
Elemental chlorine-free bleaching, 92, 227
Embodied energy, 70, 162, 175, 193, 374
Energy consumption, 3, 9, 12, 14, 17, 19, 24, 35, 51, 62, 64, 66, 92, 93, 132, 133, 158, 161, 174, 175, 187–189, 203, 204, 207, 264
Energy efficiency, 35, 47, 51, 52, 64, 65, 70, 71, 114, 159, 160, 198, 275, 276, 333
Energy generation, 146
Energy impacts, 248
Energy intensive, 96, 205
Energy recovery, 28, 53, 56, 160, 161, 235, 273, 361, 362, 364, 370
Energy reduction, 5, 18, 33, 51, 53, 75, 93, 129, 130, 202, 215, 220, 237, 317 See also Energy efficiency
Energy savings, 76
Energy use, See Energy consumption
Endocrine disrupter, 89, 374
Environmental footprint, 19 See also Carbon footprint
Environmental impact indicators, See Impact indicators
Environmental life cycle, See Life cycle, Life cycle assessment
European Organisation for Packaging and the Environment (EUROPEN), 25
Eutrophication, 11, 132, 184, 191, 248, 260, 264, 310, 334
Expanded polystyrene (EPS), 157, 166, 226, 233, 234, 238, 264, 349, 356
Extended producer responsibility (EPR), 157, 159, 164, 341–371

F

Federal Trade Commission (FTC), 122, 125, 139, 140, 144, 148, 347
Feedstock recycling, 237, 374
Food and Drug Administration (FDA), 76, 78, 165, 349
Food waste, 55, 84, 240
Forest Stewardship, 147
Forest Stewardship Council (FSC), 27, 93, 139, 147, 229, 276, 339
Foster's, 127
Functional unit, 180, 182, 184, 187–189, 198, 199, 203, 206, 253, 265, 268, 269, 306

G

General purpose polystyrene (GPPS), 233, 234, 238
Gingerbread Folk, 55
Glass, 11, 53, 54, 72, 76, 80, 83, 90, 96–101, 161, 162, 165, 174, 194, 199, 209, 213, 220–224, 230, 234, 248, 307, 341–371
Global Packaging Project (GPP), 24
Global warming, 11, 49, 56, 120, 161, 176, 184, 190, 191, 200, 201, 205, 207, 264, 308–310, 334 See also Climate change

Index 381

Green energy, *See* Renewable energy
Green marketing, 110, 112–115, 126, 127, 343
Green procurement, 13, 26, 374 *See also* Procurement
Greenwash, 110, 120–124, 133, 178, 375

H

Hazardous substances, 56, 86–91, 159, 160, 162, 275, 276, 278, 303, 361
Hazardous waste, 346
Heavy metals, 45, 82, 84, 85, 91, 99–101, 119, 162, 163, 303, 309, 312, 326, 353
Heinz, 88
Hewlett Packard, 118, 119, 325, 326
High density polyethylene (HDPE), 52, 54, 56, 76, 78, 99–101, 197, 198, 231, 232, 234, 238, 241, 242, 351
High impact polystyrene (HIPS), 233, 234, 238, 241
Hot spots, 133, 181, 263, 268, 306, 375
Human health, 28, 165, 191, 270

I

Impact assessment, *See* Life cycle impact assessment (LCIA)
Impact assessment matrix, 264
Impact assessment methods, 192, 193, 208, 265, 266 *See also* Life cycle impact assessment (LCIA)
Impact indicators, 11, 19, 79, 181, 185, 189–193, 200, 205, 207, 260, 261, 310
Indicators, 4, 6, 14, 22–24, 47, 52, 59, 78, 178, 189, 190, 193, 253–256, 264, 287, 292, 305, 326, 375 *See also* Impact indicators packaging sustainability metrics
Inks, 45–47, 78, 82–85, 87, 91, 92, 100, 101, 127, 163, 225, 228, 230, 276, 277, 323, 334, 361, 362
International Organisation for Standardisation (ISO), 24, 85, 120, 122, 137–140, 142–145, 157, 167, 174, 186, 192, 194, 265, 336–338
Inventory analysis, 187, 189
ISO standards, *See* International organisation for standardisation (ISO)

K

Kraft paper, 79, 224, 225

L

Labelling, 13, 48, 63, 70, 83, 85, 93, 94, 119–121, 132, 137, 140, 148, 157, 158, 166, 167, 185, 192, 229, 247, 289, 303, 312, 323, 325, 335–340, 348, 354
Land use, 174, 190, 191, 208, 215, 218, 248, 260, 308, 310
Leachate, 88, 143, 375
Life cycle, 4, 20, 21, 36, 44, 70, 110, 158, 321
Life cycle assessment (LCA), 9, 11, 18, 20, 22, 23, 35, 51, 73, 84, 132, 136, 171–209, 246, 253, 295–298, 322, 324, 333, 375
Life cycle assessment (LCA)-conceptual, 183, 185
Life cycle assessment (LCA)-full, 183, 185, 186, 274
Life cycle assessment (LCA)-goal and scope definition, 176, 180, 187, 189
Life cycle assessment (LCA)-sensitivity analysis, 193, 194, 202, 208, 265
Life cycle assessment (LCA)-streamlined, 49, 179–181, 183–185, 208, 253, 265, 267, 274
Life cycle assessment indicators, *See* Impact indicators
Life cycle assessment interpretation, 176, 184, 187, 193, 195
Life cycle assessment tools, 253–273, 306
Life cycle impact assessment (LCIA), 176, 187, 261
Life cycle inventory (LCI), 176, 189, 191, 204, 254, 262, 265, 268, 310
Life cycle management (LCM), 20, 21, 110, 176, 375
Life cycle map, 7, 20, 21, 87, 132, 173, 177, 180, 181, 182, 187, 188, 253, 254, 262, 263, 265, 295, 306, 322, 324, 331, 375
Life cycle thinking (LCT), 6, 20, 22, 32, 36, 51, 120, 173, 174, 177, 180, 182, 183, 208, 214, 253, 265, 375
Lightweighting, 46, 67, 70, 75, 79, 96, 117, 196, 308, 333, 350, 355, 359
Linear low density polyethylene (LLDPE), 78, 232, 234, 238, 241
Liquid paperboard, 81, 224–230
Litter, 22, 32, 53, 56, 59, 85, 86, 94, 95, 98, 100, 101, 139, 145, 146, 156, 164–166, 182, 190, 208, 242, 275, 309, 338, 346, 351
Litter reduction strategies, *See* Litter

L (cont.)
Low density polyethylene (LDPE), 78, 99–101, 175, 226, 231, 232, 234, 238, 241, 244

M
Marketing, 10, 19, 20, 26, 45, 74, 107–150, 178, 180, 184, 185, 198, 255, 258, 262, 273, 288–291, 296, 307, 319, 322, 343, 347 *See also* Green marketing
Market research, 60, 61, 77, 116, 148, 301, 314, 319
Marks & Spencer, 4, 5, 27, 77, 78, 148, 292
Material recycling, 138–142 *See also* Mechanical recycling, Recycling
Materials recovery facility (MRF), 79, 81, 216, 217, 237, 239, 375
McDonalds, 26, 28, 76, 128, 129
Mechanical recycling, 53, 76, 79–83, 237, 239, 241, 375 *See also* Recycling, Recycling rates
Merchandising packaging, *See* Secondary packaging
Metals, 213
Metrics, *See* Indicators
Mobil, 125
Mobius loop, 133, 139–141, 145, 336, 376
Moulded paper packaging (moulded pulp, moulded fibre), 52, 224–230
Municipal solid waste (MSW), 29–31, 146

N
Nestlé, 180, 181, 193, 194
Non renewable thermoplastics, 100, 214, 230–242, 246, 248, 249
Nude Food Movers, 128, 131

O
On-Pack Recycling Label, 141, 336
Orange Power, 121
Organic recycling, 54, 99, 138, 139, 142–145, 244, 337, 376 *See also* Composting, Degradable
Oxodegradable polymer, 100, 101, 242, 376
Ozone depletion, *See* Stratospheric ozone depletion

P
Packaging Impact Quick Evaluation Tool (PIQET), 180, 181, 253, 254, 260–262, 265, 268, 269
Packaging supply chain, 7, 29, 35, 58, 157, 159, 162, 198, 294, 311, 342, 376
Packaging sustainability, 24, 25, 28, 35, 46, 47, 96–101, 126, 139, 158, 162, 164, 178, 201, 289–291, 297, 303, 304, 308, 309, 311, 324–326, 376
Packaging sustainability framework, 12, 47–95, 137, 275, 277, 308, 321, 331–334
Packaging sustainability metrics, 65, 158, 190, 260, 261, 307–310
Packaging sustainability tools, 16, 20, 23, 44, 47, 159, 173, 181, 183, 196, 202, 208, 251–282, 287, 295, 296, 305, 306, 321, 326
Paper and board, 96–98, 213, 224–230, 248, 280, 317, 368 *See also* Boxboard, Corrugated board
Paperboard, *See* Boxboard
Patagonia, 134–136
Photochemical smog, 11, 191, 193, 260, 264
Photodegradable, 144, 242, 346
Phthalates, 58, 87, 89, 90, 100, 101, 163, 355, 363
Pigments, *See* Inks
Plastic bags, 55, 64, 65, 84, 116, 125, 131, 163, 165–167, 331–334, 341–371
Polycarbonate (PC), 87, 89, 163, 231, 233, 234, 238, 241
Polyethylene, 214, 225, 230, 235, 236 *See also* High density polyethylene (HDPE), Linear low density polyethylene (LLDPE), Low density polyethylene (LDPE), Polyethylene terephthalate (PET), Recycled polyethylene terephthalate (rPET)
Polyethylene terephthalate (PET), 54, 72, 76, 81, 82, 89, 99–101, 175, 199–201, 205–207, 230–242, 276, 354, 357, 362, 369, 370 *See also* Recycled polyethylene terephthalate (rPET)
Polylactic acid (PLA), 205–207, 242–249
Polymer, 89, 91, 99–101, 119, 213, 224, 225, 230–249, 280 *See also* Biodegradable polymers, Cellulose based polymers, Nonrenewable thermoplastics, Oxodegradable polymers, Polymer recycling, Starch based polymers, Thermoplastic starch (TPS) based polymers
Polymer recycling, 76, 82, 86, 139 *See also* Recycling
Polypropylene (PP), 89, 194, 214, 230–242, 331–334

Index

Polystyrene (PS), 76, 77, 116, 129, 157, 188, 205–207, 230–242, 264, 356, 357 *See also* Expanded polystyrene (EPS), General purpose polystyrene (GPPS), High impact polystyrene (HIPS)
Polyvinyl chloride (PVC), 54, 66, 76, 78, 82, 89–91, 99–101, 163, 230–242, 276, 326, 355–357, 363
Post-consumer material, 21, 52, 145, 147, 230, 237, 239, 244, 305, 376 *See also* Post-consumer recycled content (PCR)
Post consumer recycled content (PCR), 28, 52, 147, 164, 165, 242, 308, 339, 350, 357
Post-consumer waste, *See* Post consumer material
Precautionary principle, 334, 376
Pre-consumer material, 145, 230, 305, 376
Primary packaging (also sales, consumer or retail packaging), 8, 60, 62, 64, 67, 74, 76, 160, 269, 275, 376
Printing, 58, 67, 91, 96, 323 *See also* Inks
Processed chlorine-free (PCF) bleaching, 93, 147, 227, 339
Procurement, 4, 5, 57, 91, 94, 135, 136, 159, 179, 198, 259, 262, 274, 288–291, 294, 296, 307, 324–326, 343 *See also* Green procurement
Procter & Gamble, 18–19, 175, 294
Product stewardship, 157, 159, 179, 292, 327, 343

R

Recovery, 15, 22, 23, 30, 44, 47, 53, 55, 79, 80, 82, 84, 87, 96–98, 138, 145, 156–158, 161, 164, 181, 198, 201, 215, 218, 223, 237, 244, 247, 275, 276, 293, 305, 309–312, 323, 324, 334, 376 *See also* Materials recovery facility (MRF), Recycling
Recyclable packaging, 45, 65, 81, 83, 129, 134, 138, 337
Recyclate, 82, 215, 236, 240–242, 377
Recycled polyethylene terephthalate (rPET), 76–78, 198, 202
Recycling, 15, 22, 27–31, 46, 51, 53, 54, 56, 65, 69, 70, 77, 79–83, 95–101, 115, 117, 128, 129, 156–158, 160, 161, 164, 167, 175, 193, 201, 202, 242, 260, 268, 277, 293, 323, 334, 341–371, 377 *See also* Material recycling, Mechanical recycling, Materials recovery facility (MRF), Organic recycling, Recycling rates

Recycling rates, 31, 65, 80, 241
Renewable energy, 19, 23, 24, 35, 51, 86, 93, 139, 148–150, 207, 261, 276, 334, 340
Renewable packaging, 19, 23, 24, 47, 53, 57, 70, 72, 83, 96–101, 159, 276, 293, 309, 310, 334, 377 *See also* Paper and board, Renewable energy, Renewable thermoplastics
Renewable thermoplastics, 99–101, 214, 242–249 *See also* Biodegradable polymer, Cellulose based polymers, Polylactic acid (PLA), Thermoplastic starch (TPS) based polymers
Resource depletion, 190, 191, 264, 270, 310
Retail packaging, *See* Primary Packaging
Reusable plastic containers (RPCs), 202–205

S

Safe packaging, 52, 56–59, 86–95, 139, 162, 214
Sales packaging, *See* Primary packaging
Sara Lee, 307
Secondary packaging (also display or merchandising packaging), 9, 64, 67, 78, 160, 224, 226, 269, 326, 377
Sensitivity analysis, *See* Uncertainty analysis
Smog, *See* Photochemical smog
Society of the Plastic Industry (SPI), 140–142, 337, 348
Solid board, 225, 227 *See also* Boxboard
Solid waste, 12, 29–32, 73, 129, 138, 187, 193, 202–204, 260, 334, 341–371 *See also* Municipal solid waste (MSW)
Starch based polymers, 243, 244 *See also* Renewable thermoplastics
Steel, 54, 76, 80, 83, 96–98, 194, 199–201, 213, 217–220, 248, 344, 354, 357, 369
Stewardship, *See* Ecological stewardship, Forest stewardship, Forest Stewardship Council (FSC), Product stewardship
Stratospheric ozone depletion, 14, 191, 310, 326
Streamlined LCA (see Life cycle assessment - streamlined)
Supply chain, *See* Packaging supply chain, Value chain
Sustainability, 3, 23, 62, 74, 112–114, 138, 157, 174, 181, 325 *See also* Corporate sustainability, Dunphy sustainability phase model, Designing for sustainability, Packaging sustainability, Sustainable development

S (*cont.*)

Sustainable development, 3, 4, 11–15, 20, 22, 25, 26, 28, 30, 43, 44, 47, 51, 110, 125, 126, 128, 148, 157, 179, 197, 292, 296, 377 *See also* Dunphy sustainability phase model

Sustainable Packaging Alliance (SPA), 23, 47

Sustainable Packaging Coalition (SPC), 23, 24, 158, 256, 276, 277, 279–281, 303

System boundary, 182–184, 186, 188, 189, 199, 204, 206, 253, 265, 268

T

Terrestrial toxicity, 191

Tertiary packaging (also distribution, traded, transport packaging), 8, 62, 75, 198, 248, 377

Tesco, 27, 35, 48

The Consumer Goods Forum, 24, 48, 158, 308–310

Thermoplastic starch (TPS) based polymers, *See* Renewable thermoplastics, Starch based polymers

Totally chlorine-free (TCF) bleaching, 93, 139, 147, 227, 338

Toxic, 13, 19, 44, 58, 84, 87, 91, 92, 158, 159, 162, 163, 264, 270, 276, 309, 310, 312, 377 *See also* Aquatic toxicity, Bisphenol A (BPA), Heavy metals, Inks, Phthalates, Terrestrial toxicity, Volatile organic compounds (VOCs)

Traded packaging, *See* Tertiary packaging

Transport energy, 45

Transport packaging, *See* Tertiary packaging

Triple bottom line, 12, 22, 23, 47, 48, 51, 53, 133, 157, 158, 293, 298, 377

U

Uncertainty analysis, 193

Unilever, 69

V

Value chain, 18, 176, 177, 192, 378 *See also* Packaging supply chain

VIP Packaging, 4, 292, 293

Visy, 134, 135, 223, 240

Volatile organic compounds (VOCs), 58, 91, 92, 114, 264, 309, 378

W

Walmart, 23, 26, 27, 34, 52, 71, 88, 255, 260, 261, 273, 274, 292

Walmart scorecard, *See* Walmart

Waste and Resources Action Programme (WRAP), 27, 72–74, 76, 77, 141, 159, 277, 306

Waste to energy, 228, 236, 245, 261, 268, 273, 373

Water use, 11, 49, 184, 190, 191, 193, 248, 260, 310